Fourth Edition

Aircraft Rescue and Fire Fighting

Marsha Sneed
Project Manager/Editor

Contributing Authors:

Robert Lindstrom
Fire Chief
WSI Will Rogers World Airport Fire Department

William D. Stewart
Division Chief of Training
Baltimore-Washington International Airport Fire Department

Validated by the International Fire Service Training Association
Published by Fire Protection Publications, Oklahoma State University

Cover photo courtesy of Captain David Lange, Fairfax County Fire ... Fairfax, VA.

The International Fire Service Training Association

The International Fire Service Training Association (IFSTA) was established in 1934 as a "nonprofit educational association of fire fighting personnel who are dedicated to upgrading fire fighting techniques and safety through training." To carry out the mission of IFSTA, Fire Protection Publications was established as an entity of Oklahoma State University. Fire Protection Publications' primary function is to publish and disseminate training texts as proposed and validated by IFSTA. As a secondary function, Fire Protection Publications researches, acquires, produces, and markets high-quality learning and teaching aids as consistent with IFSTA's mission.

The IFSTA Validation Conference is held the second full week in July. Committees of technical experts meet and work at the conference addressing the current standards of the National Fire Protection Association and other standard-making groups as applicable. The Validation Conference brings together individuals from several related and allied fields, such as:

- Key fire department executives and training officers
- Educators from colleges and universities
- Representatives from governmental agencies
- Delegates of firefighter associations and industrial organizations

Committee members are not paid nor are they reimbursed for their expenses by IFSTA or Fire Protection Publications. They participate because of commitment to the fire service and its future through training. Being on a committee is prestigious in the fire service community, and committee members are acknowledged leaders in their fields. This unique feature provides a close relationship between the International Fire Service Training Association and fire protection agencies which helps to correlate the efforts of all concerned.

IFSTA manuals are now the official teaching texts of most of the states and provinces of North America. Additionally, numerous U.S. and Canadian government agencies as well as other English-speaking countries have officially accepted the IFSTA manuals.

ISBN 0-87939-192-8 *Library of Congress LC# 08-79391928*

Printed in the United States of America 3 4 5 6 7 8 9 10

If you need additional information concerning the International Fire Service Training Association (IFSTA) or Fire Protection Publications, contact:

Customer Service, Fire Protection Publications, Oklahoma State University
930 North Willis, Stillwater, OK 74078-8045
800-654-4055 Fax: 405-744-8204

For assistance with training materials, to recommend material for inclusion in an IFSTA manual, or to ask questions or comment on manual content, contact:

Editorial Department, Fire Protection Publications, Oklahoma State University
930 North Willis, Stillwater, OK 74078-8045
405-744-4111 Fax: 405-744-4112 E-mail: editors@osufpp.org

Table of Contents

Chapter 4 — ARFF Firefighter Safety 79

Chapter 5 — Aircraft Rescue and Fire Fighting Communications 89

Chapter 6 — Aircraft Rescue and Fire Fighting Apparatus 101

Preface

This fourth edition of IFSTA **Aircraft Rescue and Fire Fighting** presents information for airport firefighters as well as other fire and emergency service personnel who may respond to aircraft accidents/incidents. This new edition has been revised to reflect the changes in aircraft and the advances in aircraft rescue and fire fighting.

Acknowledgement and special thanks are extended to the members of the IFSTA Validation Committee who contributed their time, wisdom, and talents to this manual.

IFSTA Aircraft Rescue and Fire Fighting Validation Committee

Chair

Frederick Richards
State of New York Office of Fire Prevention and Control
Albany, NY

Secretary

Robert Lindstrom
WSI Will Rogers World Airport Fire Department
Oklahoma City, OK

Committee Members

David Covington
San Antonio Fire Department/Aviation
San Antonio, TX

George Freeman
Dallas Fire Department
Dallas, TX

Kenneth R. Gilliam
Federal Aviation Administration
Orlando, FL

Bruce Hall
JIBC, Fire and Safety Division
New Westminster, British Columbia, Canada

Richard Hall
International Fire Service Accreditation Congress
Stillwater, OK

Jeffery Hawkins
U.S. Air Force
Hickam AFB, HI

Les Omans
San Jose Fire Department
San Jose, CA

Hugh Pike
U.S. Air Force Fire Protection
Tyndall AFB, FL

Jeffrey A. Riechmann
Kern County Fire Department
Palmdale, CA

Randal L. Smith
Hobart Fire Department
Lake Station, IN

Daniel Spillman
Tallahassee Fire Department
Tallahassee, FL

William D. Stewart
BWI Airport Fire Department
Baltimore, MD

Charles R. Thomas
Air Line Pilots Association
Windsor, CA

Benjamin Warren
OSU Fire Service Training
Stillwater, OK

Frederick H. Welsh
Maryland Fire and Rescue Institute
College Park, MD

Larry Williams
Rural/Metro Corporation
Scottsdale, AZ

Special recognition is given to contributing authors William D. Stewart and Robert Lindstrom, who each wrote or revised several chapters of the manual and provided assistance throughout the production process. The Committee would like to thank John M. Olson for providing for inclusion in this manual, a comprehensive, well-written document on advanced composites/advanced aerospace materials.

The following individuals and organizations contributed information, photographs, and other assistance that made the completion of this manual possible:

Airbus Industrie
Stephen E. Allen, Sr., Associate Administrator, Division of Public Safety, Baltimore-Washington International Airport Fire and Rescue Department
American Association of Airport Executives
Aviation Emergency Training Consultants
Ballistic Recovery Systems, Inc.
Gary Berrian, United Airlines, Maintenance Training Supervisor, Washington Dulles International Airport
Nickolas Buongiorne, Firefighter, Metropolitan Washington Airports Authority Fire Department
City of Charlevoix Fire Department, Charlevoix, MI
Chemguard, Mansfield, TX
Christchurch International Airport, New Zealand
Clark County Fire Department, Las Vegas, NV
Conoco Oil Co.
Michael T. Defina, Jr., Lieutenant, Metropolitan Washington Airports Authority Fire Department
Charles F. Dusha, Logan Rogersville Fire District, Springfield, MO
Emergency One, Ocala, FL
Federal Express of Oklahoma City
Pierre Huggins, Air Line Pilots Association
International Safety Instruments
Ron Jeffers
KK Products
Brian Lowman, Firefighter, Metropolitan Washington Airports Authority Fire Department
Maryland Fire and Rescue Institute
Metro Washington Airports Authority
National Audio Visual Center
James Martin, Director, Airport News and Training Network
Chris E. Mickal, New Orleans (LA) Fire Department
National Transportation Safety Board

Jim Nilo, Fire Chief, Richmond International Airport
Oklahoma Air National Guard, Will Rogers Airport
Oklahoma City Fire Department
Oshkosh Truck Corporation, Oshkosh, WI
Riechmann Safety Services
Vespra (ONT) Fire Department
Williams Training Associates
Joel Woods, Maryland Fire and Rescue Institute

Additionally, gratitude is extended to the following members of the Fire Protection Publications Aircraft Rescue and Fire Fighting Manual Project Team whose contributions made the final publication of this manual possible:

Project Manager/Editor
Marsha Sneed, Associate Editor

Technical Reviewers
Richard Hall, International Fire Service Accreditation Congress Manager
Mike Wieder, IFSTA Projects Manager

Proofreader
Lynne M. Murnane

Production Coordinator
Don Davis, Coordinator, Publications Production

Illustrators and Layout Designers
Ann Moffat, Graphic Design Analyst
Desa Porter, Senior Graphic Designer
Ben Brock, Senior Graphic Designer

Production Assistant
Brien McDowell

Editorial Assistant
Tara Gladden

Introduction

Today's airport firefighter must be knowledgeable in the many facets of aircraft rescue and fire fighting. In addition to basic fire fighting knowledge, skills, and experience, airport firefighters must learn to use specialized techniques, tools, and equipment in order to mitigate airport emergencies. With the technological advances of recent years, aircraft are now larger and more powerful, are constructed of new advanced aerospace materials, and as a result, present new challenges to personnel responding to aircraft accidents/incidents. Effective aircraft rescue and fire fighting (ARFF) accident/incident response requires that airport firefighters be familiar with the varied aircraft at their airport as well as with the different features of the airport.

This edition of **Aircraft Rescue and Fire Fighting** provides basic information needed by firefighters involved in aircraft rescue and fire fighting. It also serves as a reference for other emergency responders likely to be involved with aircraft accidents/incidents.

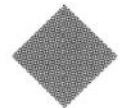

Purpose and Scope

This manual provides basic information needed by firefighters to effectively perform the various tasks involved in aircraft rescue and fire fighting. Written for all fire protection organizations, the manual discusses the use of structural and specialized aircraft fire fighting apparatus and equipment, civilian and military aircraft, and the theory and practice of aircraft fire fighting and rescue operations.

The information contained in this manual is intended to meet the requirements contained in the 2000 edition of NFPA 1003, *Standard for Airport Fire Fighter Professional Qualifications*, as published by the National Fire Protection Association. Additional material addresses the airport fire fighting apparatus covered in Chapter 7 of the 1998 edition of NFPA 1002, *Standard on Fire Apparatus Driver/Operator Professional Qualifications*. (**NOTE:** Material in this manual includes those subjects included in the training requirements of Federal Aviation Regulation [FAR] 139.319.)

Chapter 1

Qualifications for Aircraft Rescue and Fire Fighting Personnel

Job Performance Requirements

This chapter provides information that will assist the reader in meeting the following job performance requirements from NFPA 1003, *Standard for Airport Fire Fighter Professional Qualifications*, 2000 edition. Particular portions of the job performance requirements (JPRs) that are addressed in this chapter are noted in bold text.

1-3.1 For certification as an airport fire fighter, the candidate shall meet the requirements for Fire Fighter II defined in NFPA 1001, *Standard for Fire Fighter Professional Qualifications*; first responder operational level defined in Chapter 3 of NFPA 472, *Standard for Professional Competence of Responders to Hazardous Materials Incidents*; and the requirements for airport fire fighter defined in this standard.

2-1 Entry Requirements. Prior to entering training to meet the requirements of Chapter 3, the candidate shall meet the following requirements:

(1) The minimum educational requirements established by the authority having jurisdiction

(2) The age requirements established by the authority having jurisdiction

(3) The medical requirements of NFPA 1582, *Standard on Medical Requirements for Fire Fighters*

2-2 Physical Fitness Requirements. Physical fitness requirements shall be developed and validated by the authority having jurisdiction. Physical fitness requirements shall be in compliance with applicable equal opportunity regulations and other legal requirements.

All aircraft rescue and fire fighting (ARFF) personnel must have a minimum level of knowledge and skills to deal with aircraft rescue and fire fighting and haz mat operations. This chapter introduces the firefighter to requirements that are intended to identify a minimum level of competence for ARFF personnel. In addition, the chapter discusses the training needs of the airport firefighter.

Basic Requirements

The National Fire Protection Association (NFPA) identifies the job performance requirements for airport firefighters in NFPA 1003, *Standard for Airport Fire Fighter Professional Qualifications.* Other regulations pertinent to ARFF firefighters include those from the Department of Transportation (DOT), the Code of Federal Regulations (CFR), Title 14, Part 139, *Certification and Operations: Land Airports Serving Certain Air Carriers* (commonly referred to as Federal Aviation Regulation [FAR] Part 139), and the International Civil Aviation Organization (ICAO) *Airport Services Manual*, Part 1, Chapter 14. When possible, training programs also should conform to the applicable recommendations of pertinent Federal Aviation Administration

(FAA) Advisory Circulars (ACs). (**NOTE:** ACs are not requirements but rather suggested or recommended guidelines.)

ARFF personnel may respond to medical emergencies. Incidents range anywhere from a single-person medical emergency to a multiple-casualty incident. Therefore, it is important for ARFF personnel to have the appropriate emergency medical care training as specified by the authority having jurisdiction.

When dealing with a large number of victims, ARFF personnel should utilize a rapid assessment and triage system, being familiar with the parameters necessary to treat multiple victims. When working in this environment, firefighters may be at risk for exposure to blood, bodily fluids, and other potential infectious materials. People infected with diseases and viruses such as hepatitis and HIV create contamination concerns for emergency personnel. This requires ARFF personnel to know the techniques for protection from related biohazards. More information concerning emergency medical care is given in the IFSTA/Brady **Fire Service First Responder** and **Fire Service Emergency Care** manuals.

Before candidates are accepted into ARFF training programs for aircraft rescue and fire fighting, they must meet certain basic entrance requirements to be in compliance with NFPA 1003. General requirements of this standard specify that every candidate must:

- Meet the minimum educational requirements established by the authority having jurisdiction.
- Meet the age requirements established by the authority having jurisdiction.
- Meet the medical requirements of NFPA 1582, *Standard on Medical Requirements for Fire Fighters.*

In addition, prospective ARFF personnel need to be in good health and physically fit to perform adequately as airport firefighters. Before being certified as an ARFF firefighter according to NFPA 1003, the candidate must also be certified as a Fire Fighter I and II according to NFPA 1001, *Standard for Fire Fighter Professional Qualifications.*

ARFF Training Programs

Airport firefighters need to understand the causes of aircraft accidents and incidents, what factors contributed to accidents/incidents, and how to mitigate the conditions found. Through practical exercises ARFF personnel should learn to operate apparatus and equipment and use them to their fullest potential.

A comprehensive training program for ARFF personnel — whether assigned to an airport or another type of station that supports ARFF operations through a mutual aid agreement — is critically important to their effectiveness in dealing with aircraft emergencies. Only through pre-incident training can firefighters reduce the likelihood of making costly mistakes during aircraft accidents/incidents. High-quality continuing education and training enable ARFF personnel to acquire and maintain the knowledge, skills, and abilities essential for them to fulfill their mission safely (Figure 1.1). NFPA 405, *Recommended Practice for the Re-*

Figure 1.1 Firefighters perform a training evolution with a mobile ARFF trainer. *Courtesy of Maryland Fire and Rescue Institute.*

curring Proficiency Training of Aircraft Rescue and Fire-Fighting Services, contains the recommended performance criteria for maintaining proficiency and effective aircraft rescue and fire fighting services.

In many cases, airline companies and other airport agencies are willing to participate in and provide resources for ARFF training activities. There are several types of training an ARFF firefighter may receive:

- ***Cadet (recruit) training*** — Initial training received prior to being assigned to a company
- ***On-the-job training*** — Learning the job while performing the day-to-day work requirements under supervision of a more experienced firefighter
- ***In-service training*** — Proficiency training such as classroom study, hands-on skills exercises, and live fire drills conducted within the firefighter's own fire department
- ***Special courses and seminars*** — Extension programs, conferences, short courses, workshops, correspondence courses, or programs in recognized professional schools

Firefighters are not limited to any one of these types of training. In fact, firefighters are apt to be more successful if they participate in more than one of these.

General Requirements

Aircraft rescue and fire fighting personnel must meet the requirements of Fire Fighter II as defined in NFPA 1001, *Standard for Fire Fighter Professional Qualifications.* They also must meet the requirements of first responder operational level as defined in NFPA 472, *Professional Competence of Responders to Hazardous Materials Incidents,* and the job performance requirements of NFPA 1003. The requirements contained in NFPA 1003 are divided into three major duties: response, fire suppression, and rescue. The most important responsibility of a person assigned as an airport firefighter is to perform aircraft fire fighting and rescue operations. These duties, along with those specified in FAR Part 139 and in ICAO *Airport Services Manual,* Part 1, Chapter 14, require that the airport firefighter be able to demonstrate knowledge and skills in the following subjects:

- Airport familiarization
- Aircraft familiarization
- Firefighter safety
- Aircraft rescue and fire fighting communications
- Aircraft rescue and fire fighting apparatus and equipment
- Aircraft rescue tools and equipment
- Aircraft rescue and fire fighting apparatus driver/operator
- Extinguishing agents
- Aircraft rescue and fire fighting tactical operations
- Airport emergency plans
- Hazards associated with aircraft cargo

The remainder of this chapter provides a brief overview of each of these subjects. They are covered in greater detail in the remaining chapters of this manual.

Airport Familiarization

Firefighters can begin learning about their airport through training classes and by studying the overall layout of the airfield on grid maps. However, classroom training and looking at maps do not provide all the needed information. The firefighter must get out on the airport grounds and learn the airport features — control tower, runway layout, vehicle standby positions, on-airport navigation aids, instrument landing system (ILS) critical areas, airport buildings and hangars, maintenance and storage facilities, drainage systems, water distribution systems, etc. (Figure 1.2). The airport firefighter must also be able to identify airport structures and terrain features that may limit the response capabilities of ARFF vehicles or that may present a hazard to vehicles responding to accidents or incidents on the airfield. For aircraft emergencies off the airport, ARFF vehicles may need to depart the airport through other-than-normal exits or areas. Therefore, personnel must know the locations of emergency exit access points such as frangible (breakaway) fences and gates.

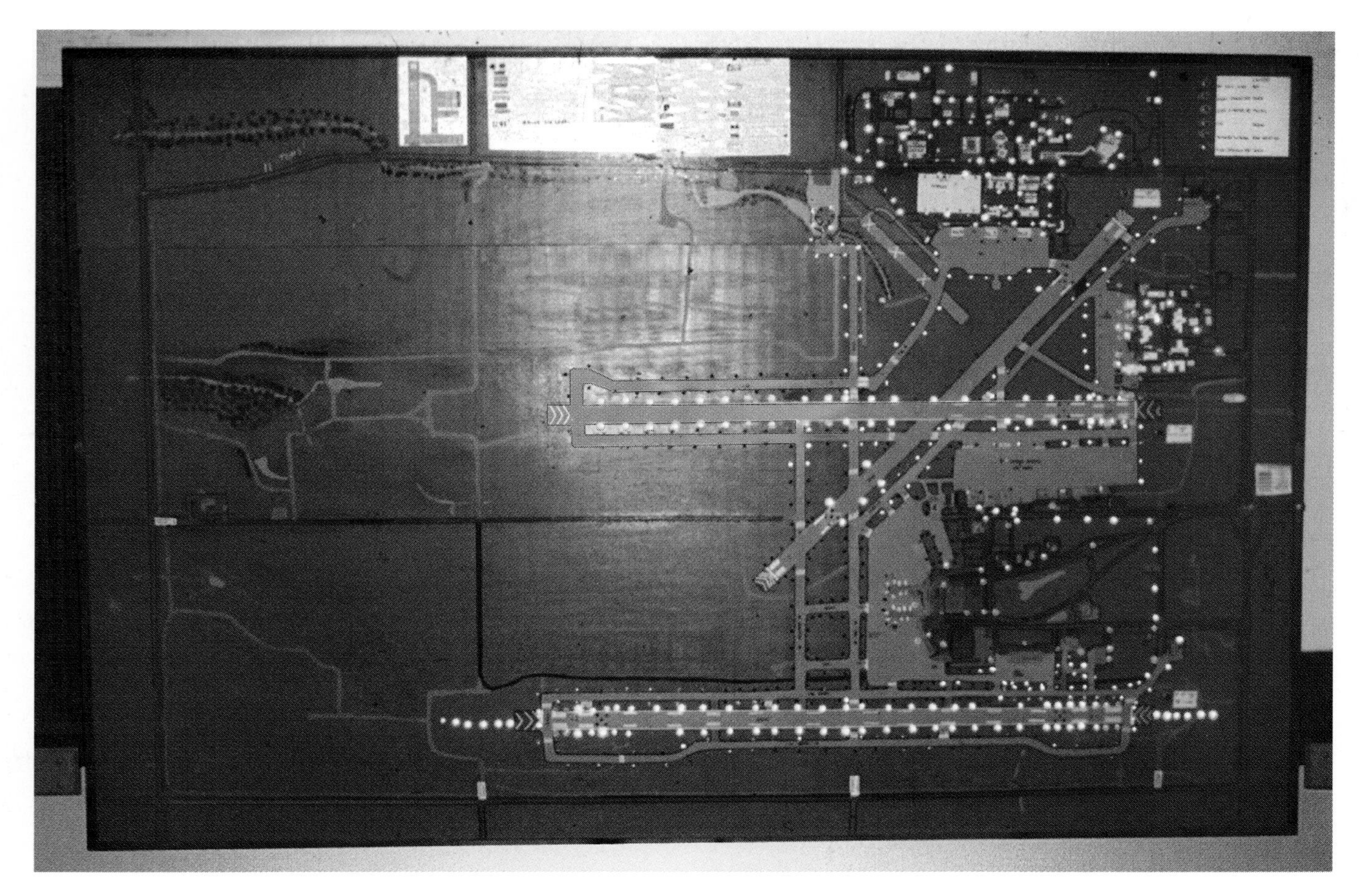

Figure 1.2 This lighted airport layout illustrates the various features of a particular airport and its surrounding area.

Airport firefighters should know the airport layout and the systems used to identify runways and taxiways. Because there are so many activities occurring simultaneously at an airport, there must be a way of managing vehicle and aircraft movements. This control is achieved by using traffic-flow procedures, airport pavement marking and signing systems, and the airport lighting color code and marking system (Figure 1.3).

All airports have target hazard areas that the airport firefighter must be able to identify. These include fuel storage and distribution systems (Figure 1.4), and dangerous goods storage and handling areas. When involved in an emergency, aircraft carrying hazardous materials/dangerous goods may be directed to designated isolation areas on the airfield. The airport firefighter must know these locations and the response routes to them.

The need for security at an airport is the responsibility of all airport employees, especially in today's environment of terrorism and vandalism. At some airports, performing security duties is part of the firefighter's job. Therefore firefighters must be familiar with the rules and regulations governing security on their airport.

Aircraft Familiarization

When confronting an aircraft accident/incident, ARFF personnel may be exposed to many hazards. This requires ARFF personnel to be familiar with aircraft design and terminology as it relates to aircraft rescue and fire fighting.

Aircraft rescue and fire fighting hazards range from fire and toxic smoke to explosions. If airport firefighters are familiar with the types of aircraft using their airport, they can plan for specific hazards. They must also be familiar with aircraft construction features and materials as they relate to forcible entry, rescue, and fire fighting operations.

The airport firefighter must be able to identify the aircrew and passenger locations and capacity

Figure 1.3 Airport firefighters should know the meanings of airport signs, lights, and pavement markings.

Figure 1.4 Knowing the location of fuel storage areas is an important part of airport familiarization.

for each type of aircraft using the airport. They must also be able to locate and operate normal entry doors, emergency exit openings, and evacuation slides of each aircraft (Figure 1.5). When normal entry is not feasible, the firefighter must be able to locate and gain entry through forcible entry points on the aircraft. For military aircraft, it is important to be familiar with canopy and seat-ejection systems as well as weapons and explosive devices.

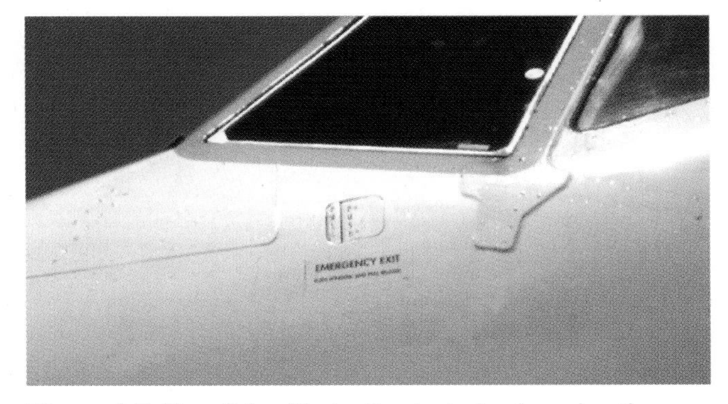

Figure 1.5 Aircraft familiarization includes knowing the locations of emergency exits on aircraft commonly encountered.

The firefighter must be familiar with the types of aircraft components and the systems found on aircraft at their airport. These include:

- Engines
- Fuel systems
- Oxygen systems
- Hydraulic systems
- Electrical systems
- Fire protection systems

- Anti-icing systems
- Auxiliary power units (APUs) (Figure 1.6)
- Aircraft radar systems
- Brake systems
- Wheel assemblies

Finally, ARFF personnel should be able to identify the flight data recorder and cockpit voice recorder found in aircraft. Both the digital flight data recorder (DFDR) and cockpit voice recorder (CVR) assist in the investigation of the probable cause of the incident and are key items for the investigation authorities. Firefighters should know the procedures to follow upon locating these devices to guard them against further damage.

Firefighter Safety

The airport firefighter faces many real and potential hazards in and around aircraft accidents/incidents. Although an aircraft incident may not always involve a fire, the firefighter must be aware of other hazards, such as engine intakes and exhausts, associated with the aircraft. ARFF personnel must understand the effects that hazards such as burning fuel, toxic smoke, aircraft wreckage, and biohazards pose to themselves, the victims of the accident, and the fire fighting equipment.

The environment in which airport firefighters must work requires that they be provided with approved personal protective equipment (PPE). As with all emergency operations training, proper training in the use of protective clothing and positive-pressure self-contained breathing apparatus is a must. Just as important as knowing how to don and use protective equipment is knowing its limitations and being able to identify common hazardous respiratory environments. The airport firefighter must be able to don protective equipment with and without SCBA. Firefighters should train in SCBA use in a low-visibility environment and should be able to demonstrate breathing techniques used under emergency conditions, including assisting other firefighters, conserving air, and using the bypass valve. The firefighter should also be able to maintain, clean, inspect, and reservice breathing apparatus.

Another very important element in providing for safe and effective aircraft rescue and fire fighting operations is the routine use of an incident management system (IMS) in all drills, exercises, and daily operations. IMS is a system of procedures for controlling personnel, structures, equipment, and communications so that all responders can work together toward a common goal in an effective and efficient manner. IMS is designed to be applicable to incidents of all sizes and types. Such organization is extremely important because the airport fire department's work often involves life-threatening situations and working with other agencies. Thus, any lapse in organization could have serious consequences for victims and firefighters alike.

Finally, injuries suffered by the victims in aircraft accidents sometimes can be extremely gruesome and horrific. The department should have in place a critical incident stress debriefing (CISD) program to handle the stress effects imposed on response personnel during and after these incidents (Figure 1.7). Also, ARFF personnel must be aware of the

Figure 1.6 An auxiliary power unit is used as an alternate power source while the aircraft is being serviced.

Figure 1.7 Dealing with the devastation of a high-impact crash can be very disturbing to emergency responders. *Courtesy of Chris E. Mickal, New Orleans Fire Department.*

assistance available to them to help cope with the results of this stress.

Aircraft Rescue and Fire Fighting Communications

The expedient and accurate handling of fire alarms or aircraft emergency notifications is a significant factor in the successful outcome of any incident. Aircraft rescue and fire fighting communications include the methods by which the telecommunications center (also known as the communications center) is notified of an emergency, the methods by which the center can notify the proper fire fighting forces, and the methods by which information is exchanged at the scene.

ARFF personnel must be able to identify the procedures for receiving single and multiple alarms. This requires the firefighter to know how to use the alarm-receiving equipment housed in the communications center. Mutual aid may be necessary for certain responses. The airport firefighter must know the procedures for notifying and requesting these resources. Mutual aid companies may use radio frequencies that are different from those the airport fire department uses, and the firefighter must be able to identify these radio frequencies.

ARFF vehicles responding to an emergency site may require clearance from the control tower to proceed into or through certain areas of the airport. ARFF personnel and telecommunicators must know the procedures for obtaining clearance from the control tower or other responsible authority for apparatus movement (Figure 1.8). Once on the scene, ARFF personnel must be able to provide an initial status report and at times communicate directly with the pilot of the emergency aircraft. They also must be able to use and understand hand signals for aircraft rescue and fire fighting and for communicating with aircrew personnel as well as with other firefighters in a high-noise environment.

Figure 1.8 Air traffic control tower.

Aircraft Rescue and Fire Fighting Apparatus and Equipment

An airport fire department may have many different types of vehicles and equipment including hoses, nozzles, and accessory fittings. The firefighter must be able to identify each hose, nozzle, and adapter, know the purpose of each, and know their location on ARFF vehicles. ARFF personnel should be trained to use an assortment of handlines, turrets, and extending/elevating waterways. They also must be familiar with how to safely use hose tools and equipment for effective fire fighting.

Airport firefighters must be totally familiar with all aspects of operating the ARFF vehicles at their airport. They must understand the primary purpose of the vehicle, type(s) of agent(s) carried, agent capacity, water capacity, agent discharge rate/range, any special features on the vehicle (Figure 1.9), personnel requirements, response and setup procedures, and response limitations for the ARFF vehicles.

ARFF vehicles are designed to deliver mass quantities of water/agent, but they have a limited capacity. Therefore, agent management is important to successful ARFF operations. When ARFF vehicles deplete their agent supply, they must be able to refill as quickly as possible. Firefighters

Figure 1.9 The forward looking infared camera allows the driver/operator to see images when operating in darkness.

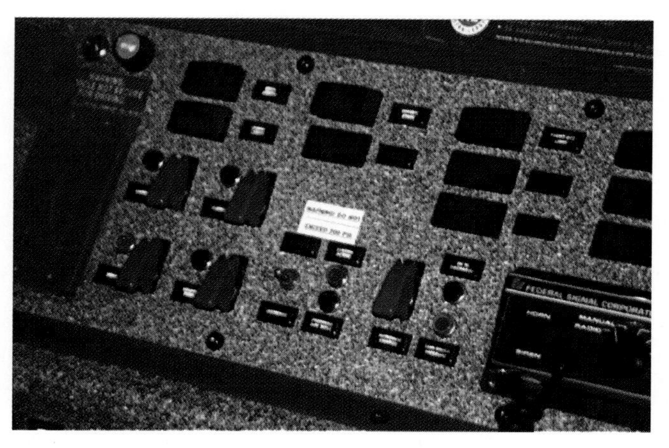
Figure 1.10 Driver/operators must know the operation of extinguishing system controls.

must be capable of establishing a resupply point using hydrants, other static water supplies, or water tenders (tankers).

Aircraft Rescue Tools and Equipment

When airport firefighters cannot gain access into an aircraft because the normal means of access is locked, blocked, or nonexistent, they must force entry. Selection and use of the appropriate tool or set of tools are imperative to gain access into an aircraft.

A variety of tools and equipment is available for gaining access into an aircraft and rescuing its occupants. Which tools are actually needed depend on the type of aircraft, its position, and the skills of the firefighter. Understanding the limitations and capabilities will assist the firefighter in selecting the right tool or piece of equipment for the job. A complete and thorough understanding of the basic types of tools used in aircraft rescue and fire fighting also ensures that the firefighter can perform such tasks efficiently and safely.

Forcible entry is a learned skill that begins with a thorough knowledge of aircraft construction and materials. Firefighters must keep up-to-date on the construction features of aircraft and the locations of doors, hatches, and windows.

Aircraft Rescue and Fire Fighting Apparatus Driver/Operator

The fire apparatus driver/operator is responsible for safely transporting firefighters, apparatus, and equipment to and from the scene of an emergency or other call for service. Once on the scene, the driver/operator must be capable of operating the apparatus properly, swiftly, and safely (Figure 1.10). The driver/operator must also ensure that the apparatus and the equipment it carries are ready at all times.

In general, driver/operators must be mature, responsible, safety-conscious adults. Because of their wide array of responsibilities, often under stressful emergency situations, driver/operators must be able to maintain a calm, "can-do" attitude under pressure. Psychological profiles, drug and sobriety testing, and background investigations may be necessary to ensure that the driver/operator is ready to accept the high level of responsibility that comes with the job.

To perform their duties properly, all driver/operators must possess certain mental and physical skills. Not every firefighter is capable of becoming a driver/operator. The required levels of these skills are usually determined by each jurisdiction. In addition, NFPA 1002, *Standard for Fire Apparatus Driver/Operator Professional Qualifications,* sets minimum qualifications for ARFF vehicle driver/operators. It requires any driver/operator who will be responsible for operating an ARFF vehicle to also meet the requirements of NFPA 1001, *Standard for Fire Fighter Professional Qualifications,* for Fire Fighter II.

Fire apparatus must always be ready to respond. Regardless of whether the truck responds to an emergency call once an hour or once a month, it must be capable of performing in the manner for which it was designed at a moment's notice. In order to ensure this, certain preventive maintenance functions must be performed on a regular basis. Most apparatus or equipment failures can be prevented by performing routine maintenance checks on a regular basis, and most fire departments require driver/operators to be able to perform routine maintenance checks and functions.

Extinguishing Agents

The airport firefighter may come into contact with fires involving many types of combustible materials, ranging from the seat coverings inside the passenger compartment, to magnesium in wheel assemblies, and to the aircraft's hydrocarbon fuels. Depending upon the material burning and the size and location of the fire, different situations may require a different type of extinguishing agent. An understanding of the classifications of fire is important to the firefighter when attacking and extinguishing a fire. Each class of fire has its own requirements for extinguishment.

Aircraft Rescue and Fire Fighting Tactical Operations

Aircraft can end up in any position during an accident. There may be fire coming from any or multiple locations within the aircraft, or fire may be impinging on the aircraft from outside. These factors must be considered in rescue and fire fighting decisions. Tactical decisions regarding approach and setup must be made during the initial response. Firefighters should apply extinguishing agent in an effort to prevent fire impingement of exposed portions of the fuselage while ensuring that egress lanes are established and maintained.

Aircraft evacuation is a priority of the airport firefighter. Evacuation may require nothing more than opening the normal entry doors; rescuing passengers may require more complex evacuation operations, including the use of evacuation slides. Based on the type of aircraft and the situation, the firefighter must identify the most appropriate areas for gaining entry into the aircraft and must protect evacuation routes.

Once the exit openings are made, they must be protected. A handline may be adequate, or turrets flowing agent to provide a foam blanket may be required to prevent ignition of fuel vapor. In any case, the firefighter must be constantly aware of the situation and be ready to protect the evacuation points on the aircraft.

Airport firefighters must know standard operating procedures as they relate to aircraft emergencies. This knowledge is necessary to select strategies and tactics for controlling the emergency and bringing it to a successful conclusion.

The success or failure of a fire fighting team often depends upon the skill and knowledge of the personnel involved in initial-attack operations. A well-trained team of firefighters with an attack plan and an adequate amount of extinguishing agent, properly applied, can contain most fires in their early stages. Failure to make a well-coordinated attack on a fire can allow the fire to gain headway and burn out of control. Losing control of the fire can result in increased damage, as well as further endangerment to firefighters and civilians alike. Other procedures with which the airport firefighter must be familiar include:

- Protecting an aircraft fuselage from fire exposure
- Providing protective streams for personnel and aircraft occupants
- Controlling runoff from fire control operations and fuel spills
- Stabilizing aircraft wreckage

Airport Emergency Plans

A plan of operation should be created so that appropriate procedures can be developed and resource needs identified before aircraft accidents/incidents occur. Such a plan addresses the need for a coordinated and structured response to emergency situations within the local jurisdiction (Figure 1.11). The plan should be as complete and detailed as necessary to ensure that all involved agencies are aware of their roles and responsibilities under various conditions.

The airport firefighters' responsibilities in understanding the airport emergency plan include:

- Identifying their duties as outlined in the plan
- Identifying and using pre-incident plans

- Knowing the various types of aircraft-related accidents/incidents possible at their airport
- Understanding the procedures used to size up an aircraft accident or incident
- Knowing how to coordinate with other organizations

The airport firefighter needs to understand the roles and responsibilities of all the agencies, groups, departments, etc., that may be involved in the emergency response to an aircraft accident/incident. Some of these organizations and agencies include the following:

- Other airport departments (administration, operations, and maintenance)
- Aircraft investigation agencies
- Police
- Emergency medical services
- Mortuary affairs
- Military
- Mutual aid agencies
- News media

Hazards Associated with Aircraft Cargo

Just as the technology explosion has changed society, so has it changed the jobs of ARFF personnel. No longer is there such a thing as a "routine incident." Aircraft cargo may be inherently hazardous, or it may become hazardous as a result of the fire. It may further complicate an aircraft fire fighting operation when exposed to fire conditions, creating an environment that is exceptionally toxic and dangerous.

Hazardous materials/dangerous goods are found in every airport. Because of this, plans and standard operating procedures for safely handling haz mat incidents must be in place. Airport firefighters must be trained in accordance with these plans and procedures (Figure 1.12). All ARFF personnel must clearly understand their role when faced with an incident involving hazardous materials/dangerous goods. Firefighters must know their limitations; they must recognize when they cannot proceed any further in an incident.

Emergency response organizations are obligated to prepare their personnel for dealing with hazardous materials/dangerous goods when they are encountered during the course of normal activities. This obligation is dictated not only by governmental regulations (such as OSHA) or by consensus standards (NFPA, for example) but also by the bounds of the moral obligation of an employer to provide a safe working atmosphere for employees.

Although informal methods of identification are helpful, it is only through positive identification of the materials involved that firefighters may fully develop a sound defensive strategy. Some of the methods available to identify hazardous materials/dangerous goods include the Department of Transportation (DOT) and the International Civil Aviation Organization (ICAO) labeling systems. ARFF per-

Figure 1.11 ARFF personnel must plan for emergencies both on and off the airport. *Courtesy of Maryland Fire and Rescue Institute.*

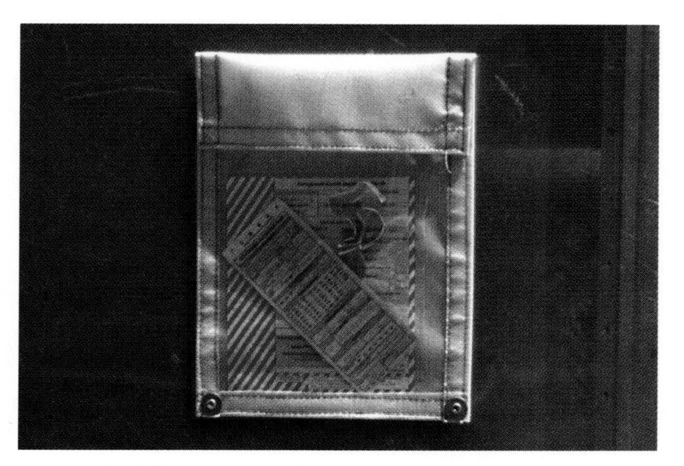

Figure 1.12 Promptly locating shipping papers on aircraft helps in quickly identifying any hazardous cargo.

sonnel must be able to identify the hazards indicated by each of these systems and the limitations of each. Firefighters must be able to use the DOT *Emergency Response Guidebook* (*ERG*) to obtain information on hazardous materials/dangerous goods for a given situation.

Emergency information centers such as the Chemical Transportation Emergency Center or CHEMTREC (in the U.S.) and the Canadian Transport Emergency Centre or CANUTEC (in Canada) are principal providers of immediate technical information assistance to firefighters. Many hazardous materials manufacturers and shippers maintain emergency phones and response teams, but most prefer to be contacted through CHEMTREC and/or CANUTEC. Both organizations operate 24 hours a day, 365 days a year. Airport firefighters must know the procedures for using CHEMTREC and other resources to obtain information concerning a hazardous material. They must be able to use the information obtained from the DOT *ERG* and CHEMTREC to identify the appropriate response, including risk assessment and rescue or evacuation requirements.

For more information on the basics of identifying hazardous materials/dangerous goods and the steps taken to mitigate an incident involving them, see the IFSTA **Hazardous Materials for First Responders** manual.

Chapter 2

Airport Familiarization

Job Performance Requirements

This chapter provides information that will assist the reader in meeting the following job performance requirements from NFPA 1003, *Standard for Airport Fire Fighter Professional Qualifications*, 2000 edition. Particular portions of the job performance requirements (JPRs) that are addressed in this chapter are noted in bold text.

3-2.1 Respond to day and night incidents or accidents on and adjacent to the airport, given an assignment, operating conditions, a location, a grid map, a vehicle, and a prescribed response time, so that the route selected and taken provides access to the site within the allotted time.

(a) *Requisite Knowledge:* **Airport familiarization including runway and taxiway designations, frangible gate locations, airport markings, lights, instrument landing system (ILS) critical isolation areas, vehicular traffic controls on airfield, bridge load limits, controlled access points, aircraft traffic patterns and taxi routes, fuel storage and distribution locations, airport and immediate local area topographic layout, drainage systems, water supplies, airport facilities**.

(b) *Requisite Skills:* Read, interpret, and take correct action related to grid maps, water distribution maps, airport markings, and lights.

3-2.2 Communicate critical incident information regarding an incident or accident on or adjacent to an aiport, given an assignment involving an incident or accident and an incident management system (IMS) protocol, so that the information provided is accurate and sufficient for the incident commander to initiate an attack plan.

(a) *Requisite Knowledge:* Incident management system protocol, the airport emergency plan, **airport** and aircraft **familiarization**, communications equipment and procedures.

(b) *Requisite Skills:* Operate communications systems effectively, communicate an accurate situation report, implement IMS protocol and airport emergency plan, recognize aircraft types.

Because fires involving aircraft may develop rapidly, the life-hazard potential in aircraft fires is tremendous. Nearby aircraft, equipment, and structures also may be exposed to the fire. Aircraft rescue and fire fighting (ARFF) units therefore must be able to respond to the scene quickly.

Many airports have continuous aircraft operations 24 hours a day, including times of inclement weather. The airport may have a fire department located on the airport itself, or it may be served by the local fire department. The local fire department also may support the airport fire department through mutual aid agreements.

Emergency response personnel must be able to find their way quickly to any point on the airport—even at night or when visibility is reduced by inclement weather. This includes using alternate response routes when the primary response routes are not available. By being thoroughly familiar with the airport, emergency personnel can respond to an accident/incident quickly, thereby increasing the effectiveness of the rescue and fire fighting effort.

Familiarity with airport layout, driving regulations, and communications procedures is important not only to ARFF personnel but also to those firefighters assigned to nearby structural stations. At times, additional resources must be called to support the airport fire department. These responding units also must know how to reach the scene quickly and safely.

ARFF personnel must be thoroughly familiar with the airport layout — particularly the runways and their numbering system, along with taxiways, roads, gates, fences, and geographical features peculiar to the airport (Figure 2.1). They must also understand how runways are used. For instance, aircraft are normally directed to take off and land into the wind. When the wind is light and not a critical factor, however, air traffic controllers may use several runways simultaneously to expedite the flow of traffic.

Fires in aircraft or associated equipment and facilities can be costly in both human and economic terms. Even if no injuries result, fire can destroy expensive property, affect airport employees' job security, affect airport suppliers, and seriously inconvenience the public. For these reasons, all personnel should practice good fire prevention techniques during all ground activities and extraordinary vigilance during some of the more hazardous operations.

Figure 2.1 ARFF personnel must become familiar with the many features and the routine activities at their airport. *Courtesy of William D. Stewart.*

Types of Airports

There are two basic types of airports: controlled and uncontrolled. Controlled airports have operating towers with air traffic controllers who manage aircraft movement both in the air and on the ground. Uncontrolled airports are those that do not have a staffed and operating control tower. Some airports may staff and operate their control tower only during specific times, such as during daylight hours, and be uncontrolled at night.

Airports are classified by various agencies such as the Federal Aviation Administration (FAA), International Civil Aviation Organization (ICAO), and the National Fire Protection Association (NFPA) in order to determine the level of fire protection needed. The FAA, for example, classifies airports that are used by air carrier aircraft with seating for more than 30 passengers into different index categories (Table 2.1). FAR Part 139.315 states that the index is determined by a combination of the following:

- The length of air carrier aircraft (expressed in groups)
- Average number of daily departures of air carrier aircraft

The average daily departures of air carrier aircraft are determined as follows:

- If there are five or more average daily departures of air carrier aircraft in a single index group servicing that airport, then the index of the group having the longest aircraft and an average of five or more daily departures is the index required for the airport.
- If there are less than five average daily departures of air carrier aircraft in a single index group servicing that airport, the next lower index from the index group with air carrier aircraft is the index required for the airport. The minimum designated index is Index A.

ICAO bases its airport categories on the longest airplanes using the airport as well as on the fuselage width of these aircraft. ARFF personnel should work with the authority having jurisdiction (AHJ) to ensure the requirements affecting fire protection for the index system used at their airport are met.

Airport Traffic Patterns

Airport firefighters must understand the air traffic patterns within the vicinity of the airport. Unless otherwise directed by an air traffic controller, all aircraft entering the airport area must do so by flying a traffic pattern. If an aircraft declares an emergency, that aircraft is given priority and may not fly a traffic pattern but rather a straight-in or modified approach.

Airport firefighters who understand how aircraft enter the traffic pattern will know the aircraft's position in relation to the airport runway during inbound emergencies. The components of a typical traffic pattern include the following legs (Figure 2.2):

Table 2.1
ARFF Requirements by FAA Airport Index

FAA Index	A	B	C	D	E
Maximum Aircraft Length (ft [m])	<90 ft (<29 m)	≥90 ft and <126 ft (≥29 m and <38 m)	≥126 ft and <159 ft (≥38 m and <48.5 m)	≥159 ft and <200 ft (≥48.5 m and <60 m)	≥200 ft (≥60 m)
Number of Fire Fighting Vehicles Required	1	1 or 2	2 or 3	3	3
Total Fire Fighting Agent Required	500 lb (227 kg) dry chemical/ Halon 1211 or 450 lb (204 kg) dry chem and 100 gal (379 kg) of water/foam	Same as "A" and 1,500 gal (5 678 L) of water/foam	Same as "A" and 3,000 gal (11 356 L) of water/foam	Same as "A" and 4,000 gal (15 142 L) of water/foam	Same as "A" and 6,000 gal (22 712 L) of water/ foam

NOTE: Exceptions are addressed in FAR 139.315.

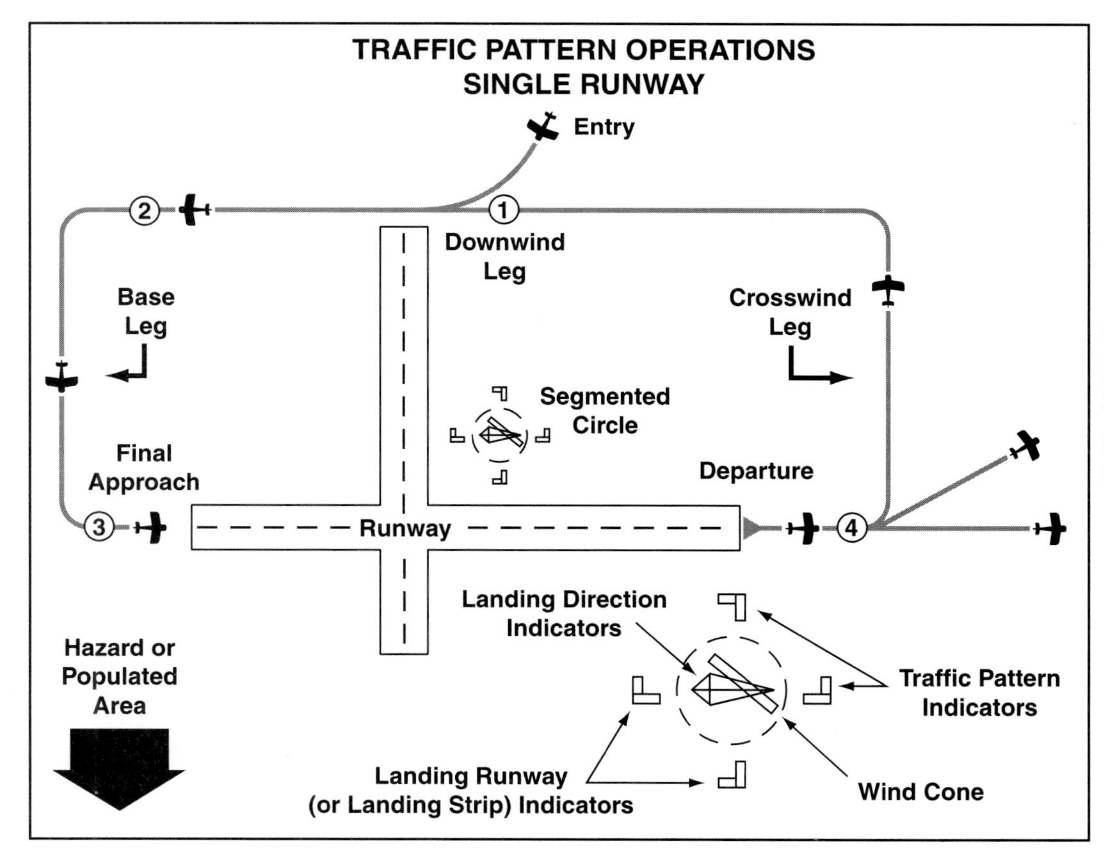

Figure 2.2 Recommended standard left-hand traffic pattern. The standard right-hand traffic pattern is the opposite.

- ***Crosswind leg*** — The flight path at right angles to the landing runway off its upwind leg.
- ***Downwind leg*** — The flight path parallel to the landing runway in the direction opposite to landing. The downwind leg normally extends between the crosswind leg and the base leg.
- ***Base leg*** — The flight path at a right angle to the landing runway off the approach end. The base leg normally extends from the downwind leg to the intersection of the extended runway line. The aircraft must make a 90-degree turn from the base leg before it can begin its final approach.
- ***Final approach*** — That portion of the landing pattern in which the aircraft is lined up with the runway and is heading straight in to land.

Once an aircraft lands on the airport, it must move along designated routes to passenger gate, cargo hangar, or maintenance areas on the airport. ARFF vehicles may use the same access routes as aircraft, so it is important that the airport firefighter know the meanings of runway and taxiway designation systems.

Runway and Taxiway Designation Systems

When an aircraft takes off and when it lands, it does so in a manner that allows it to face into the wind. Therefore, runways are laid out to take advantage of the prevailing winds at that airport. Runway numbers are taken from the nearest compass bearing (relative to magnetic north) rounded off to the nearest 10 degrees. Compass bearings start at north and run clockwise from 0 to 360 degrees. A runway with a compass heading of 340 degrees is numbered 34 for aircraft approaching from the south. The same runway is numbered 16 for aircraft approaching from the north because from that direction it has a compass bearing of 160 degrees; there is always a difference of 180 degrees between opposite ends of the same runway. When the number of the runway is 06 or 09, the number may have a bar placed underneath (06 or 09) to avoid confusion. Letters, where required, distinguish among parallel runways:

- At most airports, parallel runways are designated by a number followed by L (left) and the same number followed by R (right) (Figure 2.3). For example, one set of parallel runways might be identified as 36L and 36R from the north and as 18L and 18R from the opposite end.
- Three parallel runways are indicated in a similar fashion, but a C (for center) follows the number of the middle runway: 18L, 18C, and 18R.

Taxiways are specially designated and prepared surfaces on an airport for aircraft to taxi (travel) to and from runways, hangars, etc. In simpler terms, they are the roadways for aircraft movement on the ground. Letters, a combination of numbers and letters, or names, usually designate taxiways. Taxiway designations are not standardized and are generally determined locally.

Airport Lighting, Marking, and Signage Systems

In addition to runway numbers and taxiway identification systems, colored lights, markings, and signs

Figure 2.3 A runway properly identified with numbers and letters. *Courtesy of Air Line Pilots Association.*

are used to identify various areas, buildings, and obstructions at airports. ARFF personnel should understand the lights, markings, and signage systems used on their particular airport.

Surface Lighting

While taxiway designations may vary from airport to airport, runway and taxiway surface lighting is standard at all airports (Figure 2.4):

- *Blue lights* or reflective markers are used to outline taxiways and are usually located along the edges and are spaced 100 feet (30 m) apart.
- *White lights* are used to outline the edges of runways, with lights spaced 200 feet (60 m) apart, and to identify runway centerlines, with lights spaced 50 feet (15 m) apart. White lights are also used to denote vehicular traffic areas on ramps.
- *Green lights* are used to identify the approach end of runways and some taxiway centerlines.

NOTE: Not all taxiways are required to have centerline lighting.

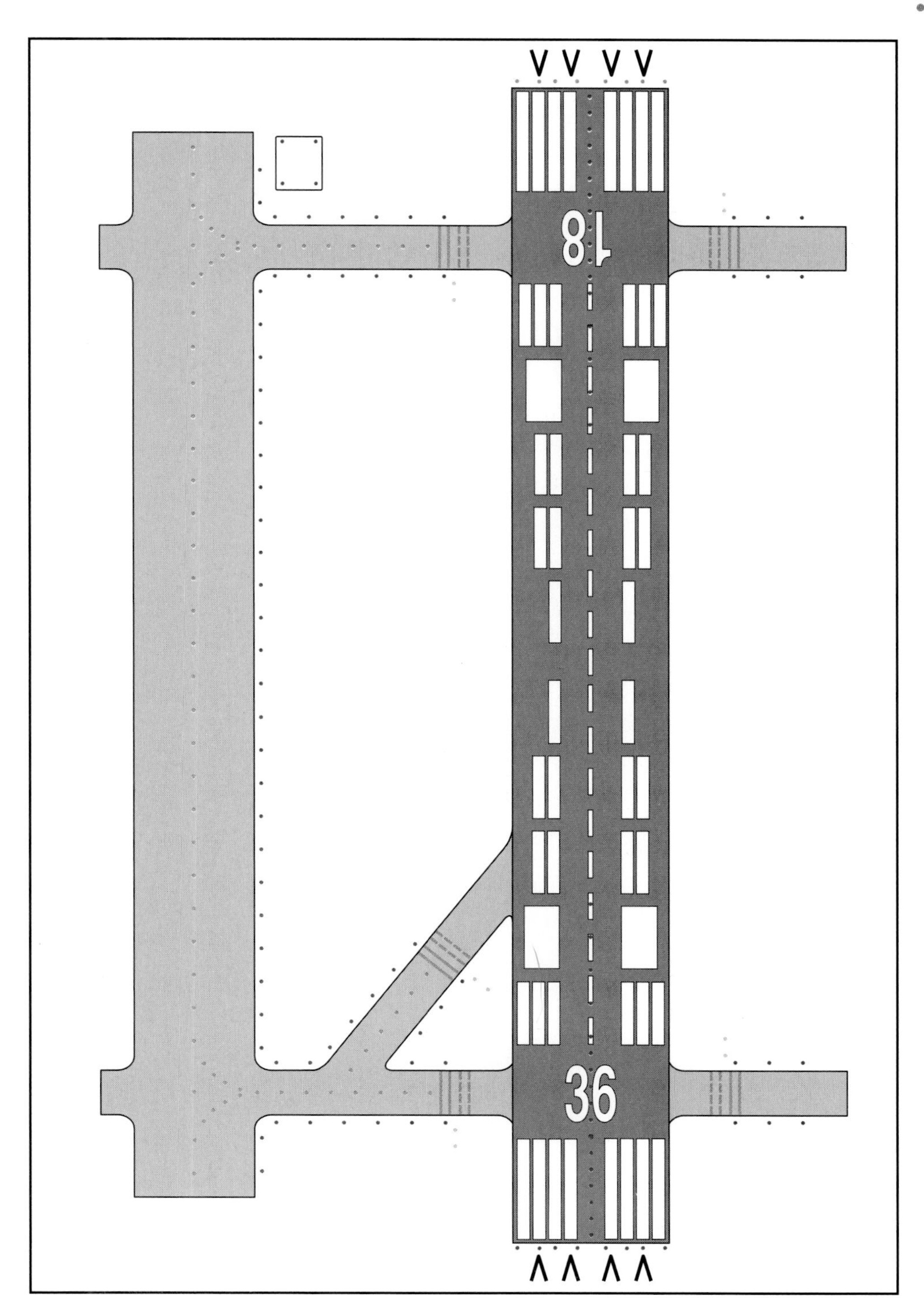

Figure 2.4 Colored lights identify different areas of runways and taxiways.

- *Red lights* are used to mark obstructions such as building structures, parked aircraft, unserviceable areas, construction work, and the departure end of the runway. Red lights may also be found at the departure end of the runway. Runway centerline lighting alternates red and white the last 3000 feet (914.4 m) and becomes all red the last 1000 feet (304.8 m).
- *Yellow or amber lights* are used to identify locations of hold bars, which are areas that can be crossed only with permission from the control tower. These lights also serve as runway edge lights for the last 2000 feet (609.6 m) at the departure end of the runway.

Markings

Colored markings are used at airports also. The three colors commonly used for airport markings are white, red, and yellow.

- *White* is used for runway identification numbers/letters, landing zone bars, and striping.
- *Red* is used to designate restricted areas such as fire lanes and no-entry areas. Permission must be granted prior to crossing a red line into a restricted area.
- *Yellow* is used for hold bars, taxiways, and Instrument Landing System (ILS) critical areas. Yellow may also mark non-load-bearing surfaces.

Hold position markings (often called hold lines or hold bars) act like stop signs for all vehicles or aircraft using taxiways (Figure 2.5). Hold position markings consist of four yellow lines — two solid and two dashed — which extend across the entire width of the taxiway. When approaching a hold position marking from the solid-line side, the vehicle or aircraft is required to stop until either visual clearance is confirmed (uncontrolled airports) or air traffic control (ATC) has approved further movement (controlled airports). When approached from the dashed-line side, the hold position marking is not applicable, and vehicles can cross immediately to the other side without obtaining clearance.

One area on the airport where hold position markings can be found is the ILS critical area. These areas are usually located on taxiways that are near the end of the runway. They are marked with hold position markings and signs to prevent ground traffic from entering these areas (Figure 2.6). No vehicle or aircraft is allowed close to the ILS system when it is in operation for an aircraft landing. This ensures that the ILS signals are not broken or otherwise obstructed.

A taxiway has a single, continuous yellow centerline along its length. A double continuous yellow line may be used to show the location of the taxiway edge.

Runways sometimes have areas that are not suitable or legal for takeoff and landing. Often referred

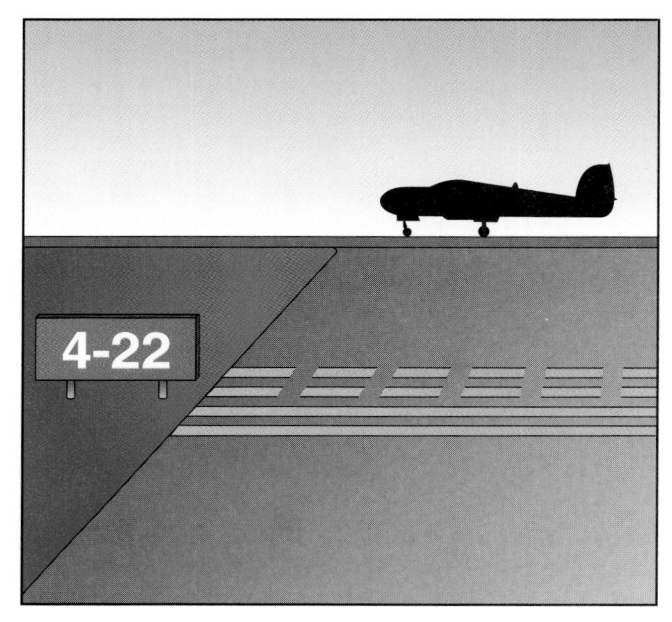

Figure 2.5 Markings that identify a holding position.

Figure 2.6 Markings that identify an ILS area.

to as "special-purpose areas," their markings indicate their purpose and associated restrictions. The most common of these areas is the displaced threshold, which is indicated by a solid white line across the runway. This line is where the runway officially begins and is followed by either the runway number or eight threshold markers. If there are white arrows pointing at the displaced threshold line, the preceding paved area can be used for taxi, takeoff, and rollout. If the area is not suitable for aircraft operations, it is marked with yellow chevrons. Extra pavement areas are often provided to allow jet blast to dissipate and for overruns during aborted takeoffs. A painted or lighted "X" on a runway or taxiway means that runway or taxiway is closed to aircraft and vehicle operations.

4-22	**TAXIWAY/RUNWAY HOLD POSITION:** Hold short of runway on taxiway.
26-8	**RUNWAY/RUNWAY HOLD POSITION:** Hold short of intersecting runway.
8-APCH	**RUNWAY APPROACH HOLD POSITION:** Hold short for aircraft on approach.
ILS	**ILS CRITICAL AREA HOLD POSITION:** Hold short of ILS approach critical area.
⊖	**NO ENTRY:** Identifies paved areas where aircraft entry is prohibited.
B	**TAXIWAY LOCATION:** Identifies taxiway on which vehicle/aircraft is located.
22	**RUNWAY LOCATION:** Identifies runway on which vehicle/aircraft is located.
	RSA/OFZ BOUNDARY: Exit boundary of runway protected areas.
	ILS CRITICAL AREA BOUNDARY: Exit boundary of ILS critical area.
J→	**TAXIWAY DIRECTION:** Defines direction and designation of intersecting taxiway(s).
↙L	**RUNWAY EXIT:** Defines direction and designation of exit taxiway from runway.
22↑	**OUTBOUND DESTINATION:** Defines directions to take-off runways.
↖MIL	**INBOUND DESTINATION:** Defines directions for arriving aircraft.
	TAXIWAY ENDING MARKER: Indicates taxiway does not continue.

Figure 2.7 Airport signs and their meanings.

Signs

The six types of signs used on airports are mandatory instruction signs, location signs, direction signs, destination signs, information signs, and runway distance remaining signs (Figure 2.7). Airports also have typical roadway signs seen on highways and roads. They are used on the airport where roads may intersect taxiways or runway approach areas. Descriptions of the signs used at airports follow:

- *Mandatory instruction signs* provide information that must be adhered to. They include identification of holding positions, runway intersections, ILS critical areas, runway approach, and entry signs.
- *Runway hold position signs* have white inscriptions on a red background.
- *Location signs* identify which runway or taxiway you are on and also identify specific locations on the airport. A location sign has a yellow inscription on a black background.
- *Direction signs* identify the direction of taxiways leading out from an intersection (Figure 2.8). They have black inscriptions on yellow backgrounds.

Figure 2.8 Taxiway direction signs.

- *Destination signs* indicate destinations such as runways, terminals, and cargo areas on the airport. Like direction signs, destination signs have black inscriptions on a yellow background.
- *Information signs* provide pilots with information such as applicable radio frequencies or noise-abatement procedures. These signs have yellow backgrounds with black inscriptions.
- *Runway distance remaining signs* indicate the distance of runway remaining. The number displayed, in thousand-foot increments (304.8 meter increments) represents how many feet (meters) of runway is left before the threshold. A white number — for example, a "4" — on a black background indicates that there is 4000 feet (1,219.2 meters) of remaining runway.

WARNING

Understand that aircraft ALWAYS have the right-of-way. ARFF apparatus drivers failing to understand or obey airport ground lighting, markings, and signs can lead to accidents on the airport as well as runway incursions.

Airport Design

The airport layout is the key factor in determining the most appropriate response route for ARFF vehicles. Determining the proper response route (primary or alternate) may require using grid maps. This section introduces the firefighter to the procedures for using grid maps. The airport firefighter must know the airport topography as well as the locations of different areas and objects on the airport including:

- Segmented circle
- Structures
- On-airport navigation aids
- Roads and bridges
- Airport ramps
- Controlled access points
- Fences and gates
- Designated isolation areas
- Water supply
- Fuel storage and distribution
- Airport drainage systems
- Terrain impassable by emergency vehicles

Grid Maps

Grid maps are important when planning emergency response routes. ARFF and airport support personnel use these maps to identify ground locations. It is important that mutual aid departments understand how to read grid maps and be familiar with the grid maps of the airports they support.

Grid maps are marked either using rectangular coordinates or with azimuthal bearings using polar coordinates. They may be standard maps, large-scale commercial maps, or modified outline maps. Whatever the type of map used, the grid map should include an area encompassing the emergency response area outside the airport.

In addition to the marked coordinates used with the grid system, these maps should include traffic patterns and control zones. One area that may be outlined on the grid map is the critical rescue and fire fighting access area (CRFFAA) (Figure 2.9). According to NFPA 402, *Guide for Aircraft Rescue and Fire Fighting Operations*, this area is the rectangular area surrounding any runway within which most accidents can be expected to occur on airports. Its width extends 500 ft (150 m) from each side of the runway centerline, and its length is 3300 ft (1000 m) beyond each runway end.

Standard map symbols indicating streams, marshes, wooded areas, or any type of ground surface may also be used. By inspecting the mapped area, ARFF personnel can determine pertinent terrain features that should be included on the map. The addition of landmarks, bodies of water, roads, bridges, drainage systems, and other features not ordinarily shown on maps is necessary so that a complete description of the accident/incident scene can be given to responding emergency personnel.

A map locator system using a grid map is the link that ties together the operations of all groups having some responsibility in the event of aircraft

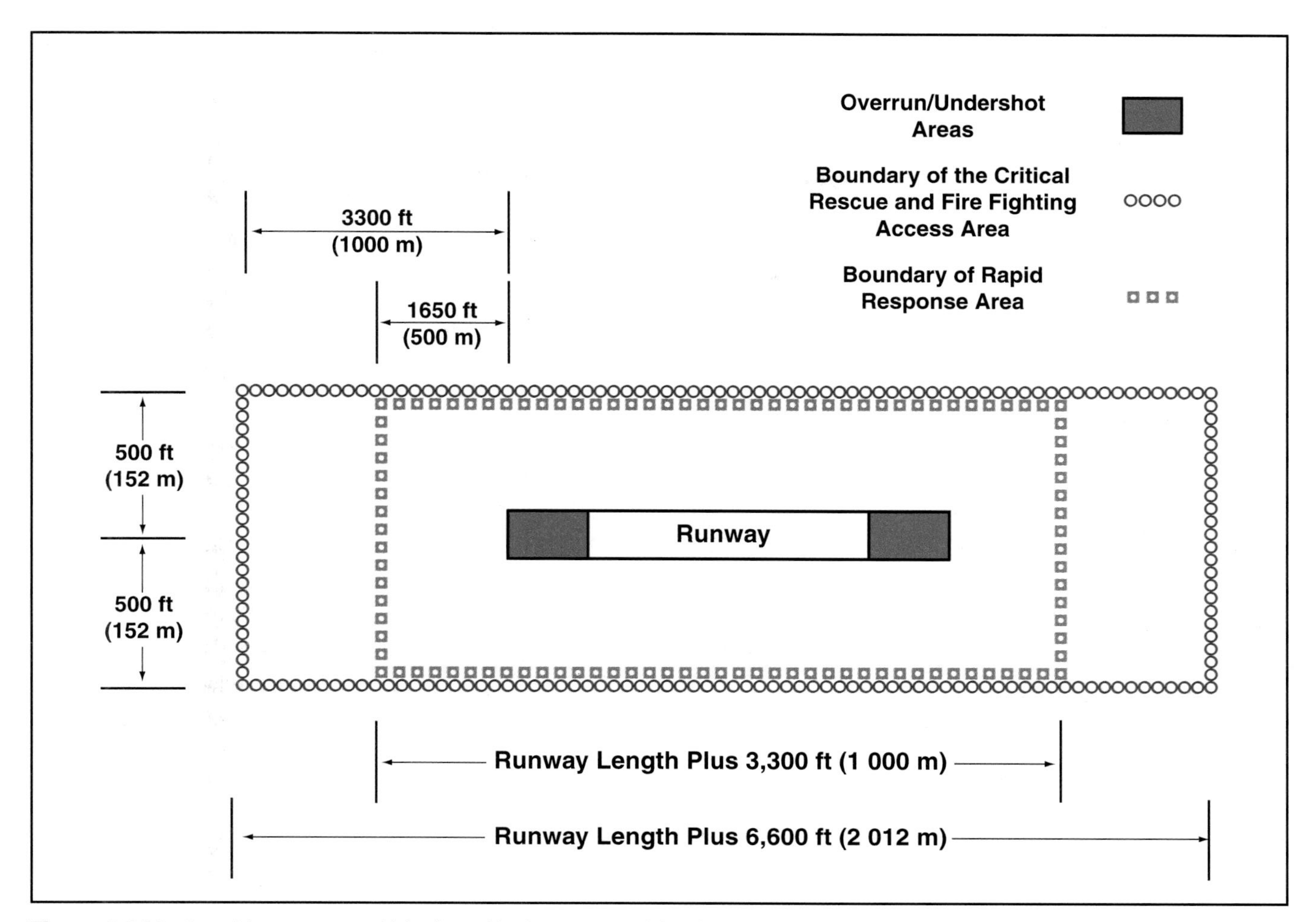

Figure 2.9 Most accidents occur within the critical rescue and fire fighting access area of a runway.

accidents/incidents. Complete, up-to-date copies of such maps must be furnished to control tower personnel; to all emergency response units, including ambulance and emergency medical personnel; and to all others with legitimate interests.

In an emergency situation, the location of the emergency scene should be described in terms of the grid system. The incident location can be quickly identified by using the numbered and lettered grid coordinates. It is also helpful for the description to include identifying landmarks. All information possible should be included in the description of the accident/incident location so that ARFF personnel may correctly identify and quickly reach the location.

Other maps that may be available to the airport firefighter include utility maps (water distribution, electric service, gas line distributions), drainage maps, structural locator maps, and fuel spill directional maps.

Airport Topography

Topography is composed of the features of the earth's surface, both natural and constructed, and the relationships among them. The airport firefighter must be knowledgeable of the topographic layout of both the airport and its immediate surrounding area. Airport topography is important when determining response routes for apparatus and fuel drainage direction when a spill has occurred. Terrain may be impassable during inclement weather (Figure 2.10). Normally dry areas may be converted to mud by heavy rains; snow may pile beyond the capabilities of snow-removal crews; or water may collect in low-lying areas blocking access to some points on the airport.

Topography is also an important aspect when predicting fire spread if an aircraft crashes and catches fire in an area that has varying elevations. The topography in the immediate area of a fire affects both its intensity and its rate and direction of spread. Wind channeling through various topographical features also affects the fire spread. The effects of terrain and wind are discussed further in Chapter 10, "Aircraft Rescue and Fire Fighting Tactical Operations."

For more information on how fire is affected by weather and topography, see IFSTA's **Fundamentals of Wildland Fire Fighting** manual.

Airport Structures

ARFF personnel may be required to respond to structural alarms as well as aircraft emergencies, so they should be cross-trained in structural fire fighting. Likewise, structural firefighters stationed near airports should be cross-trained in aircraft rescue and fire fighting techniques. The following sections include the hazards of which airport firefighters should be aware.

Terminals

The occupant load and fuel load within a terminal varies depending on the volume of air traffic at the time. Some of the major concerns for ARFF personnel responding to an emergency at an airport terminal follow:

- ***Life safety***. Large crowds of people unfamiliar with exit locations may occupy terminal buildings. The exits may egress onto restricted airport operations areas, which pose additional hazards.
- ***Jetways***. Because jetways connect aircraft to the terminal, they can provide a means for smoke and flame to spread from one area to the other (Figure 2.11).
- ***Baggage handling and storage areas***. These areas, usually located on lower levels, may be loaded with baggage and cargo, which could make it difficult to extend handlines and conduct other fire suppression operations.

Aircraft Maintenance Facilities

Aircraft maintenance facilities conduct a variety of operations that are a concern to fire safety personnel. These include the following:

- Maintenance and repair of aircraft fuel tanks and systems
- Use of flammable and hazardous chemicals for painting and striping operations
- Repair of aircraft electrical, avionics, and radar systems
- Heavy aircraft maintenance that includes disassembling large parts of the aircraft and its interior, using cleaning fluids, and reassembling the aircraft with various sealant, glues, and paints
- Welding, cutting, and grinding operations used to fabricate or repair aircraft parts or assemblies
- Offices, parts rooms, or record keeping areas in the aircraft hangar used by maintenance personnel
- Storage of hazardous materials to be used in aircraft maintenance operations

All these maintenance activities present potential fire and safety hazards in a facility. The operating maintenance areas need to be a part of the firefighter's airport familiarization and fire inspection duties.

Figure 2.10 An area may become more difficult to access due to a change in weather conditions. *Courtesy of Michael T. Defina, Jr., Metro Washington Airports Authority Fire Department.*

Figure 2.11 Jetways are a hazard for spreading smoke and fire. *Courtesy of Michael T. Defina, Jr., Metro Washington Airports Authority Fire Department.*

Smoking should be prohibited inside aircraft hangars or near any flammable liquid or hazardous material associated with the maintenance activities. Inspection duties should be thorough in an effort to eliminate all potential ignition sources.

Fire extinguishers of sufficient size and number should be available in aircraft maintenance facilities to extinguish fires in their incipient stages. The local AHJ usually dictates the size and number of fire extinguishers. Aircraft in hangars may contain substantial amounts of fuel. More information about fire extinguishers in aircraft hangers can be found in NFPA 10, *Standard for Portable Fire Extinguishers*, and NFPA 409, *Standard on Aircraft Hangars*.

Figure 2.12 The air traffic control tower can present entry and egress challenges to emergency responders. *Courtesy of Michael T. Defina, Jr., Metro Washington Airports Authority Fire Department.*

Other Facilities and Airport Activities

ARFF personnel should be familiar with the other airport buildings and the activities and concerns associated with them. These structures and situations likely to be encountered and their challenges are described in the following list:

- Utility structures and vaults
  - — Confined spaces
  - — Large electrical loads
  - — Diesel engines
  - — High-voltage switching equipment
- Air-freight facilities
  - — Combustible and flammable storage
  - — Stored hazardous materials
- Control tower (Figure 2.12)
  - — Probability of electrical problems due to the high concentration of electrical/electronic equipment
  - — Forcible entry difficulties posed by security measures
  - — Limited means of emergency egress due to tower height
- Passenger transportation systems (Figure 2.13)
  - — Subways
  - — Monorails
  - — Electrically driven sidewalks
  - — Horizontal elevators

Figure 2.13 ARFF personnel should be prepared to deal with emergencies involving the passenger transportation systems at their airport. *Courtesy of Michael T. Defina, Jr., Metro Washington Airports Authority Fire Department.*

- Multilevel parking structures (Figure 2.14)
  - Low overhead clearances
  - Weight restrictions
  - Limited water supply
  - Alternate-fuel vehicles (those using compressed natural gas [CNG], for example)
- Hotels/motels, stores, and restaurants
  - May be attached to the terminal
  - May be located inside the terminal
  - May be located on airport property away from the terminal area

On-Airport Navigation Aids

Navigational aids (NAVAIDs) are visual or electronic devices, either airborne or on the ground, which provide point-to-point guidance information or position data to aircraft in flight. Airport firefighters do not need to know the details of how navigational aids work. They should, however, be able to identify navigation aids and know their locations on the airport (Figure 2.15). The presence of ARFF vehicles within the operating locations of some navigational aids may interfere with their signals; therefore, ARFF vehicles should respond via routes that do not hamper the operation of these devices (ILS systems). Also, firefighters within some of these operational areas may be harmed by the radio waves produced by this equipment. Figure 2.16 describes some of the navigational aids that may be found on an airport and shows their typical placement on the airport.

WARNING

Several navigational aids pose an electrical hazard to firefighters. The airfield lighting system and navigational aids operate on high-voltage electrical systems.

Roads and Bridges

Roads on the airport are used for normal vehicle transportation on and around the airport. Service roads are used to reach the ends of runways and other remote parts of the airport. Some roads may not be suitable for ARFF vehicle use. Some may become impassable in adverse weather. Some of these roads may have bridges along them. ARFF crews should know the load limits of all bridges on the airport and in the local vicinity and plan alternative routes to areas serviced by bridges with low load limits.

Airport Ramps

Ramps/aprons tend to be the most congested areas on the airport. The following list includes functions, equipment, and vehicles that make up the bulk of the activity in these areas (Figure 2.17):

- Pedestrian traffic
- Fueling operations
- Baggage handling

Figure 2.14 Weight restrictions and low overhead clearances may make it difficult for ARFF vehicles to respond to an accident/incident at a parking garage. *Courtesy of Michael T. Defina, Jr., Metro Washington Airports Authority Fire Department.*

Figure 2.15 Airport emergency responders must be able to identify the various navigational aids at the airport. *Courtesy of William D. Stewart.*

- Service-vehicle movements
- High-voltage electrical feed to aircraft from mobile ground power units (GPUs)
- Aircraft maintenance operations
- Hazardous materials/dangerous goods being shipped and/or transferred

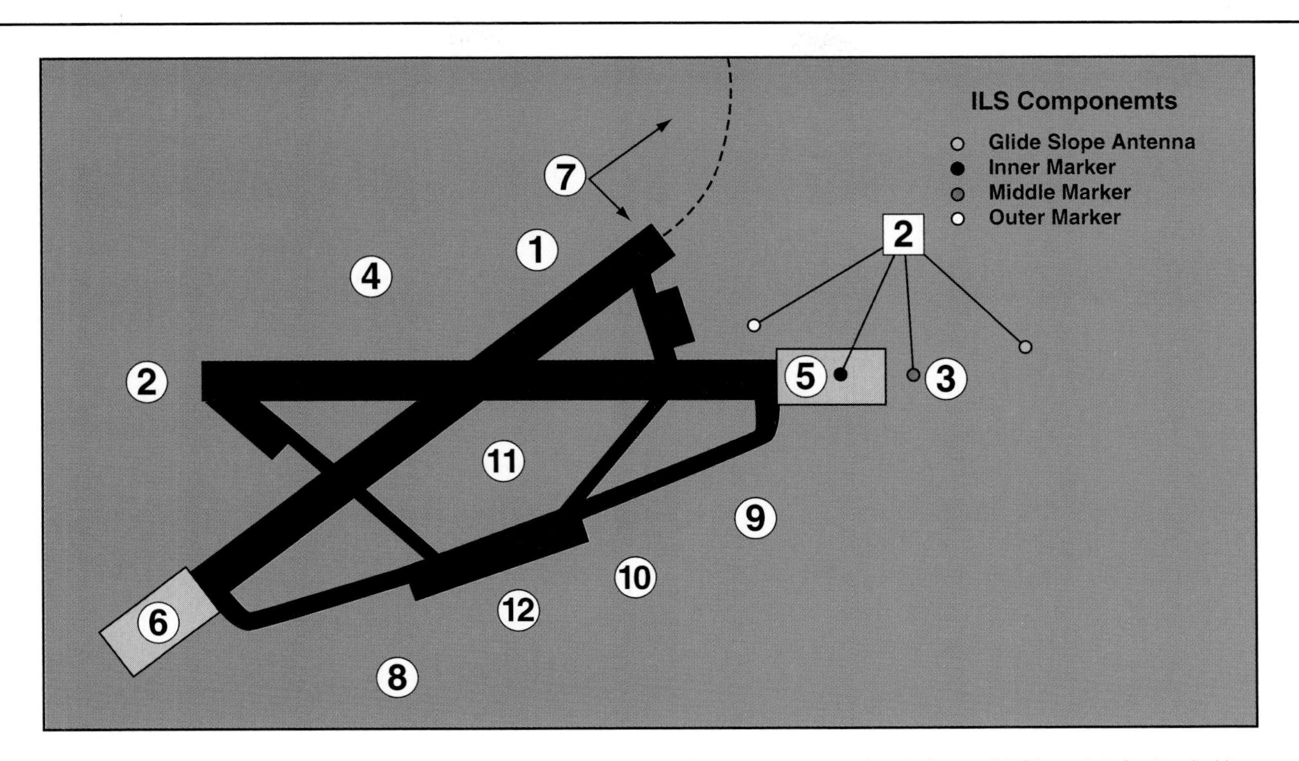

1. **Visual Approach Light Systems**. Precision approach path indicators (PAPI) or visual approach slope indicators (VASI) consist of red and white lights located adjacent to the runway to provide the pilot with a visual descent path.

2. **Instrument Landing System (ILS)**. The instrument landing system provides pilots with electronic guidance for aircraft alignment, descent grade, and position until visual contact confirms the runway alignment and location.

3. **Non-Directional Beacon (NDB)**. The non-directional beacon transmits radio signals by which a pilot, using aircraft instruments, can determine his or her location from the signaling station. An NDB is usually mounted on a 35 foot (11m) pole.

4. **Very High Frequency Omnirange (VOR)**. The standard very high frequency omnirange located on an airport is known as a TVOR. TVORs radiate azimuth information for nonprecision instrument approach procedures.

5. **Approach Lighting Systems (ALS)**. Approach lighting systems are configurations of lights positioned symmetrically along the extended runway centerline. They provide visual guidance to landing aircraft by radiating light beams in a directional pattern by which the pilot aligns the aircraft with the extended centerline of the runway on the final approach for a precision landing.

6. **Omnidirectional Approach Lighting Systems (ODALS)**. Omnidirectional approach lighting systems are configured so as to radiate flashing light beams in all directions. These systems are positioned at the approach end of runways where aircraft conduct non-precision landings.

7. **Lead-In Lighting Systems (LDIN)**. Lead-in lights are flashing lights installed at or near ground level to designate the desired course an aircraft should fly to an ALS or to a runway threshold.

8. **Airport Rotating Beacon**. Airport rotating beacons indicate the location of an airport by projecting beams of light spaced 180 degrees apart. Alternating white/green flashes identify a lighted civil airport; white/white flashes an unlighted civil airport.

9. **Airport Surveillance Radar (ASR)**. Airport surveillance radar scans 360 degrees of the airport to provide air traffic controllers with the location of all aircraft within 60 nautical miles of the airport.

10. **Airport Surface Detection Equipment (ASDE)**. Airport surface detection equipment is used to compensate for the loss of line of sight to surface traffic during periods of reduced visibility.

11. **Automatic Weather Observation Stations (AWOS)**. Automatic recording instruments measure cloud height, visibility, wind speed and direction, temperature, dew point, etc.

12. **Airport Traffic Control Tower (ATCT)**. Air traffic controllers control flight operations within the airport's designated airspace and the operation of aircraft and vehicles on the movement area.

Figure 2.16 Airport navigational aids.

Figure 2.17 Much congestion and activity is commonplace at airport ramps. *Courtesy of Michael T. Defina, Jr., Metro Washington Airports Authority Fire Department.*

Firefighters should stay clear of aircraft on ramp and apron areas. Do not park behind an aircraft at a terminal gate. Try to check with ramp personnel to confirm where the best and safest place to park is. Try to leave someone with the fire department vehicle when parked in questionable areas. Yield to aircraft pushing back from gates, unless waved on by one of the ground crew. A pushback is when a passenger aircraft backs away from the jetway or terminal area to taxi to the departure runway. At some airports, aircraft may use engine thrust reversers to back up, but normally the aircraft will be pushed back with a tug. Aircraft pushing back will have their red anticollision lights illuminated, located on the top and bottom of the aircraft fuselage. Air stairs, jetways, and wheel chocks will be secured prior to moving the aircraft. Aircraft doors, hatches, cargo doors will be closed. Baggage-handling equipment, fueling vehicle, and other ground support equipment will have completed loading operations and moved away from the aircraft. A tug and tow-bar will be attached to the aircraft nosewheel, and the tug driver will be seated, with inter-phone connected. One or more wing-walkers with wands or lights may be in place.

Firefighters need to be aware of foreign object debris (FOD) on airport ramps and other driving surfaces. This is loose debris, trash, and other objects on the airport that can be sucked up and into the intakes of jet engines, causing considerable damage. Firefighters should always be vigilant for FOD and take the time to pick up any and dispose of it properly. When driving from unpaved areas onto aircraft movement areas, personnel should always stop and check vehicle tires for rocks, mud, and other objects stuck in the tire tread.

ARFF personnel should be able to monitor ground activities through direct observation. For this reason, airport fire stations should be in a central location where there is a good view of the flight line, taxiways, apron/ramp, and hangar areas. To facilitate surveillance, some airport fire stations incorporate an observation tower for monitoring ground activities. If properly constructed, a fire station observation tower may be an ideal location for the fire department communications/dispatch center.

Although it is not always practical for ARFF personnel to constantly observe ground operations from the airport fire station, when feasible, they should visually monitor the following:

- Taxiing operations, integrity of aircraft landing gear, ground operation of engines, and aircraft maintenance on the flight line
- Fueling/defueling operations
- Roads, taxiways, and fire lanes that may be blocked by aircraft or other vehicles
- Current weather conditions that might affect the movement of emergency vehicles and the takeoff/landing patterns of aircraft.

Observing these activities allows firefighters to gain an awareness of what is happening on their airport and also assists in an effective fire prevention program.

Controlled Access Points

Controlled access points are areas where access is limited in order to eliminate unnecessary or unauthorized traffic (Figure 2.18). A solid red hold line, a red and white dashed line, or a mandatory sign may identify these points. These areas may be staffed by a security guard at the designated entry-control point as part of the security identification display area (SIDA). The entry control point may be the only way to enter a controlled area. Controlled access points are also used for controlling entry into areas designated as isolation areas, ILS areas, munitions areas, and fuel storage areas.

Fences and Gates

Airport facilities require protection from vandals and any unauthorized individuals. For security purposes, most airports provide perimeter fences to keep people and animals from inadvertently entering the airport as well as keeping them from entering restricted areas of the airport.

While serving their intended purposes, these fences also pose a barrier to ARFF vehicles trying to leave the airport using other-than-normal exit points. Frangible (breakaway) fences and gates are strategically located to allow rapid access for ARFF vehicles to areas outside the airport boundaries. By knowing the exact locations of frangible fences and gates, ARFF crews can reduce vehicle response times to areas outside the airport boundaries. If the airport fire department does not keep keys to the gates, security personnel should carry them. If time permits, security may be notified to unlock and open the gates. If this is not possible, ARFF vehicles may be used to strike sections of the fence or a gate designed to break away or collapse when struck with the ARFF vehicle's bumper. Firefighters must also know whether these areas are accessible year-round and during inclement weather.

Designated Isolation Areas

The isolation area is a predetermined area designed for temporary parking of aircraft experiencing problems with hazardous cargo. It can also be used for handling dangerous circumstances such as a hijacked aircraft, bomb threat, or terrorist attack. The location is selected because of its distance from the major facilities and other aircraft traffic.

Figure 2.18 Gates at a controlled access area of the airport. *Courtesy of Michael T. Defina, Jr., Metro Washington Airports Authority Fire Department.*

Water Supply

The availability of water for fire suppression should be identified. There are generally two sources of water for airport fire protection: fixed systems and mobile supplies. Common fixed systems include wells, plus surface-level and elevated storage tanks. Distribution of water from fixed systems is achieved through domestic water supply mains unless the system is designed otherwise. Hydrants located along aircraft operational areas may be located underground. Common mobile water supply systems include fire apparatus and water tankers/tenders. For airport locations known to be deficient in water supply, arrangements for the automatic dispatch of water tankers/tenders need to be made in advance.

Fuel Storage and Distribution

Having a working knowledge of the airport fuel storage and distribution facilities is important for ARFF personnel (Figure 2.19). These facilities include the following:

- Fuel storage tanks or supply pipeline
- Fuel distribution systems or loading areas
- Ramp areas where aircraft are fueled

Studying the location of these facilities on a map of the airport and learning the function and operation of shutoff valves and switches from a technical manual is important. Still, all firefighters assigned to the airport should periodically visit each of these sites and become thoroughly familiar with their location, functions, and operation. The following

Figure 2.19 A typical "fuel farm." *Courtesy of Michael T. Defina, Jr., Metro Washington Airports Authority Fire Department.*

sections discuss fueling operations and the associated hazards that ARFF personnel may experience on a regular basis.

Fueling Operations

A major area of concern at airports involves fueling operations. This activity is a constant hazard and represents the number one fire prevention consideration at airports. Fuel is delivered to airports by tank truck, railcar, or pipelines. It is also stored in aboveground or belowground bulk storage tanks. (**NOTE:** See *Storage Tank Emergencies* [1996] by Hildebrand, Noll, and Donahue for more information on emergencies involving fuel storage tanks.) The fuel is loaded onto aircraft by one of three methods. At larger airport facilities, the fuel is transferred by underground piping, which terminates at a sub-surface fuel hydrant located at each gate. A fuel service truck connects to the underground system and pumps the fuel into the aircraft (Figure 2.20).

Figure 2.20 An aircraft refueling from an underground fuel system. *Courtesy of William D. Stewart.*

Figure 2.21 An aircraft refueling from a tank truck. *Courtesy of William D. Stewart.*

The most common method of aircraft fuel delivery is by tank truck (Figure 2.21). These tankers transport fuel from a storage location and pump their contents into the aircraft. Common capacities of these tankers can range from 500 to 10,000 gallons (2 000 to 40 000 liters). In order to load the tank truck or transfer fuel into the aircraft, fueling personnel must hold open a "dead man device" or hold tension on a "dead man rope" that holds open a spring-loaded valve (Figure 2.22). The term *dead man* is used to denote the release mechanism of the valve during an emergency or the incapacitation of fueling personnel. This release shuts down the fueling operation. Fuel trucks are also required to have emergency shutoff switches at both ends of the vehicle. The third method is referred to as a fueling island. This operation is similar to an automobile gas station, where small aircraft can taxi up and get fuel.

During aircraft fueling operations, metal cables are used to equalize static electrical charges between the fueling operation (such as a vehicle or loading rack) and the aircraft. Grounding to a static ground electrode in the pavement is not required by NFPA standards but may still be requested by the carrier or required by other standards or the airport or military regulations.

There are two general methods of loading fuel onto the aircraft. Larger aircraft receive fuel through single-point fueling connections located under the wing (Figure 2.23) or in the side of the fuselage. As the name implies, all onboard fuel tanks can be filled from this single location. These aircraft may

Figure 2.22 The "dead man device" holds open a spring-loaded valve which serves as an emergency shutoff. *Courtesy of Michael T. Defina, Jr., Metro Washington Airports Authority Fire Department.*

Figure 2.23 With single-point refueling, all onboard fuel tanks are filled from one location. *Courtesy of Michael T. Defina, Jr., Metro Washington Airports Authority Fire Department.*

also have over-the-wing fueling locations that fill directly into the individual tanks. Once fueled, personnel can transfer fuel throughout the aircraft by using transfer pumps, which are built into the aircraft. Smaller aircraft predominantly use the over-the-wing method of fueling which is accomplished by using a handheld fuel nozzle.

Because of the demand for "on-time" performance and the necessity for all-weather flying, servicing crews must perform their duties quickly and at all hours of the day and night. This increases the risk of fuel handlers cutting corners on safety procedures. Some examples of poor fueling practices include circumventing safety shut-off devices, operating poorly maintained vehicles and equipment, and overfilling aircraft or truck tanks.

In addition to poor safety practices, fuel vapor poses another hazard during fueling operations. When fuel is transferred into an aircraft tank, the incoming fuel forces vapors out through tank vents that are usually found at the wingtips. So, an explosive vapor-air mixture can be formed in the vicinity of any fueling operation. As the ambient temperature increases, so does the amount of vapor generated by the fuel. Fuel vapors are not only invisible but also heavier than air and may be moved by the wind or may settle onto the ground and into depressions.

Ignition Sources

In any aircraft area, there are numerous ignition sources that may ignite fuel vapors. These sources include static electricity (such as that caused by low-conductivity liquids, refueling vehicles, and clothing), adverse weather conditions (lightning), electromagnetic energy (radar), portable communication equipment, and open flames.

Static electricity. Controlling static electricity is extremely important during fueling operations. Aircraft, similar to any vehicle with rubber tires, has the ability to build up a static charge when moving or at rest. Static charges also build up when air flows over aircraft surfaces. The generation of static charges is greater when the humidity is low (dry air) or when the air carries particles such as dust, dry snow, or ice crystals. Certain aircraft service operations such as fueling and fuel filtering also produce static charges. The degree of static buildup depends on the fuel type; the amount of fuel; the velocity of the fuel moving through piping, hoses, and filters; and the presence of impurities in the fuel. Bonding equalizes the electrostatic potential between the fueling vehicle and the aircraft or loading facility. Bonding should only be done with properly maintained equipment connected to unpainted metallic surfaces.

Certain fabrics are notorious for accumulating a static charge. Fuel handlers should avoid wearing materials made of polyester, nylon, silk, or wool.

Electromagnetic energy. Transferring fuel is hazardous in close proximity of electromagnetic energy created by operating radar sources. Portable and mobile radio equipment and cellular telephones should not be used around aircraft when fueling operations are being conducted.

Open flames. Open flames should be strictly controlled or prohibited in aircraft operation areas or within 50 feet (15 m) of any aircraft fueling operation. The most commonly encountered open-flame hazard is smoking near aircraft or aircraft fueling operations. Other hazards include welding or other hot-work maintenance operation.

For more information about fueling operations, see NFPA 407, *Standard for Aircraft Fuel Servicing.*

General Considerations

In addition, aircraft batteries, battery chargers, or other electrical equipment should not be connected, disconnected, or operated during fuel servicing. De-fueling operations can be just as hazardous as fueling. Radios and electronic flash equipment should not be operated within 10 feet (3 m) of fueling equipment or of the fueling points or vents of the aircraft. Ground power units (GPU) should be located as far as practical from aircraft fueling points and tank vents in order to reduce the danger of igniting any flammable vapors released during fueling operations (Figure 2.24).

Fire Extinguishers Required for Fueling Operations

Fire extinguishers of appropriate size and type (with a minimum rating of 20-B) should be readily accessible within the area of fueling operations.

Figure 2.24 A GPU attached to an aircraft. *Courtesy of William D. Stewart.*

They must also be installed at emergency remote-control stations of airport fixed fuel systems. If extinguishers are not permanently located in the area but are brought to the servicing area prior to the refueling operation, they should be stationed upwind within 100 feet (30 m) of the aircraft being serviced. Multipurpose dry chemical (monoammonium phosphate) fire extinguishers are no longer recommended for use due to their corrosive effects on aircraft materials.

For normal ramp area protection, extinguishers may be located approximately midway between gate positions. In these situations, the distance between extinguishers should be less than 100 feet (30 m), and the extinguishers should have at least a 20-B rating.

Additional information concerning portable extinguishers for aircraft servicing ramps and aprons may be found in local publications or NFPA 10, *Standard for Portable Fire Extinguishers.*

Airport Drainage Systems

The airport's drainage system is designed to control the flow of fuel that may be spilled on a ramp and to minimize resulting possible damage. The system may be equipped with drain inlets with connecting piping or open-grate trenches. The aircraft fueling ramps must slope away from terminal buildings, hangars, loading walkways, or other structures. Fuel is not permitted to go directly into the sewage system and must flow through an approved fuel/water separator. The final separator/interceptor for the entire airport drainage system should be designed to allow disposal of combustible or flammable liquids into a safely located, approved containment facility.

Not all airports have this type of system. Airport firefighters must know the type of drainage system design at their airport. This information is critical when planning, confining, and containing fuel spills.

Chapter 3

Aircraft Familiarization

Job Performance Requirements

This chapter provides information that will assist the reader in meeting the following job performance requirements from NFPA 1003, *Standard for Airport Fire Fighter Professional Qualifications*, 2000 edition. Particular portions of the job performance requirements (JPRs) that are addressed in this chapter are noted in bold text.

3-1.1.1 General Knowledge Requirements. Fundamental aircraft fire-fighting techniques, including the approach, positioning, initial attack, and selection, application, and management of the extinguishing agents; limitations of various sized hand lines; use of proximity protective personal equipment (PrPPE); fire behavior; fire-fighting techniques in oxygen-enriched atmospheres; reaction of aircraft materials to heat and flame; **critical components and hazards of civil aircraft construction and systems related to ARFF operations; special hazards associated with military aircraft systems;** a national defense area and limitations within that area; characteristics of different aircraft fuels; **hazardous areas in and around aircraft;** aircraft fueling systems (hydrant/vehicle); **aircraft egress/ingress (hatches, doors, and evacuation chutes);** hazards associated with aircraft cargo, including dangerous goods; hazardous areas, including entry control points, crash scene perimeters, and requirements for operations within the hot, warm, and cold zones; and critical stress management policies and procedures.

3-1.1.2 General Skills Requirements. Don PrPPE; operate hatches, doors, and evacuation chutes; approach, position, and initially attack an aircraft fire; select, apply, and manage extinguishing agents; **shut down aircraft systems, including engine, electrical, hydraulic, and fuel systems;** operate aircraft extinguishing systems, including cargo area extinguishing systems.

3-2.2 Communicate critical incident information regarding an incident or accident on or adjacent to an airport, given an assignment involving an incident or accident and an incident management system (IMS) protocol, so that the information provided is accurate and sufficient for the incident commander to initiate an attack plan.

(a) *Requisite Knowledge:* Incident management system protocol, the airport emergency plan, airport and **aircraft familiarization,** communications equipment and procedures.

(b) *Requisite Skills:* Operate communications systems effectively, communicate an accurate situation report, implement IMS protocol and airport emergency plan, **recognize aircraft types.**

3-3.5 Attack a fire on the interior of an aircraft while operating as a member of a team, given PrPPE, an assignment, an ARFF vehicle hand line, and appropriate agent, so that team integrity is maintained, the attack line is deployed for advancement, ladders are correctly placed when used, access is gained into the fire area, effective water application practices are used, the fire is approached, attack techniques facilitate suppression given the level of the fire, hidden fires are located and controlled, correct body posture is maintained, hazards are avoided or managed, and the fire is brought under control.

(a) *Requisite Knowledge:* **Techniques for accessing the aircraft interior according to the aircraft type,** methods for advancing hand lines from an ARFF vehicle, precautions to be followed when advancing hose lines to a fire, observable results that a fire stream has been applied, dangerous structural conditions created by fire, principles of exposure protection, potential long-term consequences of exposure to products of combustion, physical states of matter in which fuels are found, common types of accidents or injuries and their causes, and role of the backup team in fire attack situations, attack and control techniques, techniques for exposing hidden fires.

(b) *Requisite Skills:* Deploy ARFF hand line on an interior aircraft fire; gain access to aircraft interior; open, close, and adjust nozzle flow and patterns; apply agent using direct, indirect, and combination attacks; advance charged and uncharged hose lines up ladders and down interior and exterior stairways; locate and suppress interior fires.

3-3.6 Attack an engine or auxiliary power unit/emergency power unit (APU/EPU) fire on an aircraft while operating as a member of a team, given PrPPE, an assignment, ARFF vehicle hand line or turret, and appropriate agent, so that the fire is extinguished and the engine or APU/EPU is secured.

(a) *Requisite Knowledge*: **Techniques for accessing the aircraft engines and APU/EPUs,** methods for advancing hand line from an ARFF vehicle, methods for operating turrets, **methods for securing engine and APU/EPU operation**.

(b) *Requisite Skills:* Deploy and operate ARFF hand line, operate turrets, **gain access to aircraft engine and APU/EPU, secure engine and APU**.

3-3.8 Ventilate an aircraft through available doors and hatches while operating as a member of a team, given PrPPE, an assignment, tools, and mechanical ventilation devices, so that a sufficient opening is created, all ventilation barriers are removed, the heat and other products of combustion are released.

(a) *Requisite Knowledge:* **Aircraft access points,** principles, advantages, limitations, and effects of mechanical ventilation; the methods of heat transfer; the principles of thermal layering within an aircraft on fire; the techniques and safety precautions for venting aircraft.

(b) *Requisite Skills:* **Operate doors, hatches,** and forcible entry tools; operate mechanical ventilation devices.

3-4.1 Gain access into and out of an aircraft through normal entry points and emergency hatches and assist in the evacuation process while operating as a member of a team, given PrPPE and an assignment, so that passenger evacuation and rescue can be accomplished.

(a) *Requisite Knowledge:* **Aircraft familiarization, including materials used in construction, aircraft terminology, automatic explosive devices, hazardous areas in and around aircraft, aircraft egress/ingress (hatches, doors, and evacuation chutes), military aircraft systems and associated hazards;** capabilities and limitations of manual and power rescue tools and specialized high-reach devices.

(b) *Requisite Skills:* Operate power saws and cutting tools, hydraulic devices, pneumatic devices, and pulling devices; operate specialized ladders and high-reach devices.

This chapter provides information to familiarize aircraft rescue and fire fighting (ARFF) personnel with aircraft classifications and aircraft systems, along with immediate and long-term operational hazards they pose. To enhance personal safety, ARFF crew members must exercise extreme caution when working in and around aircraft. One of the most important aspects of ARFF operations is aircraft familiarization training because rescuing the occupants is the number one priority. Enhanced knowledge of the aircraft that operate in and around your airport helps ensure that rescue operations can be performed in the quickest, safest manner. In-depth knowledge of egress systems allows rescue crews to assist or perform the evacuation process, thus increasing the chance of passenger and crew survival.

Types of Aircraft

Generally, aircraft are categorized according to their intended purpose. Depending on their uses, some aircraft may be included in more than one category. Aircraft such as the DC-10 may be configured as a commercial transport (passenger aircraft), as a cargo aircraft, or as a refueling tanker by the military. Naturally, the hazards around the aircraft stay the same, but the hazards found inside the aircraft can vary dramatically. Classifications of aircraft generally include the following:

- Commercial transport
- Commuter/regional
- Cargo, including combination aircraft (combi-aircraft)
- General aviation

- Business/corporate aviation
- Military aviation
- Rotary-wing (helicopters)
- Other

Commercial Transport

Those aircraft used for commercial transport of passengers are generally of large-frame construction and can be categorized as either narrow- or wide-bodied. However, newer designs of aircraft known as *new large aircraft* may soon be seen in use.

- ***Narrow body*** (Figure 3.1). These aircraft have two or sometimes three jet engines and carry up to 13,000 gallons (52 000 L) of jet fuel. Narrow-body aircraft cabins are designed with a single aisle usually 18 to 20 inches wide and may seat up to 235 persons if arranged in a single (coach) class configuration (Figure 3.2). Most cabin doors are plug-type in design and swing out and forward. Some aircraft doors incorporate pneumatic emergency opening systems that assist in opening the door if jammed during a low-impact crash. In accordance with Federal Aviation Regulation (FAR) 121.310 any aircraft with a doorsill height of 6 feet (2 m) or more off the ground when the aircraft's wheels are extended must be equipped with an emergency escape slide. ***Some narrow-body aircraft escape slides cannot be de-armed from the outside and will automatically deploy when the doors are opened from the outside.*** Over-the-wing escape hatches are provided and may contain an escape slide, which activates when the hatch is opened from the inside. Cargo and luggage are usually bulk-loaded into two to three cargo compartments that are found along the bottom of the fuselage with access being provided on the right side of the aircraft.
- ***Wide body*** (Figure 3.3). These aircraft have two, three, or four jet engines and may carry over 58,000 gallons (220 000 L) of jet fuel. Wide-body aircraft cabins have dual aisles creating a center section of seats, allowing the aircraft to carry

Figure 3.1 Boeing 757, a single-aisle commercial aircraft. *Courtesy of William D. Stewart.*

Figure 3.3 Boeing 747, a wide-body commercial aircraft. *Courtesy of William D. Stewart.*

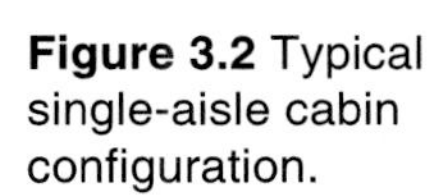

Figure 3.2 Typical single-aisle cabin configuration.

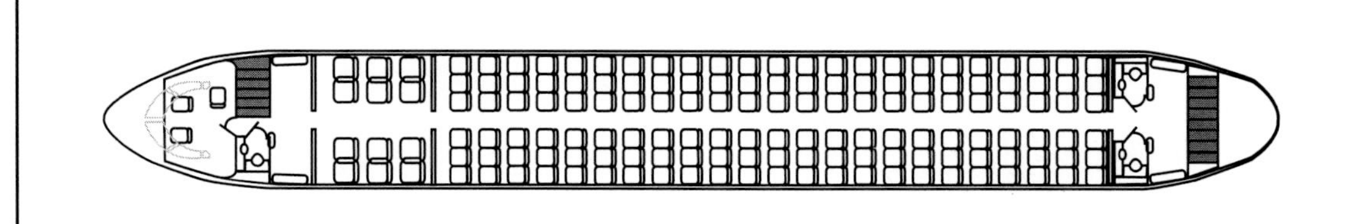

NARROW-BODY MIXED-CLASS CABIN CONFIGURATION

over 500 passengers (Figure 3.4). Doors are generally power-assisted and may contain a pneumatic or spring-tension emergency operation system. Some wide-body aircraft doors open up into the overhead area, while others swing out and forward. Almost all escape slides can be de-armed from the outside. Most slides are doublewide in design and expand beyond the outside door opening when inflated. In wide-body aircraft, over-the-wing escape doors are more common than hatches. Most of the luggage and cargo is preloaded into containers or onto pallets before being unit-loaded into lower cargo compartments. Both fire detection and fire extinguishing systems can be found in wide-body aircraft cargo compartments.

- ***New large aircraft (NLA)*** (Figure 3.5)**.** With the availability of lightweight, strong composite material components, aircraft manufacturers are developing a new breed of large aircraft. Referred to as NLA, or very large aircraft (VLA) these aircraft will dwarf today's commercial aircraft and may incorporate a passenger capacity of up to 900 passengers. Cabin configuration will incorporate a double-deck seating configuration and pose numerous rescue concerns for responding ARFF personnel. This large-capacity aircraft is made possible by the increased use of lightweight composite materials. Airports may need to be redesigned to accommodate aircraft of this size.

Commuter/Regional Aircraft

Aircraft used for the commercial transport of passengers on short routes, typically to and from hub airports to smaller airports, are referred to as *commuter/regional aircraft.* Twin engine turboprop aircraft (Figure 3.6) are commonly used for this class with current trends moving towards the use of jet-powered aircraft in this role (Figure 3.7). Most

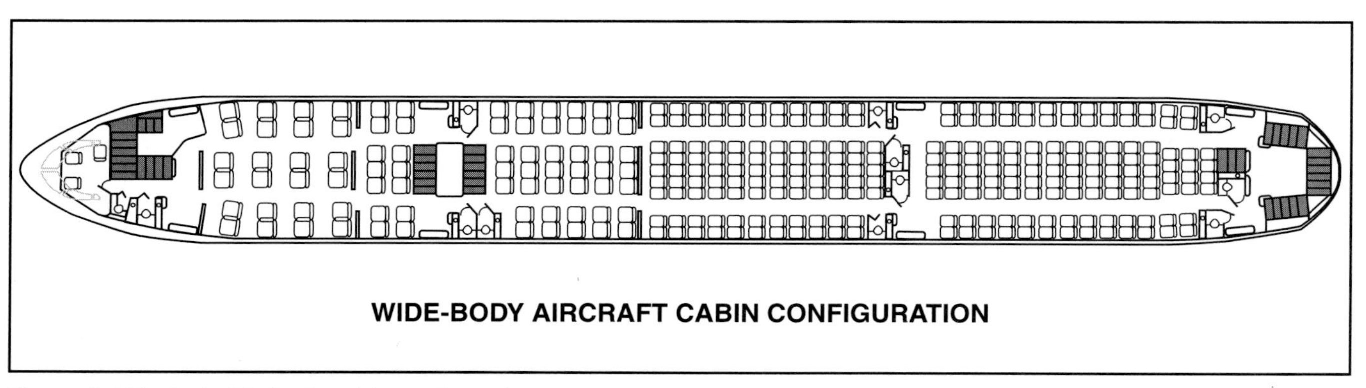

Figure 3.4 Typical wide-body cabin configuration.

Figure 3.5 New large aircraft (NLA). *Courtesy of Airbus Industrie.*

Figure 3.6 Saab 340, a commuter-style aircraft. *Courtesy of William D. Stewart.*

are pressurized and can carry from 19 to 100 passengers. The interiors can be cramped and congested and can present a difficult work environment under emergency conditions (Figure 3.8). They tend to have a limited number of egress locations and often have only one door. Sometimes the passenger cabin may be accessed through the rear cargo area. There is usually no flight attendant on aircraft with less than 30 passengers.

Cargo Aircraft

These aircraft are used primarily for the transport of cargo and can include any of the previously discussed aircraft. They are commonly referred to as *freighters* (Figure 3.9). Many freighters are former passenger aircraft modified to carry cargo pallets or containers and can contain significant amounts of dangerous goods. *Combi-aircraft* (combination aircraft) are those aircraft carrying passengers and cargo on the main deck and additional cargo below the deck. Some cargo aircraft are used as freighters during the week and converted to passenger aircraft for weekend excursions.

ARFF personnel must be aware that except for the two forward entry doors, all other doors and exit hatches may be disabled or blocked off as part of the modifications to an all-cargo configuration. Large-frame aircraft have large, hydraulically operated cargo doors located forward or aft of the wing on the left side of the aircraft. Although most have cargo doors that can be operated manually in an emergency, electrical power is needed to open these doors under normal conditions. Containers and pallets are loaded sequentially into numbered or lettered positions from the front of the aircraft back. On narrow-body cargo freighters, the lower compartments are usually bulk-loaded with packages no heavier than 70 pounds each. Once the aircraft is loaded, it is often impossible for personnel to move through the cargo compartment.

General Aviation

A wide variety of aircraft include those used primarily for pleasure or training. Most of these aircraft are small, light, and non-pressurized (Figure 3.10). They are typically powered by single or twin

Figure 3.7 Canadair Regional Jet, a regional aircraft. *Courtesy of William D. Stewart.*

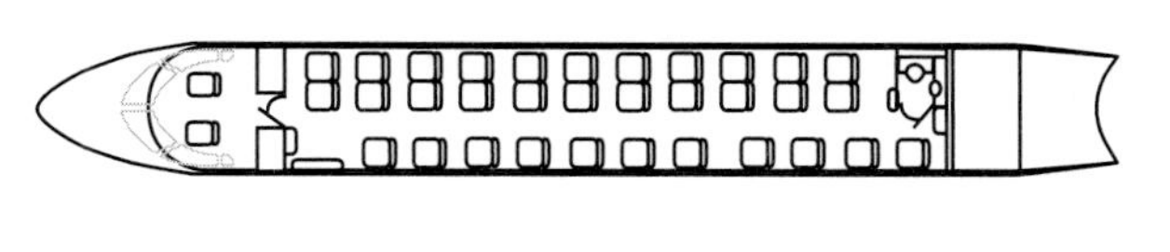

Figure 3.8 Typical commuter/regional aircraft cabin configuration.

Figure 3.9 A typical wide-body cargo aircraft. *Courtesy of William D. Stewart.*

Figure 3.10 A small general aviation aircraft. *Courtesy of William D. Stewart.*

internal-combustion engines and present a fire fighting and rescue problem similar to a highway vehicle accident. General aviation aircraft usually carry one to ten passengers and up to 90 gallons (360 L) of aviation gasoline (AVGAS). Some general aviation aircraft may be larger and carry up to 500 gallons [2 000 L] of fuel. According to National Transportation Safety Board (NTSB) statistics, a majority of aviation accidents involve this type of aircraft and account for most of the aircraft-related fatalities.

Business/Corporate Aviation

Aircraft primarily used for business-related transport can range from smaller, light, non-pressurized aircraft to large "commercial-type" jets (such as a Boeing 737) and include many different models and manufacturers. Typically, they are pressurized aircraft that generally accommodate six to nineteen passengers (Figure 3.11). They are often powered by twin jet engines that operate on jet fuel (Figure 3.12). Many have custom-designed interiors that differ greatly from normal configurations. This type combined with general aviation aircraft account for the largest variety of aircraft styles and configurations.

Figure 3.11 A business/corporate aircraft. *Courtesy of Jeff Riechmann, Riechmann Safety Services.*

Figure 3.12 A twin-engine business/corporate aircraft. *Courtesy of William D. Stewart.*

Military Aviation

Aircraft used by all branches of the armed services make up the military aircraft classification. Many civilian models have been converted to function as military aircraft in a variety of missions. Each type of aircraft is identified by a letter prefix that indicates its function. For example, the A-10 (Figure 3.13) is an attack aircraft, and the F-16 (Figure 3.14) is a fighter aircraft. Some of the common types are discussed in the "Military Aircraft" section later in the chapter.

Aircraft types range from single-engine fighters to large, multiengine transports and bombers. Because of the high altitude, high speed, complex instrumentation, and armament required by the military, this type of aircraft presents additional hazards for emergency responders. Although the crew is often limited to a few persons, the aircraft may have armament, liquid oxygen, high-powered radar, extensive composite material construction, and explosive ejection devices.

Figure 3.13 The A-10 aircraft.

Figure 3.14 The F-16 aircraft.

Fire Fighting Aircraft

In addition to medevac and high-angle rescue roles, aircraft can be used in a variety of roles in the support of fire fighting operations. These roles include the use of fixed-wing aircraft like the Canadair® 415 (Figure 3.15) for transporting smokejumpers relatively short distances. Fixed-wing air tankers can carry 800 to 3,000 gallons (3 200 L to 12 000 L) of fire fighting agent that can be dropped on a fire.

Rotary-wing aircraft can carry 100 to 1,000 gallons (400 L to 4 000 L) of agent in slung buckets suspended from the aircraft or in tanks mounted on the underside of the aircraft and can carry up to 3,000 gallons (12 000 L) of agent. Rotary-wing aircraft can also be used to transport firefighters and cargo, to serve as infrared imaging platforms, and as a tool for conducting backfiring operations. The firefighter needs to be aware that the helicopter supporting backfiring operations carries flammable "ping pong balls" in the cargo area or has a torch carrying jellied gas slung under the helicopter.

Various light aircraft, like the Aero Commander®, are used by the Air Tactical Group Supervisor to coordinate all aircraft operations over a fire.

Rotary Wing (Helicopters)

Rotary-wing aircraft or helicopters can range in size from small, single-seat models to large transports capable of carrying up to 50 passengers. Some helicopters like the Sikorsky® Skycrane® (Figure 3.16) may be equipped to carry loads weighing more than 10 tons (9 100 kg). Because most helicopters are not as rigidly constructed as "fixed-wing" aircraft, if involved in an accident, they tend to collapse, trapping occupants.

Figure 3.15 A Canadair 415. *Courtesy of Roger Ward.*

Helicopters may have piston engines or gas turbines with fuel capacities ranging from 70 to 1,000 gallons (280 L to 4 000 L). The internal fuel tanks are usually located under the cargo floor and may have rubber bladders, while auxiliary fuel tanks may be either located inside the main cabin in the aft section or attached to the outside of the aircraft. The main rotor(s) serves the same purpose as the wings and propeller on a fixed-wing aircraft — that is, to provide lift and directional motion (Figure 3.17). The helicopter tail rotor, if the helicopter is so equipped, provides directional control. Helicopters are constructed of materials similar to those used for fixed-wing aircraft such as aluminum, titanium, magnesium, and a variety of composite materials.

Other Types of Aircraft

Airports are home to many different types of aircraft or aviation-type activities not included in the

Figure 3.16 A Sikorsky Sky Crane. *Courtesy of William D. Stewart.*

Figure 3.17 A business/corporate helicopter. *Courtesy of William D. Stewart.*

categories previously listed above. It is extremely important for ARFF personnel to be familiar with the aircraft that operate in and around their airport. This helps to ensure a safer working environment when rescue operations become necessary. The following list includes some types of aircraft that might be part of any airport:

- Vintage/antique aircraft
- Lighter-than-air crafts (blimps, hot-air balloons) (Figure 3.18)
- Tilt-rotor aircraft
- Ultralight aircraft
- Experimental/amateur aircraft
- Agricultural spraying (crop-dusting) aircraft (Figure 3.19)
- Skydiver transport aircraft
- Aerobatics aircraft
- Medical evacuation/transport aircraft

Figure 3.18 The Goodyear blimp.

Figure 3.19 A typical crop duster.

Major Components of Aircraft

For airport firefighters to understand aircraft and the possible emergencies they may encounter, they must be familiar with the terminology of the major components involved in aircraft construction. This information will assist the firefighter in sizing up the situation when performing aircraft rescue and fire fighting operations. The following sections discuss the major components of both fixed-wing and rotary-wing aircraft.

Fixed-Wing Aircraft Components

Features and components of a fixed-wing aircraft include the fuselage, wings, and tail section (Figure 3.20).

- ***Fuselage.*** The main body of an aircraft to which the wings and tail are attached is referred to as the *fuselage.* Approximately 85 percent of the fuselage is constructed of aluminum. Depending on the type of aircraft, the aircraft skin varies in thickness as it forms and covers the various sections along the structure. The fuselage houses the crew, passengers, cargo, and additional fuel storage. Most of the aircraft systems are found in areas of the fuselage.
- ***Wings.*** The wings are designed to develop the major portion of the lift required for flight. They, too, are generally constructed of aluminum and carry a majority of fuel. Some military aircraft may have weapons and additional fuel tanks attached to the wings.
- ***Engines.*** The engines produce thrust that propels the aircraft. They can be either internal-combustion reciprocating or gas turbine. Turbine engines vary in size and thrust-producing capability depending on the type and use of the aircraft.
- ***Landing gear.*** The landing gear provides a mechanism for supporting the aircraft while on the ground and is commonly either tricycle or conventional design. Tricycle gear consists of a single strut under the nose and two main struts extending from under the wings or out of the fuselage. Conventional gear consists of a tail wheel and two main struts under each wing. The

nose gear or tail wheel is used for steering while the main gear is equipped with brake systems.

- ***Tail (empennage).*** The aircraft tail section includes the vertical and horizontal stabilizers, rudders, and elevators. Generally, the tail section houses the auxiliary power unit (APU), which provides electrical power to operate the essential systems while the aircraft is parked at the gate. Some aircraft are equipped with rear stairs or a tail-cone jettison system that is designed to provide additional means of egress.

Other components of the aircraft with which the airport firefighter must be familiar include:

- ***Cockpit.*** The cockpit, also referred to as the *flight deck* on airliners, is the fuselage compartment occupied by the flight crew. The cockpit in certain military (fighter, attack, bomber, and training) aircraft may be equipped with ejection seats. The canopy is a transparent enclosure over the cockpit of various types of aircraft. It is usually constructed of special plastics for durability during flight.
- ***Nacelle.*** The nacelle is the housing around an externally mounted aircraft engine. It can be constructed of aluminum or composite materials. In the event of an engine fire, fuel often pools in the bottom of the nacelle. This pooling can create a hazardous situation if the nacelle is opened during the extinguishment phase of fire fighting operations.
- ***Flight control surface.*** This is the general term used for devices that enable the pilot to control the direction of flight, altitude, and attitude of the aircraft. These control surfaces include: ailerons, elevator, rudder, flaps and slats, spoilers, and speed brakes.
- ***Ailerons.*** The ailerons are attached to the trailing edge of the wings. They are the movable, hinged, rear portion of the aircraft wing that controls the rolling (banking) motion of the aircraft.
- ***Elevator.*** The elevator is the hinged, movable control surface found along the rear of the horizontal stabilizer. It is attached to the control wheel or stick and is used to control the up-and-down pitch motion of the aircraft.
- ***Rudder.*** The rudder is the hinged, movable control surface attached to the rear part of the vertical

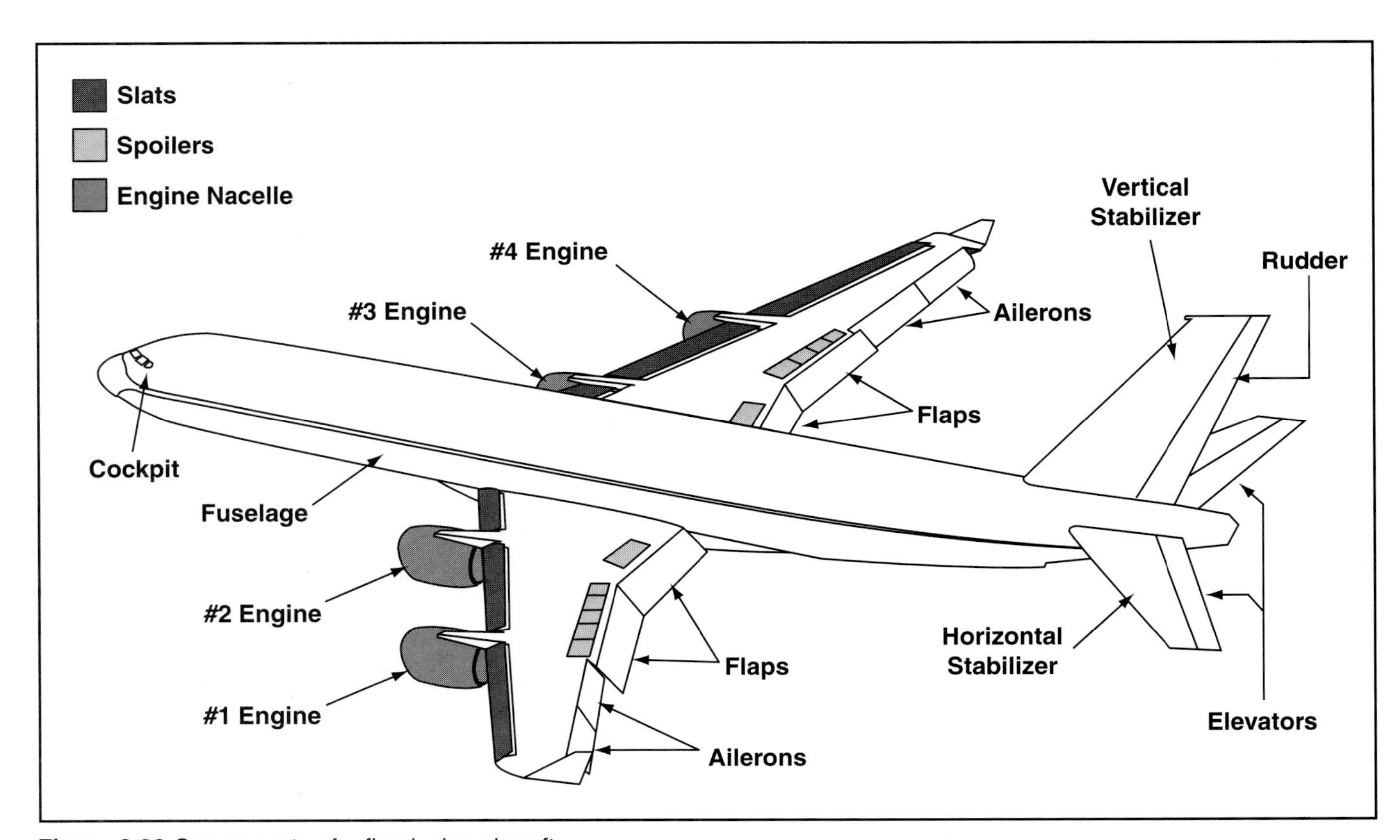

Figure 3.20 Components of a fixed-wing aircraft.

stabilizer and is used to control the yaw or turning motion of the aircraft.

- ***Flaps and slats.*** Flaps are airfoils that extend from the leading edge and/or trailing edge of a wing. Slats are airfoils that extend only from the leading edge of a wing. These devices are used to improve the aerodynamic performance of the aircraft during takeoff and landing.
- ***Spoilers* and *speed brakes.*** Spoilers are movable panels located on the upper surface of a wing and raise up into the airflow to increase drag and decrease lift. Speed brakes are aerodynamic devices located on the wing or along the rear or underside of the fuselage that can be extended to help slow the aircraft.

Rotary-Wing Aircraft Components

The main sections of a rotary-wing aircraft include the following:

- ***Fuselage.*** The fuselage is similar to that of fixed-wing aircraft; it houses the same components as the fixed-wing aircraft does.
- ***Main rotor*(s).** The main rotor(s) provide lift and propulsion for the helicopter to fly. Some helicopters are designed with two main rotors, and others have one main rotor and one tail rotor. Depending on the helicopter type and use, the main rotor may consist of two to seven rotor blades.
- ***Tail rotor.*** The tail rotor provides the helicopter with directional control. It counteracts the torque produced by the main rotor. Some newly designed helicopters can operate by using engine exhaust to provide control of the aircraft, thus eliminating the need for a tail rotor.
- ***Landing gear.*** The landing gear on helicopters is used to support the aircraft when it is not in flight. The two types of landing gear assemblies are conventional and skid support.
  - —*Conventional gear* consists of main gear and either a nose or tail gear. This gear may or may not retract depending on the type of helicopter. Retractable landing gear is housed in pontoons that provide flotation support for helicopters that land in water.
  - —*Skids* are used on smaller helicopters in the place of conventional landing gear. The skids are permanently mounted to the exterior and resemble platforms. Because they are without wheels, helicopters with skids often "hover taxi" to move along taxiways or the parking ramp.

Engine Types and Applications

The two different types of engines used to power aircraft are the internal-combustion reciprocating engine and the gas turbine, or jet engine. This section examines these types of engines and the hazards they may pose to responding firefighters.

Internal-Combustion Reciprocating Engines

In early aviation history, all aircraft were powered by the internal-combustion reciprocating engine. The cylinders may be configured around a central crankshaft (radial engine) (Figure 3.21) or in a horizontally-opposed arrangement much like the engine found in an automobile. Power from the engine is transmitted through the crankshaft to the propeller. Reciprocating engines use aviation gasoline (AVGAS) as their fuel. Unlike most automotive engines, they are air-cooled to eliminate the weight of the heavy engine blocks typical of liquid-cooled engines. These aircraft engines use relatively large amounts of oil and often carry a large oil tank adjacent to the engine. An accessory section drives the

Figure 3.21 Radial-style internal-combustion engine. *Courtesy of Jeff Riechmann, Riechmann Safety Services.*

pumps for the fuel, oil, and hydraulic systems and the generators for the electrical system.

Most aircraft with this type of engine are used primarily for general aviation. The fuselage is usually made entirely of lightweight metal or of a metal frame with a fabric covering. An aircraft powered by reciprocating engines can be rated up to 400 horsepower, weigh up to 3,500 pounds (1 588 kg), and may carry from one to six passengers (Figure 3.22).

Also powered by this type of engine are larger aircraft, including twin-engine and four-engine types that are used for general, commercial, and military aviation. The total number of passengers carried is often limited, but the aircraft may be configured to carry as many as 90 passengers.

Spinning propellers and hot engine parts produce hazards from these engines. A magneto is another hazard that poses an immediate risk during extrication. At least two magnetos are found on all internal combustion engines and are designed to produce sparks that keep the engine operating. If while performing extrication, rescuers were to bump or rotate the propeller, unspent fuel remaining in the engine cylinders could be ignited by the magneto causing the engine to restart and the propeller to rotate (Figure 3.23).

WARNING

Disconnecting the battery does not prevent the magneto from functioning, so personnel must exercise caution when working in the area of the propeller. A safety zone should be established around the engine, keeping all personnel clear of the engine.

Figure 3.22 A twin-engine general aviation aircraft. *Courtesy of William D. Stewart.*

Gas Turbine Engines

There are four main types of gas turbine engines: turbojet, turbofan, turboprop (Figure 3.24), and turboshaft. In all types of jet engines, air is drawn in through the front, compressed, mixed with fuel and ignited, and then exhausted out the back. Engine power is generated by the rapid expansion of the fuel/air mixture when ignited and is used for one of two purposes:

- To drive the aircraft by expelling high-speed exhaust gases
- To drive a propeller or rotor

Gas turbine engines use jet fuel to operate and can be damaged if AVGAS is mistakenly added.

The four major components of all gas turbine engines are the compressor section, combustion section, turbine and exhaust section, and accessory section. Air is drawn into the compressor section at the front of the engine where it is compressed and accelerated by rotating blades. Some air is bled from the compressor section and used to pressurize and air-condition the cabin and, when necessary, hot air can be ducted to the leading edge of the wing and engines to keep them from accumulating ice.

The compressed air enters the combustion section, which is divided into a number of chambers, where it is mixed with atomized fuel and ignited. This action results in the expansion of heated gases and the production of high-pressure, high-velocity exhaust gas. At this point, the superheated gas is directed through turbine blades at the rear of the

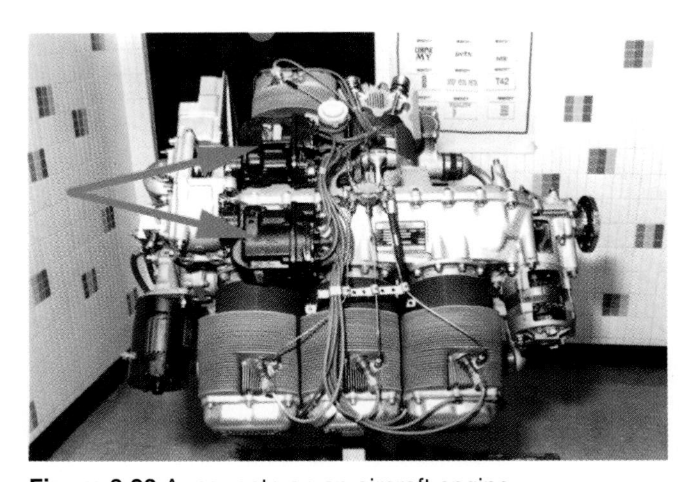

Figure 3.23 A magneto on an aircraft engine.

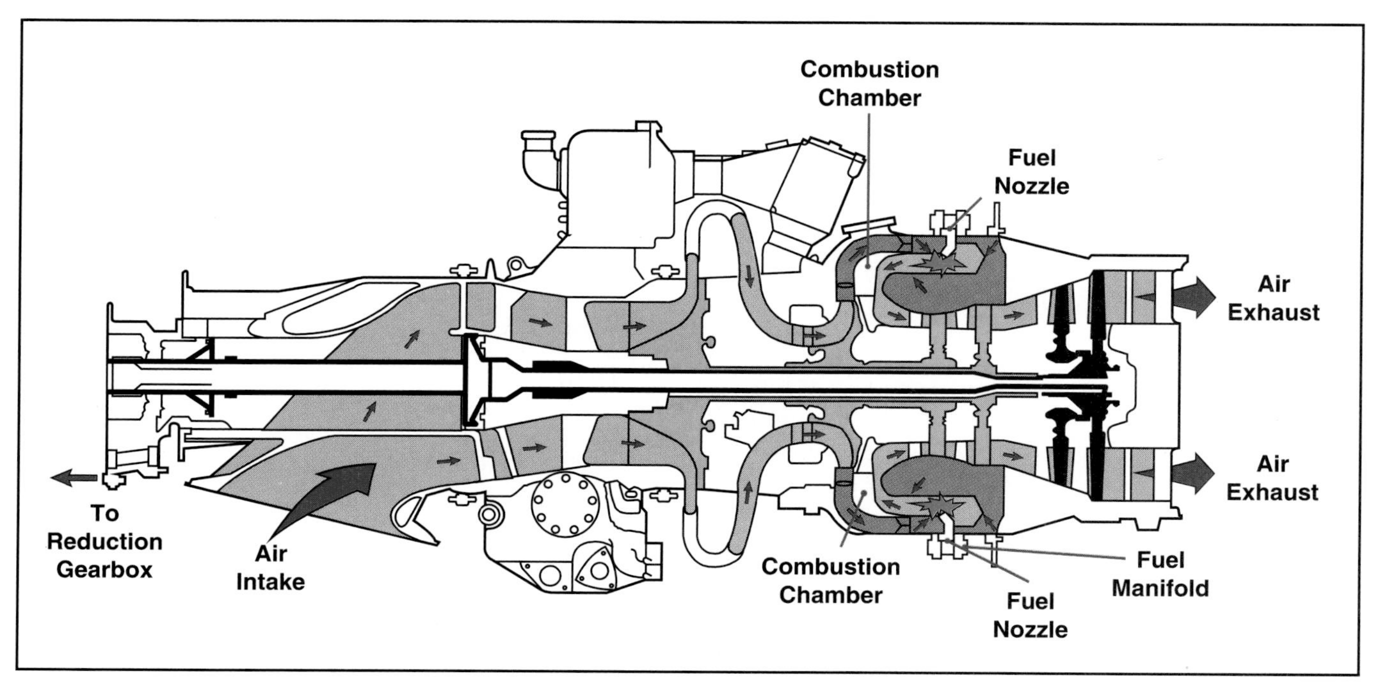

Figure 3.24 The internal workings of a turboprop engine.

engine. The turbines are attached to a common shaft with the compressor blades. In this arrangement, the high-speed gases cause the turbines to rotate which in turn drives the compressor section. Components of other aircraft systems that support the engine or are powered by it are contained in the accessory section. These accessories include the fuel control unit and fuel pump, hydraulic pump, oil pump and cooler, and electrical generator.

Turbojet

The turbojet engine is the simplest of the gas turbine engines (Figure 3.25). It functions as detailed in the previous paragraph.

Figure 3.25 An aircraft equipped with turbojet engines. *Courtesy of William D. Stewart.*

Turbofan

The turbofan engine is the gas turbine engine most commonly found on aircraft today, especially on large jetliners. It contains an additional component that the turbojet does not have — a large fan at the front of the engine (Figure 3.26). This fan helps increase the engine's thrust by increasing the total airflow of the engine. Latest technological advances have developed an engine capable of delivering over 100,000 pounds of thrust.

Figure 3.26 The latest style of a very large turbofan engine. *Courtesy of William D. Stewart.*

Turboprop

The turboprop engine is widely used for small- and medium-sized commuter and cargo aircraft. Instead of the fan previously discussed, the turboprop consists of a propeller that is driven by a small turbojet engine. Turboprop engines are easily distinguished from piston engines by the turboprop's streamlined engine nacelle and a single or dual exhaust port that is much larger in diameter than those on piston engines (Figure 3.27). Turboprop engines are used on a variety of aircraft having one, two, or four engines.

Figure 3.27 A typical twin-engine turboprop aircraft. *Courtesy of William D. Stewart.*

Turboshaft

The most common application for turboshaft engines can be found in helicopters. The turboshaft engine is basically the same as the turboprop; however the output shaft is not connected to a propeller. Instead, the power turbine is connected, either directly or through a gearbox, to a shaft that drives the helicopter's main and tail rotors.

Engine Additions and Variations

Additional components may be added to the basic gas turbine engine to redirect engine exhaust gas streams, to increase engine thrust, and to slow aircraft speed when landing. These include:

- Exhaust nozzles that rotate to redirect the exhaust gas stream downward to enable vertical takeoff and landing. The Harrier attack jet uses this type of engine exhaust system (Figure 3.28). A unique variation of this principle is found incorporated in the design of tilt-rotor aircraft. These have turboprop engines driving very-large-diameter propellers. The entire engine nacelle pivots from vertical, for helicopter-like takeoff and landing, to horizontal, for aircraft-like high-speed flight.
- An afterburner (augmentor) to provide additional thrust for short periods, improving takeoff and climb capability, and enhancing the performance of military fighter aircraft. This is accomplished by injecting and burning raw fuel in the superheated exhaust stream behind the turbine section.
- Thrust-reversal system incorporated into the exhaust section. Thrust reversers consist of internal or external doors and vanes that operate to deflect jet exhaust forward to assist in slowing the aircraft during its landing rollout. These devices can be hydraulically powered, or on some aircraft, pneumatically actuated.

Figure 3.28 Variable pitch exhaust nozzles allow this aircraft to hover.

Aircraft Hazards

ARFF personnel must possess a thorough knowledge of the aircraft types that use their airfield. Familiarization training must include the dangers these aircraft and their systems pose during normal and emergency operations. Injuries that firefighters receive while conducting on-scene operations have a great impact upon occupant safety and evacuation. Some of the hazards ARFF personnel may face include:

- Pinching and limb-severing hazards
- Propeller dangers
- Helicopter hazards
- Jet-engine dangers

Pinching Hazards

As stated previously, the hydraulic system on an aircraft is tasked with a wide variety of functions to safely operate an aircraft. Flight controls, braking systems, thrust reversing systems, landing gear bay doors, cargo door operation, and other necessary systems rely heavily on hydraulic pressure. When the surface systems and doors operate, personnel must exercise extreme caution.

> **WARNING**
>
> **Pressures in excess of 3,000 psi (21,000 kPa) generate enough energy to sever fingers, hands, and arms. Guard against leaning against moving parts; stay clear of ALL moving parts when working around an aircraft.**

Propeller Dangers

It is very difficult to see a propeller when it is rotating at high speed. Personnel must be especially careful when approaching a rotating propeller from the front as the aircraft could suddenly move forward without warning. When approaching a propeller, personnel should remain at least 15 feet (5 m) from it (Figure 3.29).

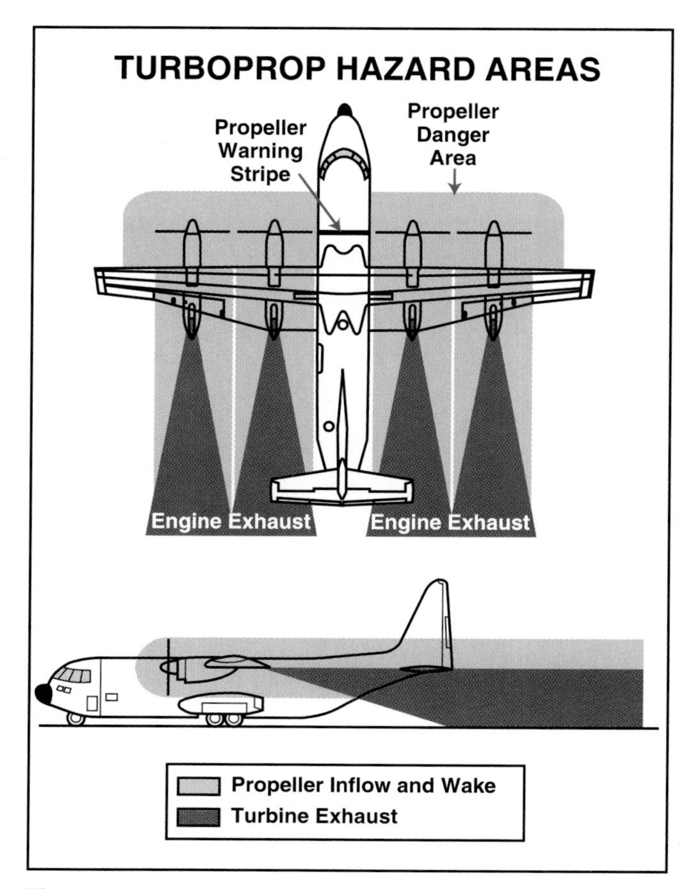

Figure 3.29 Personnel should be aware of danger areas when approaching aircraft.

> **WARNING**
>
> **Even if a propeller has stopped, do not move it under any conditions. Piston engines that have recently stopped can sometimes cycle, violently rotate, or restart if the propeller is moved.**

Helicopter Hazards

The helicopter must be approached with caution. The rotors present the greatest hazard and should be avoided at all times. In gusty wind conditions, the main rotor may dip to within 4 feet (1.3 m) of the ground. Because the pilot is most familiar with rotor behavior under various conditions, he or she should decide when it is safe for personnel to approach the helicopter. Therefore, before personnel attempt to approach a helicopter, they should wait until the pilot has them in sight and signals when it is safe to approach the aircraft (Figure 3.30). The tail rotor rotates at very high speed and is also difficult to see, so personnel should ***never*** approach a helicopter from the rear. Personnel should approach and leave the helicopter in a crouched position and *always* within view of the pilot. On uneven ground, personnel should always approach and leave on the downhill side — ***never*** on the uphill side.

When approaching a helicopter, personnel must carry all equipment such as shovels, axes, or tools horizontally and below waist level – *never* upright or over the shoulder. Any loose articles of clothing must be properly secured before approaching or leaving the helicopter. Personnel should make sure that any gear or cargo is secure and should *never* throw anything in the vicinity of a helicopter.

When landing, helicopters must have sufficient clearance of all ground cover within 100 feet (33 m) of the site selected for the landing zone. Clear access into the landing zone must avoid elevated high-voltage lines and cables, towers, and structures along with clearance over trees. The operating areas should be kept clear of all personnel, cargo,

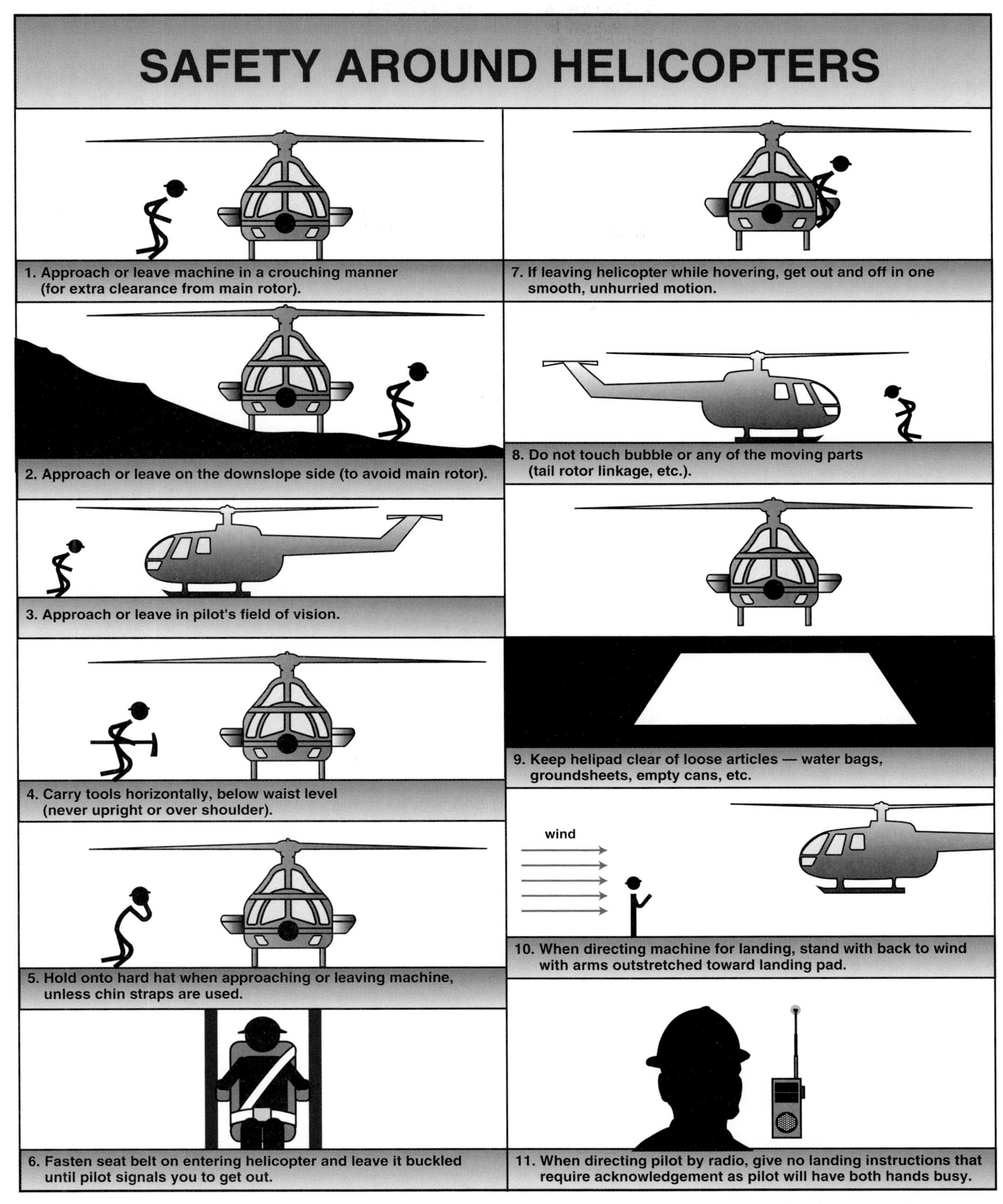

Figure 3.30 To avoid personal injury when working around helicopters, personnel must observe all safety precautions.

personal belongings, or other loose articles that may be blown around by the rotor downdraft while the helicopter is approaching or leaving.

Jet-Engine Dangers

Some extreme hazards exist for those working in or around jet engines. All aircraft engines are noisy, but noise from jets is excessive, and hearing protection is needed. Jet-engine exhaust or blast is superheated and may approach velocities well over 800 miles per hour (1,287 km/h). The exhaust from a jet engine may blow loose objects considerable distances, so personnel should avoid exhaust areas when jet engines are operating. The same awareness to operating jet engines must be practiced when operating an emergency response vehicle. Jet blast can easily upset any vehicle that is driven too close to the rear of an operating jet engine (Figure 3.31).

The suction generated by running jet engines is another severe hazard. To ensure a safe distance, personnel should not approach the front of the engine and should stay at least 30 feet (10 m) away from the front and sides of the engine. It is important to communicate to the pilot, with hand signals or radio, prior to inspecting any system on or under an operating aircraft. When a number of jet engines are operating in a given area, it is often difficult for ground personnel to tell which engines are operating and which ones are not, especially if the personnel are wearing hearing protection. Therefore, they should assume that all engines are operating and be aware of the hazard areas (Figure 3.32). The suction of a jet engine also poses a hazard

Figure 3.31 The exhaust from a jet engine is strong enough to blow away a vehicle.

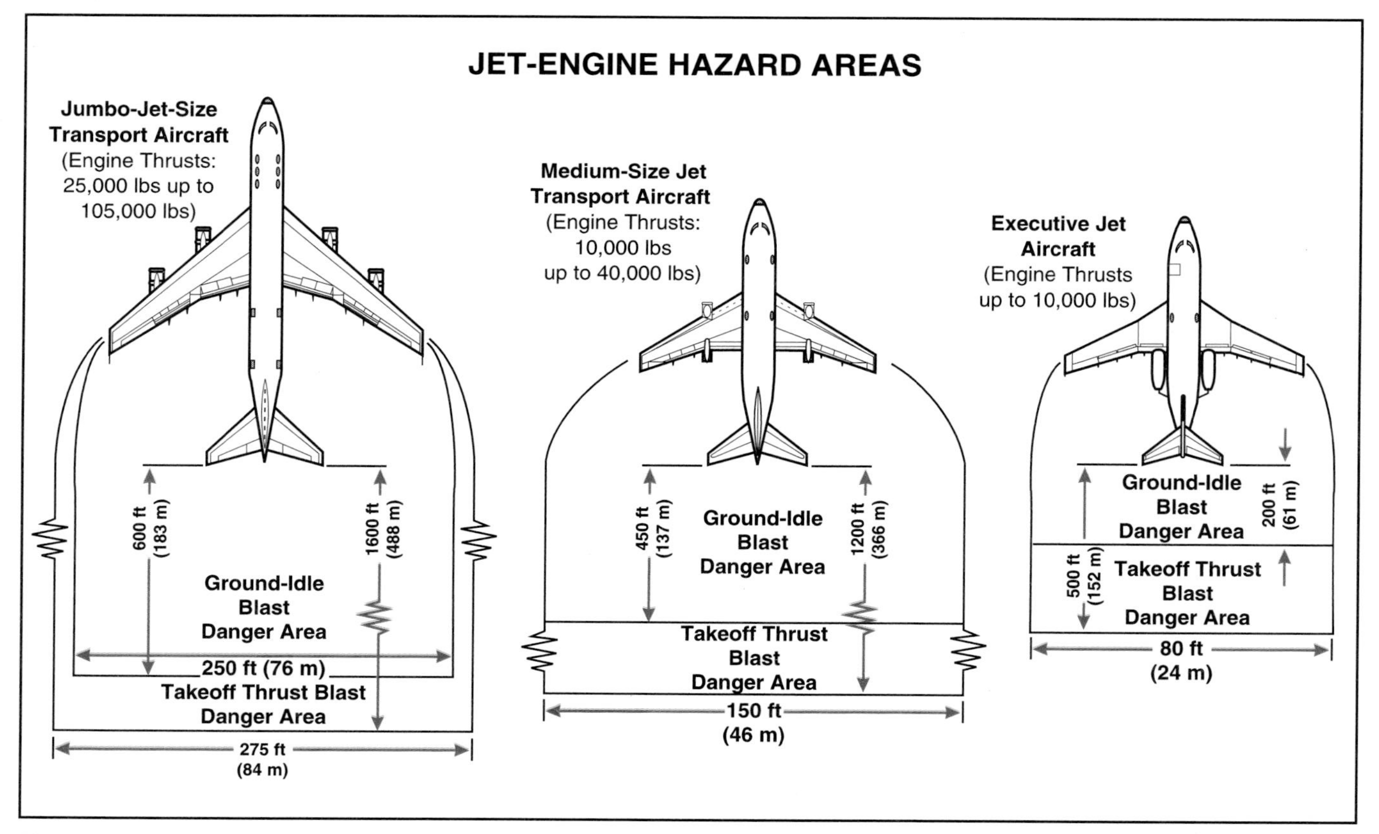

Figure 3.32 Both intake and exhaust hazards are present around jet engines.

to the engine itself. Airport firefighters always should be on the lookout for objects that could be drawn into jet engines. Called FOD or foreign object debris, these materials can cause significant damage to engines.

CAUTION: After an accident, a jet engine may continue to run if the fuel is not shut off. Even after shutdown, jet engines retain sufficient heat to ignite spilled flammable materials for up to 20 minutes. Also, the rotation of the engine may draw in vapors from spilled fuel and ignite them. When possible, cordon off the area around the engine, and establish a safety zone.

Ballistic Recovery Systems

A growing trend in lightweight sport aircraft is the use of ballistic recovery parachutes (Figure 3.33). These systems use an ejection device to quickly deploy a parachute during catastrophic emergencies. These systems present an extreme hazard to ARFF personnel.

WARNING

Ballistic recovery parachutes can be fired vertically or horizontally and are not always readily visible on the aircraft or location from where they are fired.

Figure 3.33 A ballistic recovery system relies on a small ballistic charge to deploy a recovery parachute. *Courtesy of Ballistic Recovery Systems, Inc.*

Other Aircraft Components

Aircraft Lighting

When responding to an aircraft emergency at night, aircraft lighting may be the only means by which personnel can designate their location and setup relating to the aircraft. A red light can be found at the left wingtip while a green light can be found at the right wingtip. A white light(s) is found at the tail section on the end of the fuselage. Lights designed to illuminate the logo found on the sides of the vertical stabilizer are referred to as "logo lights." Landing lights consist of high-intensity spotlights that can be found on the wings and landing gear. Rotating or flashing red anti-collision lights are also used to indicate that aircraft engines are operating. They can be found on the top of the vertical stabilizer or on the top and underside of the fuselage.

Cargo Compartments

Classes of Cargo Compartments

ARFF firefighters should know the various classes of cargo compartments on aircraft at their airports. Cargo compartments are different from stowage areas on aircraft. This section discusses cargo areas other than those considered stowage compartments by Federal Aviation Administration (FAA) requirements. Stowage compartments such as overhead storage areas used for the storage of carry-on articles and baggage are not considered as cargo compartments.

Each class of cargo compartment is usually larger than the preceding class with Class A being the smallest cargo compartment and Class E comprising the entire main deck of a cargo aircraft. Following are the classes of compartments as defined by FAA requirements:

Class A. A compartment in which the presence of a fire would be easily discovered by a crewmember while at his or her station, and where all compartments are easily accessible in flight. These compartments can be located between the flight deck and the passenger cabins. They can also be found adjacent to the galley or at the back of the aircraft.

Class B. A compartment with a separate, approved smoke or fire detection system to give warning to the pilot or flight engineer station and with sufficient access in flight to enable a crewmember to effectively reach any part of the compartment with a hand-held fire extinguisher. Class B cargo compartments are usually located remote from the flight deck. On combi-aircraft, Class B cargo compartments can be located either in front of or behind the passenger cabins.

Class C. Class C compartments differ from Class B compartments primarily in that built-in extinguishing systems are required for control of fires in lieu of crewmember accessibility. Smoke or fire detection systems must be provided. Class C compartments are usually found under the passenger cabin floor in wide-bodied aircraft. Class C and upgraded Class D compartments are the types of cargo compartments usually found on modern passenger aircraft. Class C and upgraded Class D compartments are also found under the main deck floor on cargo-only aircraft.

Class D. Prior to industry changes, Class D compartments were originally designed without fire detection or fire extinguishing systems. Any fire occurring was supposed to be inhibited by very low airflow within the compartment. Class D compartments are no longer an option for new aircraft. Current aircraft that have Class D compartments under previous definitions must be upgraded to comply with the requirements of Class C compartments if the aircraft is used for passenger transport or Class E compartments if the aircraft is used only for cargo transport.

Class E. A cargo compartment used only for the carriage of cargo. Typically, a Class E compartment is the entire cabin of an all-cargo airplane. A smoke or fire detection system is required. In lieu of providing extinguishment, means must be provided to shut off the flow of ventilating air to or within a Class E compartment. In addition, procedures such as depressurizing a pressurized airplane are stipulated in the event that a fire occurs. This is not something that can be done on a passenger-carrying aircraft.

Gaining Access to Cargo Compartments

Most cargo doors are hinged at the top of their opening and swing out and up. A few open up and into the compartment. Most older narrow-body aircraft cargo doors open manually. Newer narrow-body and almost all wide-body aircraft cargo doors open electrically and hydraulically. Mechanically operated cargo doors can usually be opened manually by releasing a latching handle that releases the door locks and inserting a ¼- or ½-inch ratchet drive into a ¼- or ½-inch socket hole and rotating the drive. Pneumatic drivers cannot be used because they turn too fast and jam up the mechanism. The socket hole is usually found in the vicinity of the cargo door. Large cargo doors may also have mechanical locking devices which when opened, relieve the pressure in the compartment.

Aircraft Construction and Structural Materials

In an effort to help ensure personnel safety, airport firefighters should have a thorough knowledge of aircraft construction, the materials used, and the hazards they may pose during and after fire fighting operations. Fire fighting operations may be affected by the inherent properties of the materials and by the manner in which these components are assembled. Materials commonly used in late-model aircraft construction include aluminum, steel, magnesium, titanium, wood, and plastic. These materials are often used in any number of combinations. In an effort to reduce the overall weight of an aircraft, manufacturers have incorporated the use of composite materials and metal alloys. ARFF personnel should be aware that when the surfaces of different materials are uniformly painted, the variation in construction may not be apparent without close investigation.

Aluminum and Aluminum Alloys

Currently, aluminum comprises 85 percent of aircraft construction (Figure 3.34). Due to its lightweight characteristics, along with the ability to be molded into a variety of shapes, it is an ideal material for aircraft construction. This lightweight material also can be molded and used in sheets for skin surfaces, or it can be formed into honeycomb sheets which are often used to form walls and floor sections. One disadvantage to using aluminum for

Figure 3.34 Typical aircraft construction is composed of at least 85 percent aluminum. *Courtesy of Airbus Industrie.*

aircraft construction is that it does not withstand heat well, and it melts at relatively low temperatures (approximately 1,200°F [649°C]). Aluminum alloys are created by mixing components of different types together in a molding process that produces stronger, yet lighter construction materials. These alloys can be found molded into landing gear parts, structural and load-bearing members, as well as parts of the door operating assemblies.

Steel

In certain parts of the aircraft, such as in the engine and landing gear, high strength is required and/or heat tolerance is critical. Steel is used in these components, even though the weight per volume is much higher than other structural materials.

Magnesium and Magnesium Alloys

Because they are both strong and lightweight, magnesium and magnesium alloys are used for the landing gear, wheels of some older aircraft, engine-mounting brackets, crankcase sections, cover plates, and other engine parts. Magnesium and its alloys are generally used in areas where forcible entry will not be required. Unless ground into a dust or into small particles, magnesium is difficult to ignite; however, once ignited, it burns intensely and is very difficult to extinguish.

Figure 3.35 Landing gear may be composed of either magnesium or titanium. *Courtesy of William D. Stewart.*

Titanium

Titanium is a metallic element used to reinforce skin surfaces to protect them from impinging exhaust flames or heat. Titanium is used for internal engine parts such as turbine blades, auxiliary power unit enclosures, along with landing gear parts (Figure 3.35). Like magnesium, titanium is a combustible metal that burns with intensity and makes extinguishment difficult.

Advanced Aerospace (Composite) Materials

As mentioned before, a host of lightweight composite or advanced aerospace materials are currently being used in modern aircraft construction. The percentage of composite materials will only increase as manufacturers develop more and more ways to incorporate their use (Figure 3.36). Much of the success of NLA's is based on the use of composite materials. The most comprehensive study to date has been completed and is added as an appendix to this manual. Included within Appendix B, "Advanced Composites/Advanced Aerospace Materials (AC/AAM): Mishap Risk Control and Mishap Response," by John M. Olson, are established guidelines that departments may want to incorporate into their standard operating procedures in an effort to enhance ARFF personnel safety. The department must emphasize that personnel should be familiar with this study and take the necessary precautions when faced with aircraft emergency involving advanced aerospace materials.

Wood

Some older aircraft have a considerable amount of wood in structural areas such as wing spars and bulkheads; bulkheads of some aircraft are made almost entirely of wood. However, the most

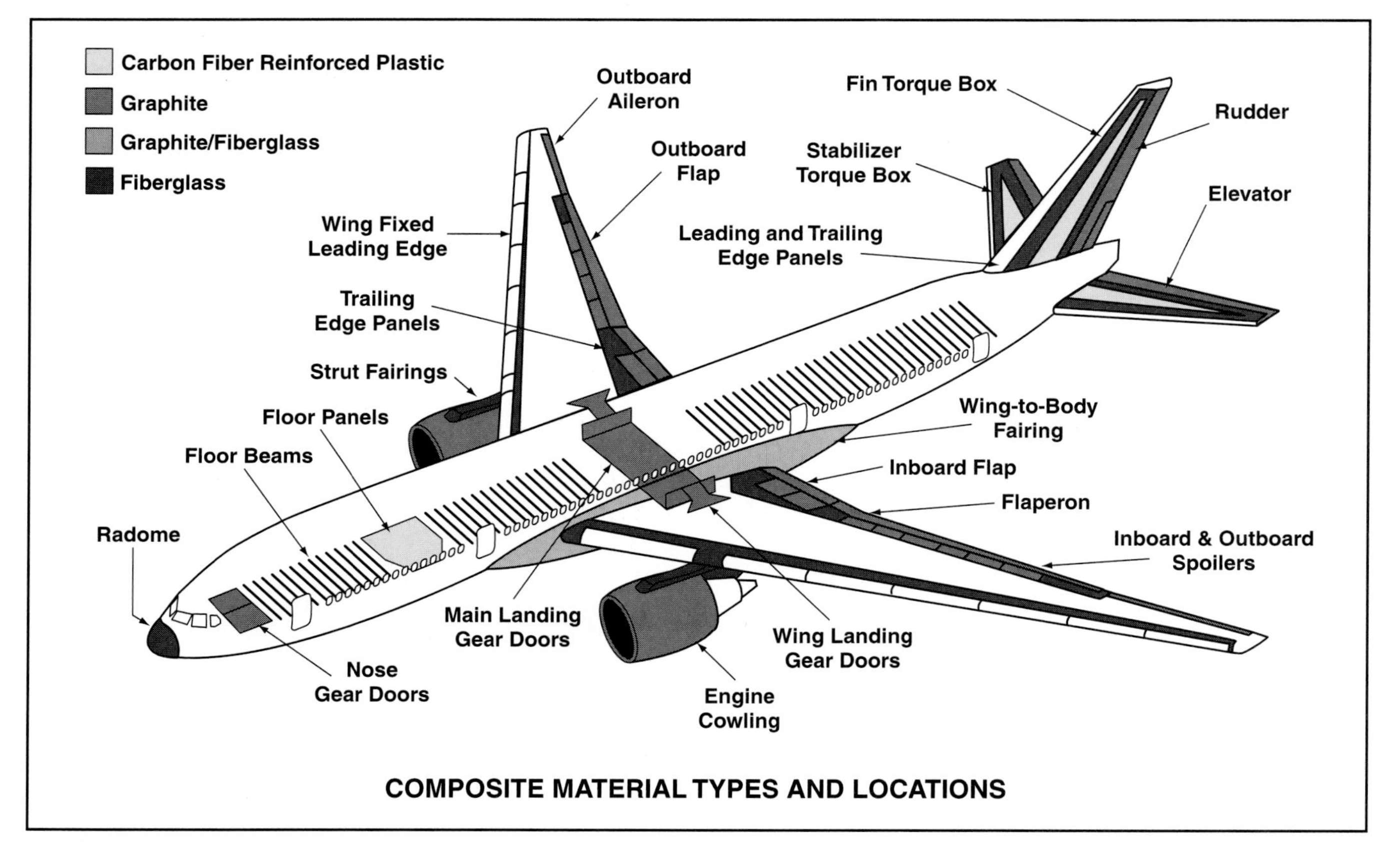

Figure 3.36 Composite material types and locations.

common construction technique is to combine tubular steel framing with wooden components. Because there is a high probability that an aviation-type fuel is present, aqueous film forming foam (AFFF) should be used in fire fighting operations even though these are Class "A" materials. Corporate-style aircraft incorporate elaborate wood fixtures for interior furnishings. These furnishings are often structural framework, fabrics, and laminated wood products with epoxy-type finishes that when exposed to fire produce a wide variety of toxic vapors.

Aircraft Systems

Hazards or potential hazards are created by such aircraft systems as fuel, hydraulic, electrical, oxygen, flight-control, landing-gear, and egress or escape systems. When planning strategies for aircraft accidents/incidents, rescue personnel should carefully consider each of these potential hazard areas and develop tactics, SOPs, etc. that address and attempt to eliminate and control the hazards while performing a rescue.

Standardized Coding

In an effort to assist aircraft mechanics, aircraft manufacturers have developed a color-coded labeling system that addresses all tubing, piping, and cabling found in an aircraft. ARFF personnel can use this system to assist them when performing extrication operations. Aircraft of all types contain varying amounts of tubing, hose, and other conduits that may be of the same size and appearance, so it is often difficult to distinguish among them. Therefore, a standardized coding system has been designed to simplify their identification and reduce the risk of misidentification.

The coding is presented in three different forms: colors, labels, and symbols to make it easier to accurately identify tubing, hose, and piping (Figure 3.37). Color is used because it may be identified from a distance. Labels are necessary for the color-blind and for situations in which fire, heat, and smoke may obscure or alter the color. Finally, symbols not only aid in helping confirm the colors and labels, but also they are more readily recognized by those who do not read English. Being aware of these

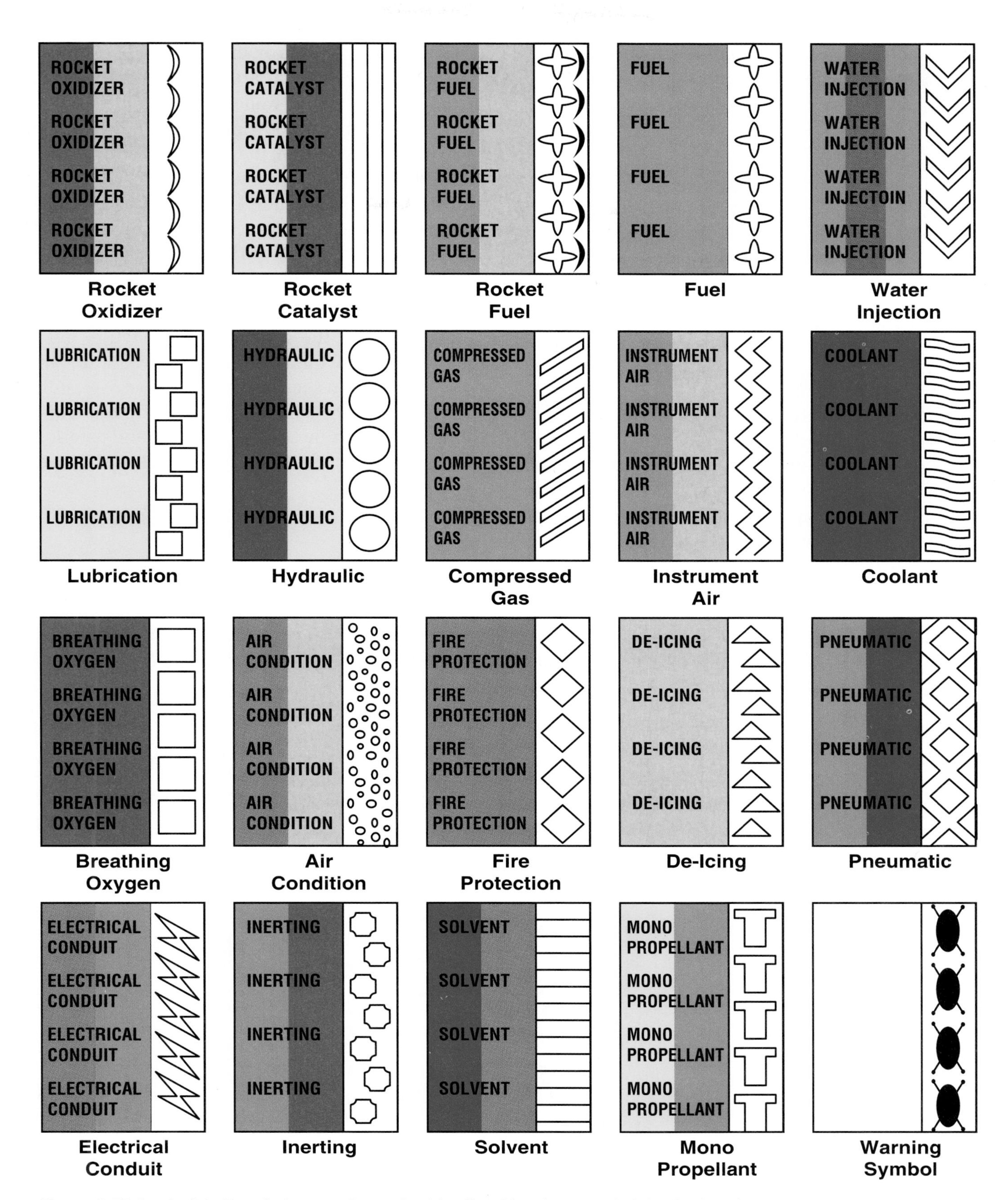

Figure 3.37 A color/label/symbol system is used to identify tubing, hose, and piping in aircraft.

codes allows ARFF personnel to proceed quickly but cautiously when encountering these conduits during aircraft extrication procedures.

Fuel Systems

The largest system in the aircraft is the fuel system. The components of the fuel system — tanks, lines, control valves, and pumps — are located throughout the aircraft. Therefore, the fuel system presents the greatest hazard in an aircraft accident. The fuel system consists of two major parts: the tanks and the distribution system.

Fuel Tanks

Depending on the type and use of aircraft, fuel tanks may be found and constructed as separate units or as an integral part of the aircraft. Small general aviation aircraft have tanks that are generally found in the wings and that are constructed of aluminum, composite, or rubber bladders. Business-style, commuter, and commercial aircraft also use the wings to store fuel by incorporating integral tanks. Integral tanks are formed by sealing the inside structure of the wing with specialized epoxy. In addition to the wing area, these aircraft use the center fuselage section between the wings to store fuel (Figure 3.38). Additional tanks are sometimes installed forward or aft of the center fuselage tank. Some aircraft that are designed to fly great distances use double-walled fuselage tanks. Because these tanks are outside the box structure of the wing center section, they do not have any substantial structural protection. Other areas for additional tanks include center fuselage pods, wingtips (Figure 3.39), tail (horizontal or vertical stabilizers), or tailcones. Military aircraft also use auxiliary tanks or fuel pods, which are mounted externally, to extend their flying range.

Regardless of the fuel-tank construction, fuel may be released if the aircraft is damaged. Although the damage may seem insignificant and remote from the aircraft cabin, ARFF personnel should thoroughly examine both the interior and exterior of the aircraft for fuel leaks. Even minor damage

Figure 3.39 Some aircraft may be equipped with tip tanks. *Courtesy of William D. Stewart.*

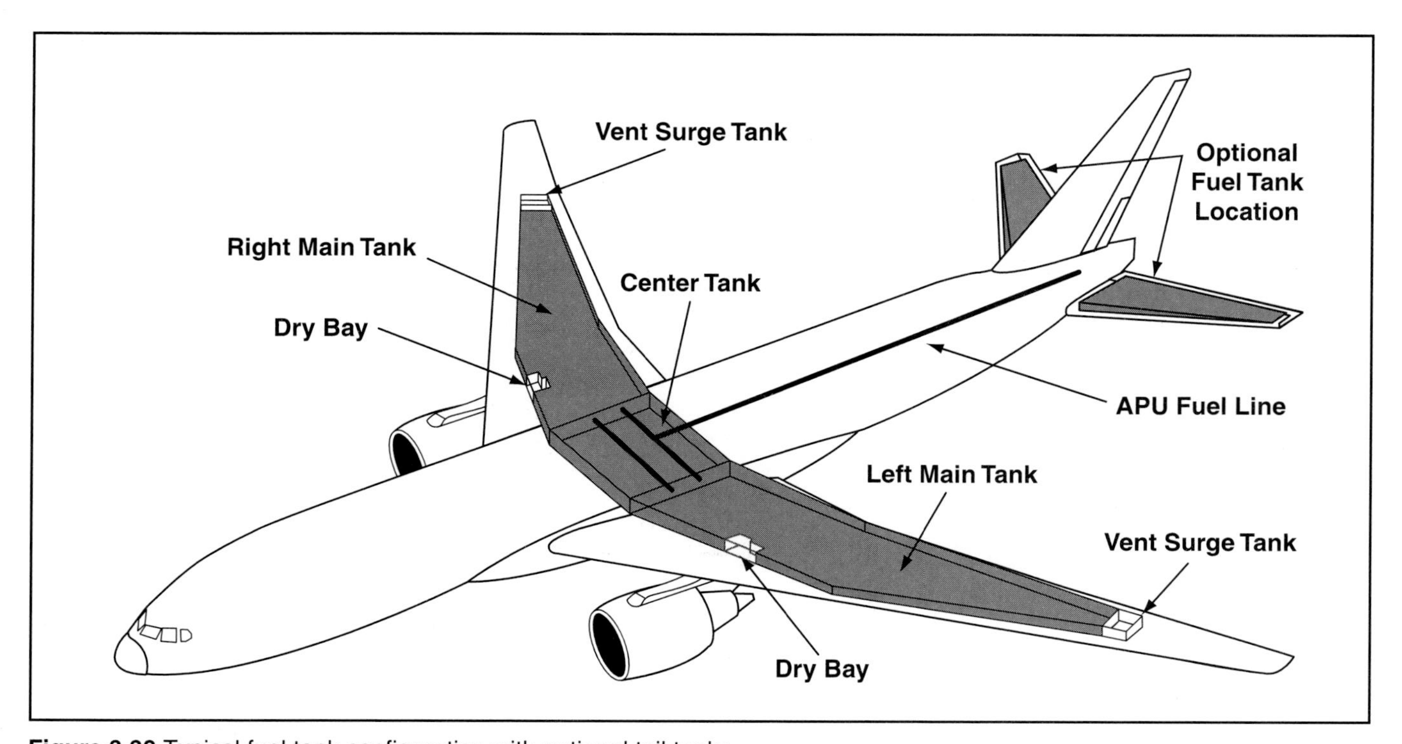

Figure 3.38 Typical fuel tank configuration with optional tail tanks.

can be critical because leaking or seeping fuel may pool in low-lying sections of the fuselage.

Civilian and military aircraft, both fixed-wing types and helicopters, utilize auxiliary fuel tanks. On military aircraft, tanks may be jettisoned in flight to improve speed and maneuverability. The fuel capacity of auxiliary tanks can vary from 30 gallons (120 L) in a small civilian aircraft to 2,000 gallons (8 000 L) per tank in large military aircraft. Helicopter auxiliary tanks are normally located inside or outside the cabin. During flight operations, the fuel within auxiliary tanks is usually consumed first. Crash-resistant fuel tanks with self-sealing fittings and automatic shutoffs are in limited use. Although the technology for fuel tanks has continued to advance, these improvements have not been widely adopted. Some military aircraft have open-celled foam blocks that are cut to fit and placed in the tanks. While these vapor-suppressing blocks are primarily to protect against explosion after projectiles, such as incendiary bullets, have penetrated the vapor space, they are also effective in suppressing fire after a crash.

Fuel tanks may be filled individually through service openings on the top side of the wings (gravity refueling) or filled through a single point or multiple fueling points on the underside of the wings (pressure refueling) or the side of the fuselage (Figure 3.40). In pressure refueling, a system of valves directs the fuel to the tanks needing to be filled. Sensing devices within the individual tanks automatically stop the flow to a particular tank when it is full or filled to the required level. Dipstick fuel-quantity gages are also located on the bottom of some wings.

Figure 3.40 Typical pressure refueling. *Courtesy of William D. Stewart.*

Over a period of time, the epoxy sealing the fuel tanks may develop a leak. Normal repair procedures involve maintenance personnel entering the confined spaces of fuel tanks through access ports found on either the top side or the bottom side of the wing. Due to the confinement of these spaces, rescue workers should meet with aircraft maintenance personnel and develop a response plan in which to effectively handle any emergency that may occur during these maintenance operations. As a precautionary measure, the airport fire department should be notified by the aircraft maintenance division when this procedure is being performed.

Fuel Distribution

Commercial aircraft fuel capacity can range from 3,000 gallons to over 58,000 gallons (12 000 L to 220 000 L) as found on a Boeing 747-400. Fuel is distributed from the aircraft's fuel tanks to its engines through fuel lines, control valves, and pumps located throughout the aircraft. Aircraft with engines located in the tail section may have fuel lines routed through interior walls, through the roof, or between the main cabin floor and the cargo area of the aircraft. Because most aircraft incorporate an auxiliary power unit, the fuel line is often between the main cabin floor and the cargo area (Figure 3.41).

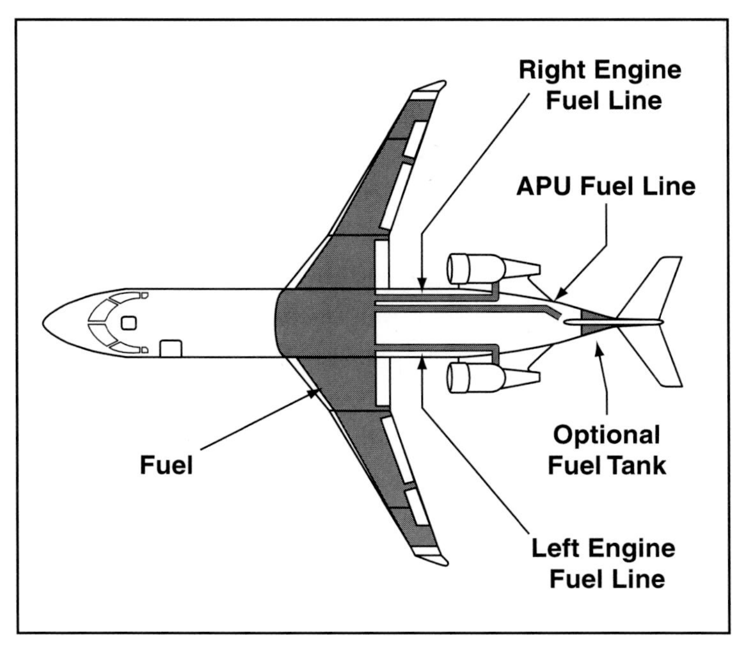

Figure 3.41 Fuel lines for the engines and APU may be routed through the area between the floor and the cargo hold.

Fuel lines vary in sizes from ⅛ inch (3 mm) to 4 inches (100 mm) in diameter. They are constructed of metal, rubber, or combinations of materials and are often shrouded to control developing leaks. The fuel flow within these lines is controlled by pumps that are capable of producing pressures from 4 to 40 psi (28 kPa to 280 kPa). Fuel system leaks can be controlled by shutting down fuel pumps, which is best accomplished by securing aircraft power and fuel controls in the flight deck area.

Temperature changes cause fuel in the tanks to expand and contract. In order to reduce pressure buildup caused by expansion, fuels tanks are equipped with vents and vent tanks, which hold residual and released fuel vapors. Under normal conditions, the small amounts of fuel that escape evaporate quickly, so such venting is usually not hazardous. Quite often, fueling personnel overfill the main tanks so that when expansion does occur, the fuel vents into the vent tanks, continuing through the overflow tube and onto the aircraft parking area. Heating of fuel cells exposed to direct fire or radiant heat can also cause expansion of fuel which may release fuel vapors from vents. There are also other causes for fuel spills under aircraft that are not hazardous under normal conditions. For example, minor spillage may occur when aircraft engines are shut down or when petcocks are opened to drain water and sediment from fuel tanks. These small amounts of fuel usually do not represent a significant fire hazard.

Two basic types of fuel that rescue workers encounter include AVGAS and jet fuel. These fuels can be mixed in a variety of ways depending on the use of aircraft, and are covered in greater detail in Chapter 9, "Extinguishing Agents." The level of hazard also varies depending on the fuel, how it is mixed, and the scenario in which it has been released.

Hydraulic Systems

It takes a great deal of power to operate the control surfaces on an aircraft, not to mention extending and retracting the landing gear. Aircraft manufacturers have developed a high-pressure hydraulic system to accomplish these tasks. The hydraulic system of an aircraft consists of a hydraulic fluid reservoir, electric or engine-driven pumps, appliances, various hydraulic accumulators, and tubing that interconnects the system (Figure 3.42). The hydraulic fluid is supplied to a pressure pump that moves the fluid throughout the hydraulic system and to accumulators where some of the fluid is stored under pressure. This stored fluid may then be used to supply hydraulic pressure to critical aircraft systems such as landing gear, nose-gear steering, brakes, and wing flaps. The accumulator may store this fluid under pressure for a considerable period of time even after the engines have been stopped. Most modern aircraft hydraulic systems operate at a pressure of 3,000 psi (21,000 kPa) or higher and may carry as much as 185 gallons (740 L) of hydraulic fluid.

Of the three types of hydraulic fluid produced, synthetic hydraulic fluids are the most widely used. Old vintage aircraft relied on either a hydrocarbon-based fluid or a vegetable-based fluid. Synthetic fluid is most popular because it presents a significantly reduced flammability hazard: Its flash point is twice that of non-synthetic fluid, and once on fire, the flame-spread rate is slower. The most common synthetic hydraulic fluid is a phosphate-ester-based material.

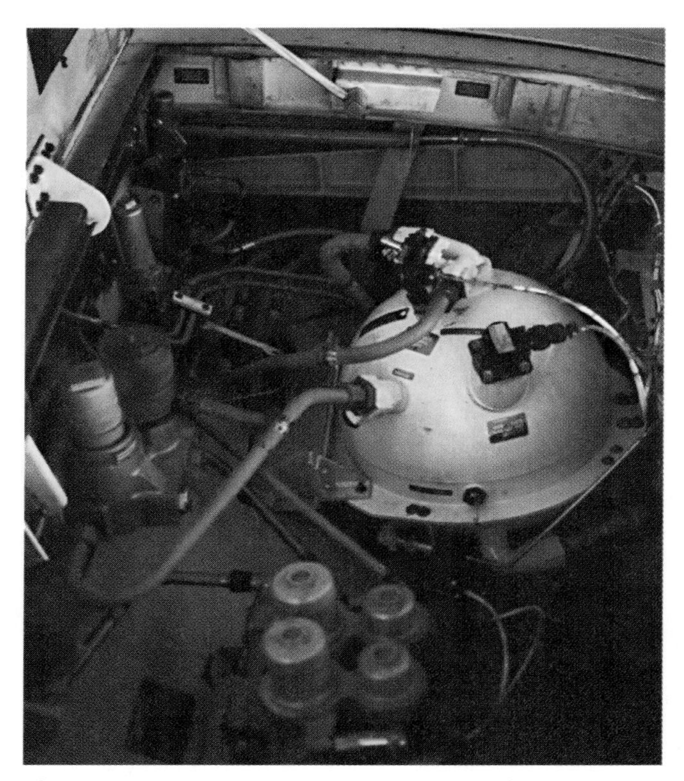

Figure 3.42 Hydraulic system components. *Courtesy of William D. Stewart.*

There are drawbacks associated with synthetic fluids. During rescue or fire suppression operations, ARFF personnel must exercise extreme caution to avoid cutting pressurized hydraulic lines. When released in the form of a fine mist, synthetic fluid is extremely flammable. If sprayed on hot brakes or hot engine components, the fluid may ignite. A hydraulic fire produces a "torch effect" or, if confined, may explode. Personnel also must guard against hydraulic fluid contacting skin, eyes, and protective clothing because this type of fluid can cause severe skin and eye irritation as well as erode protective clothing surfaces.

WARNING

When dealing with aircraft emergencies around operating aircraft, keep personnel aware and clear of areas containing hydraulically and pneumatically operated parts. Because of the pressure and force needed to move these parts personnel can easily injure or sever body parts if caught or pinched between moving surfaces.

Wheel Assemblies

As stated earlier, the landing gear is designed to support the weight of the aircraft when it is on the ground. The landing gear contains the wheel assembly, which consists of rims, brakes, and tires. Wheel assemblies or *bogies* contain rims that in older style aircraft may be made of magnesium and in newer aircraft, titanium or aluminum alloys. Most all aircraft rims are equipped with fusible plugs (Figure 3.43). These plugs are designed to melt, automatically deflating the tires when the rim reaches a predetermined temperature. Aircraft brakes are designed to slow and stop the aircraft after landing, during an aborted takeoff, or while taxiing. Aircraft brake systems can be very complex. There may be as many as three independent sources of hydraulic power for the brakes on a large jet aircraft as well as two separate anti-skid systems and an auto-brake system. The brake assemblies are made of magnesium, beryllium, or asbestos in older aircraft and carbon composite in newer aircraft. Aircraft brakes frequently become overheated due to the combined effects of the aircraft weight, the landing speed, and the extra braking power required for short runways (Figure 3.44). Brakes and wheels may reach their maximum temperatures 20 to 30 minutes after the aircraft comes to a stop.

WARNING

When dealing with a landing gear emergency such as a hot brake or gear fire, always approach landing gear either forward or aft of the gear assembly. If heated beyond limits, landing gear assemblies and tires may explode, sending debris and pieces out from the sides of the assembly. These pieces can travel with enough velocity to kill or severely injure a responding firefighter. These pieces can also travel with enough velocity to strike and puncture fuel cells located in the wings.

Figure 3.43 Fusible plugs are designed to melt and deflate the tire. *Courtesy of William D. Stewart.*

Figure 3.44 A landing gear tire fire. *Courtesy of Airport News and Training Network.*

Large aircraft tires may have pressures exceeding 200 psi (1,400 kPa). They are usually filled with nitrogen (an inert gas) due to the tremendous amount of heat generated during takeoffs and landings. Procedures for handling aircraft emergencies involving wheel assemblies are covered in Chapter 10, "Aircraft Rescue and Fire Fighting Tactical Operations."

Electrical Systems

An aircraft relies on an electrical system to supply current for lights, electronic equipment, hydraulic pumps, fuel pumps, armament systems, warning systems, and other devices (Figure 3.45). Aircraft electrical systems use both AC and DC current to supply electrical power because some equipment operates more efficiently on one type than on the other. Light aircraft operate on 12- or 24-volt DC systems; large aircraft operate on 24/28-volt DC and 110/115-volt AC.

Aircraft Batteries

Aircraft batteries are divided into two general types: lead acid and nickel cadmium. Fundamentally, there is no difference in the operation of aircraft and automobile batteries. Both have the same type of plates immersed in an electrolyte solution and operate on the same basic principles. The aircraft battery, however, requires a great deal more care because of the unusual conditions under which it operates. Aircraft batteries are built so that they will not leak when the airplane is upside down. Voltage is usually 12 to 30 volts. To save weight, aircraft batteries have an exceedingly small capacity – only one-third that of the average automobile battery.

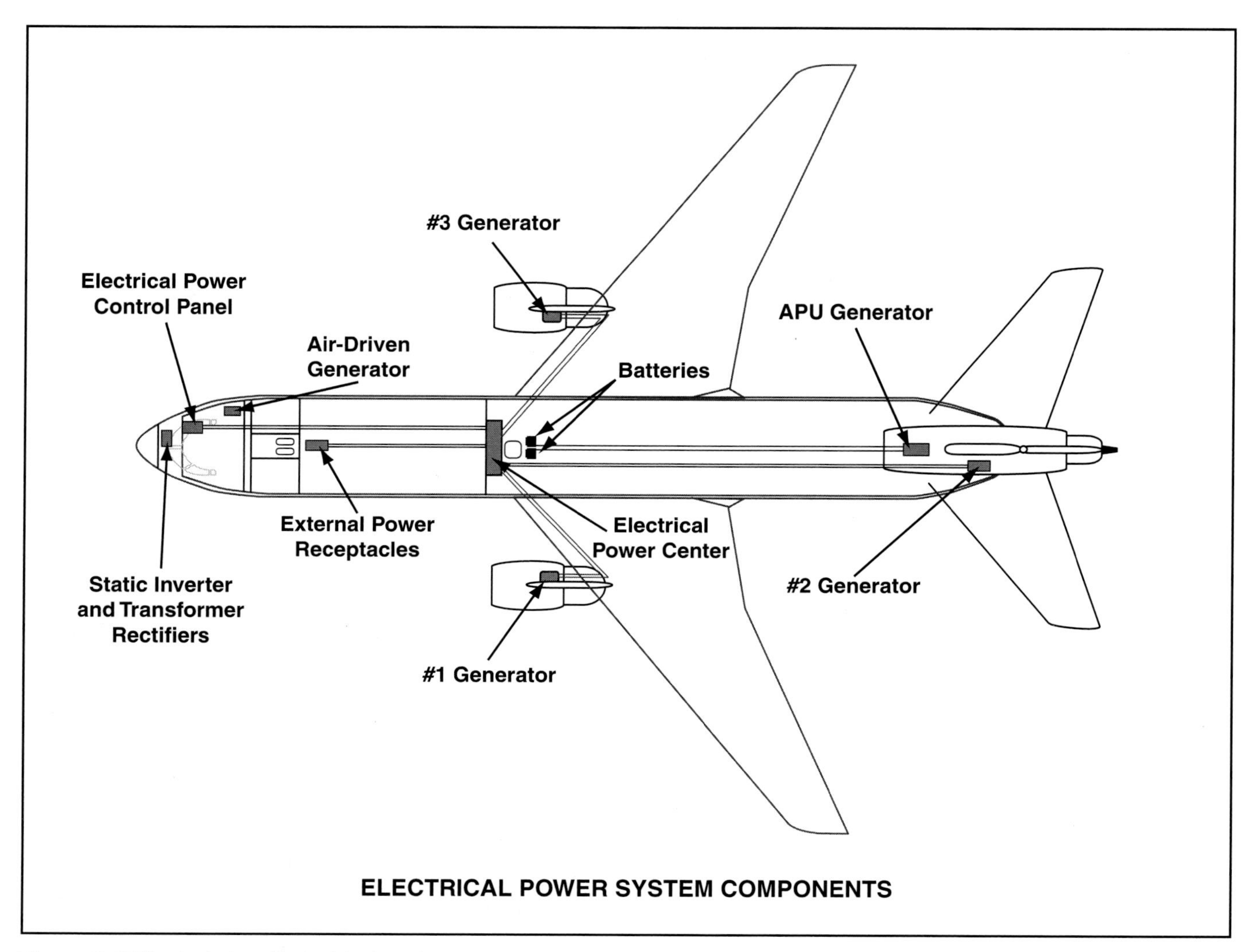

Figure 3.45 Typical aircraft electrical system.

Both lead-acid and nickel-cadmium batteries produce hydrogen gas when charging. Hydrogen gas is highly flammable and presents serious fire and explosion hazards to ARFF personnel.

CAUTION: The electrolyte used in nickel-cadmium batteries is a strong alkaline solution of potassium hydroxide, which is caustic and corrosive. The electrolyte used in lead-acid batteries is sulfuric acid and distilled water. Serious burns will result if either electrolyte contacts the skin.

Most commercial and military aircraft batteries are equipped with quick-disconnect terminals. Usually, a quarter-turn terminal device on the battery cable terminal will connect or disconnect the battery (Figure 3.46). On large aircraft, there are at least one and sometimes two or three batteries usually located in an avionics compartment. The location of the avionics compartment may vary. Some aircraft have multiple battery locations. On helicopters, general aviation, commuter, and military aircraft, the batteries could be located in a number of places. Again, depending on the type and use of aircraft the number and location will vary. ARFF personnel assigned to airport duty should familiarize themselves with the locations of the batteries and electrical system shutoffs on the types of aircraft common to that facility. The battery(ies) will usually be near the ground power connection. Sometimes the battery location can be identified by a compartment drain and/or vent on the bottom of the aircraft. Military aircraft will often have the battery compartment marked. It is important to remember that all aircraft shutdown functions must be accomplished prior to de-energizing the electrical system. Functions such as normal cargo-door operation, cockpit shutdown procedures, and emergency shutdown procedures cannot be completed without electrical power on the devices that perform these functions.

Figure 3.46 A 28-volt nickel-cadmium battery with a quarter-turn quick disconnect. *Courtesy of William D. Stewart.*

Auxiliary and Emergency Power Systems

Auxiliary Power Unit (APU)

An auxiliary power unit (APU) is a small jet engine with a generator attached which is used while the aircraft is on the ground and at the gate to operate systems instead of running one of the engines (Figure 3.47). Running engines would create hazards to ground maintenance personnel while servicing the aircraft. These small turbine engines provide pneumatic air and AC electrical power to start the aircraft engines, power the cockpit, recharge the batteries, light the cabin, and maintain comfortable cabin temperatures. While the aircraft is airborne, the APU can sometimes be used as a backup electrical power source. Found on all commercial aircraft,

Figure 3.47 Typical APU usually found in the tail section of most commercial aircraft. *Courtesy of William D. Stewart.*

some commuter and corporate, the unit is generally located in the tail section of the aircraft. External APU controls on large aircraft usually are found on the nose gear, belly, tail, or main gear compartment (Figure 3.48).

Because the APU is a small jet engine, it can create a noise hazard and an exhaust hazard to anyone who walks past the exhaust port while it is running. Also, because the APU operates on jet fuel, there is always the possibility that it will catch fire. Almost all aircraft are equipped with some type of system designed to shut down and extinguish an APU fire. Many newer aircraft incorporate an automatic system that shuts down the unit if a fault, overheat or fire is detected. Manual controls are located in the cockpit and on an external fire protection panel to shut the unit down and discharge the APU fire extinguishing bottle. The airport firefighter should be familiar with the APU locations, internal and external shutdowns, and the battery that supplies the APU for the aircraft that operate at their airport.

The APU may often use two air intakes to function. One is used for the operation of the unit while the other is used to cool the compartment. Personnel should be familiar with the intake that cools the compartment. The firefighter might be able to use this intake to discharge extinguishing agent into the APU compartment. When battling an APU fire, rescuers must exercise extreme caution when accessing the unit because the access doors are often found below the unit. Pooling fuel may be held within the voids of these access doors and could spill out when the doors are opened.

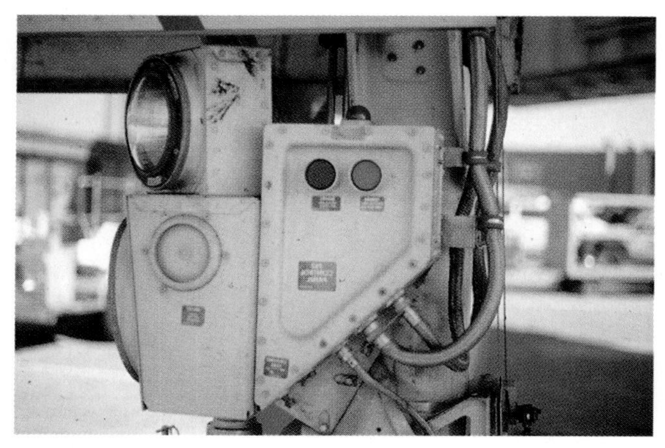

Figure 3.48 External APU shutdown and fire extinguishing system control panel. *Courtesy of William D. Stewart.*

Ground Power Unit (GPU)

GPUs can be mobile (on carts, trailers, or trucks), fixed-mounted in buildings, or bridge-mounted on jetways that connect the aircraft to the terminal building, and they are used to provide onboard electrical power while the engines or APU are not operating. GPUs can be used to produce either AC or DC power and come in diesel- or gas-fueled models (Figure 3.49).

WARNING

Disconnecting the GPU from the aircraft prior to the power being shut off can cause electrocution or arcing. Arcing could provide an ignition source for flammable vapors that have collected in the area.

Airport firefighters should be familiar with the shutdown and disconnection procedures of GPUs at their airport.

Emergency Power Unit (EPU)

EPUs fill the need for a highly reliable and quickly responsive means of obtaining emergency electrical power (for restarting engines) and hydraulic power (for flight-control operation) aboard airborne aircraft. There are basically three types of EPUs: ram-air-turbine (RAT), jet-fuel, and the monopropellant.

The airport firefighter should be familiar with the general location of the ram-air-turbine because it may deploy when the electrical system is de-energized (Figure 3.50). Deployment could result

Figure 3.49 Aircraft not equipped with an APU rely on an external power unit. *Courtesy of William D. Stewart.*

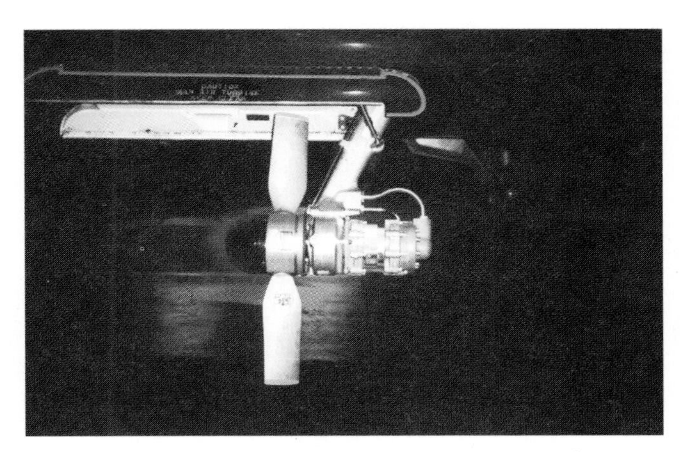

Figure 3.50 During an electrical emergency, aircraft power may need to be supplied by a ram-air-turbine (RAT). *Courtesy of William D. Stewart.*

in striking other rescue workers. The jet-fuel EPU has the same hazards as the jet-fuel APU. On the other hand, monopropellant EPUs, used in aircraft such as the U.S. Air Force F-16 fighter and the space shuttle, are extremely hazardous because they are powered by a toxic and caustic fuel called *hydrazine*. This fuel is classified as hypergolic, which means that it ignites spontaneously on contact with an oxidizer (for example, hydrazine with an oxidizer). The F-16 uses UDMH-70, which is 70 percent hydrazine and 30 percent water.

Hydrazine is a clear, oily liquid that has a smell similar to ammonia. It poses a health hazard in both the liquid and vapor forms. Liquid hydrazine can cause severe local damage or burns if it comes in contact with the eyes or skin. It can penetrate the skin to cause systemic effects similar to those produced when swallowed or inhaled. If inhaled, the vapor causes local irritation of the eyes and respiratory tract and the following systemic effects:

- Short-term exposure effects involve the central nervous system with symptoms including tremors.
- High concentrations can cause convulsions and possible death.
- Repeated or prolonged exposures may cause toxic damage to the liver and kidneys, as well as anemia (blood deficient of red blood cells).

WARNING

Wear full personal protective equipment (PPE) at all times when dealing with hydrazine emergencies as it may be absorbed through the skin. Even short exposures may have serious effects on the nervous and respiratory systems.

Oxygen Systems

All aircraft intended for high-altitude operations use an oxygen-supply system to provide life support for crew members and passengers. Oxygen is normally stored in either a gaseous or liquid state; however, some commercial aircraft are equipped with a system for chemically generating oxygen for passengers. Chemically generated oxygen systems, when activated, produce substantial amounts of heat because of the exothermic chemical reaction. This heat is normally contained within the oxygen-generating unit but may ignite combustibles if in direct contact. Once the reaction is started, it is impossible to stop until the unit has exhausted its chemical. These units are often located in the seat backs or in overhead compartments.

On older passenger aircraft, the crew and passenger oxygen supply is stored in pressurized cylinders within the fuselage (Figure 3.51). Small first-aid-type cylinders can be found in a variety of locations throughout the cabin. Ejection seat systems of military aircraft have a small emergency

Figure 3.51 Some older aircraft utilize oxygen cylinders to supply oxygen in an emergency. *Coutesy of William D. Stewart.*

oxygen cylinder attached to the seat. Some medical transport helicopters and most military fighter, bomber and attack aircraft employ the use of liquid oxygen cylinders; a regulating system converts the liquid oxygen into usable oxygen.

In most cases, oxygen cylinders aboard aircraft are painted green; however, this coloring system is not used universally. ARFF personnel should not depend on the color to identify cylinders following an accident/incident.

Liquid oxygen (LOX) is light blue and transparent, with a boiling point of -297°F (-147°C). LOX may produce burns similar to but more severe than frostbite if it is allowed to contact the skin. Like gaseous oxygen, LOX is not flammable by itself, but it will support combustion. LOX readily forms combustible and explosive mixtures when it comes in contact with most substances — especially materials such as oil, grease, cloth, wood, paper, acetylene, gasoline, kerosene, powdered metal, and asphalt.

WARNING

Do not disturb asphalt onto which LOX has been spilled because it is explosively unstable and extremely shock-sensitive. Until the LOX has dissipated, merely walking on the spill or dropping something onto it may cause a violent reaction.

Oxygen systems on aircraft may present severe hazards to firefighters during emergency operations. As long as an oxygen-enriched environment is present, fire will burn with greater intensity. There is danger of an explosion if liquid oxygen mixes with flammable/combustible materials. An explosion or deflagration may also result if an oxygen storage tank or liquid oxygen container ruptures because of heat expansion or impact.

Cylinders that have shifted or have been displaced from their mountings by the impact of a crash should not be disturbed unless doing so is necessary to perform a rescue. The area should be isolated, and the containers should be protected from fire or unnecessary manipulation until they can be disposed of properly. If shutting off the cylinder lessens the intensity of the fire, firefighters should do so if they can accomplish this safely.

In fires involving LOX, the flow of oxygen and/or fuel should be stopped. The smothering and blanketing agents normally used in aircraft fire fighting are ineffective if the fire is being supplied with liquid oxygen. One acceptable method of stopping a liquid oxygen leak is to spray the leak with water fog. The super-cold LOX immediately converts the water to ice, which forms a plug, sealing the leak.

Radar Systems

Radar energy, much like microwaves, can present an ignition source, as well as a health hazard. Because of this, most airborne radar systems are operated on the ground only before takeoff and just after landing. Because the radar system is in the nose, personnel should never approach the nose of an aircraft if it is suspected that the radar system is on; it can have serious health-related effects causing cellular damage to humans. If the aircraft engines and power have been shut down, then so has the radar. Certain military command and control aircraft and surveillance aircraft have very powerful radar systems, which may be obvious by the presence of large external radar antennas and devices.

Fire Protection Systems

Many modern aircraft are equipped with fire protection systems that may be activated by the flight crew or ground crew to extinguish fires in engines, APUs, and cargo compartments. The quantity of extinguishing agent and configuration of the system is specifically designed for each aircraft type. A typical fire suppression system consists of pressurized containers, tubing to deliver the agent, nozzles, and appliances for actuating and controlling agent discharge. After a crash, these systems may or may not be usable, but ARFF personnel should be familiar with their location and operation because they may assist in securing aircraft systems. Remember, once the battery has been disconnected and all electrical power removed, the fire suppression system will not operate.

Handheld extinguishers for use on interior fires are located in the cockpit and throughout the cabin. Detection and suppression units are used in lavatories aboard some aircraft. A smoke detector sounds an audible alarm to alert the cabin crew, and a small heat-activated extinguisher bottle is installed to protect the lavatory trash bin.

Miscellaneous Systems and Components

Anti-icing Systems

Many aircraft are equipped with electric and/or pneumatic anti-icing systems. Electric components are typically used to heat cockpit windows, propellers, and items such as probes, ports, and drain masts along the fuselage. High-temperature bleed air from the engine exhaust is used to heat engine inlets and the leading edge of the wings.

Pressurized Cylinders

There can be many pressurized cylinders located throughout any size aircraft (Figure 3.52). Some of these, such as oxygen cylinders, have pressure-relief valves. Other cylinders used for hydraulic fluids, fire extinguishing systems, rain repellent, pneumatic systems may explode during aircraft fire fighting operations if heated due to external heat sources.

Pitot Tubes

Two to four pitot tubes are usually located on both sides of the forward fuselage, just below the cockpit windows, of transport-type aircraft. Pitot tubes measure air pressure for use in certain cockpit instrument displays. Because these L-shaped probes protrude from the side of the fuselage and are heated, they pose a hazard for personnel operating near them. Due to the vital functions performed in conjunction with these tubes, personnel should not touch or handle these devices during training.

Figure 3.52 Small, pressurized cylinders, which may be filled with a variety of gases or liquids, may be found throughout the aircraft. *Courtesy of the National Audio Visual Center.*

Antennas

Aircraft are equipped with multiple antennas for communications and navigation. VHF, UHF, and satellite communications, global positioning, and onboard air-phones are some of the systems connected to these antennas. They protrude from the top and underside of the fuselage.

Flight Deck Emergency Shutdown Procedures

It may be necessary for ARFF personnel to conduct emergency shutdown procedures on an aircraft. These procedures vary from a simple step to a complicated sequence of procedures. A common first step in any aircraft is to move the throttles(s) to the IDLE or OFF position. Accomplishing this may require lifting the throttle(s) past a detented position. The next step may be to activate the fire protection system. Shutting off the battery switch(es) should be the last step in cockpit shutdown procedures.

On almost all commercial transport-type aircraft and some commuter aircraft, the shutoff procedure involves activating T- or L-shaped engine and APU fire shutoff handles. Pulling these handles simultaneously shuts off the engine's fuel, hydraulic, pneumatic, and electrical connections while arming the fire suppression system. An extinguisher bottle discharge button is usually located adjacent to each fire shutoff handle for activating the suppression system. The T- or L-shaped handles are usually located around the throttles or, in some cases, on the cockpit overhead panel (Figures 3.53 a and b). Some aircraft also have APU shutdown and bottle discharge buttons on an external fire protection panel located on the nose landing gear or in the main wheel well. When operating the fire protection systems, the batteries must be on to provide electrical power to the system. Once engine shutdown is complete, battery shutdown and disconnect procedures can be accomplished if access is possible. Smaller general aviation aircraft may require fuel switches or fuel cutoffs to be de-

Figures 3.53 a and b Commercial aircraft engine shutdowns (fire T-handles) are usually centrally located and easy to find. *Courtesy of William D. Stewart.*

Figure 3.54 Locating and operating throttles and fuel control levers may need to be accomplished to shut down general aviation aircraft. *Courtesy of William D. Stewart.*

activated in addition to retarding the throttle(s) to shut down the aircraft (Figure 3.54).

Military aircraft often require personnel to follow a highly detailed set of procedures to accomplish aircraft shutdown. If any ARFF personnel are unfamiliar with these procedures, it is recommended that they stay clear of the cockpit to avoid injury. Many of the larger military aircraft resemble commercial aircraft of the same type, so shutdown procedures often are the same.

It is crucial that rescue workers become familiar with the aircraft that operate in and around the airport. It is also important to be familiar with the operation of cockpit windows and hatches that can be opened to aid in ventilation, extrication, and egress if necessary.

Ingress/Egress Systems

Aircraft are generally designed to be evacuated in 90 seconds or less in the event of an emergency. A main cabin door is provided for normal enplaning and deplaning operations while service doors are provided for catering and cleaning operations. These cabin doors are the primary means of egress, with secondary means consisting of over-/under-wing hatches, tail-cone jettison systems, rear air-stairs or stairs that lower at the rear of the aircraft, and roof hatches.

Airline flight crews identify cabin doors on an aircraft by reference to a number and a left or right designator. For example, a door may be referred to as "L1" or "1 Left," meaning left side, first door closest to the front of the aircraft. Remember that the left side refers to the pilot's left as he or she is seated in the cockpit. "R2" or "2 Right" would refer to the right side, second door back from the cockpit (Figure 3.55). These designations become very important when communicating with the flight crew or assigning personnel to assist evacuation operations. Because the design of aircraft egress systems varies so much, ARFF personnel must make it a point to become familiar with the aircraft that frequent their airport and know the operation of all the various emergency egress systems.

Aircraft Doors

Primary egress from aircraft is through the doors normally used for servicing or for routine entry or exiting. These doors may be located on both sides or on just one side of the fuselage and are usually operated by simple mechanisms. All doors have an exterior latch release that disconnects the locking device and permits the door to swing open, pivot open, swing down, or fall free from the aircraft (Figure 3.56).

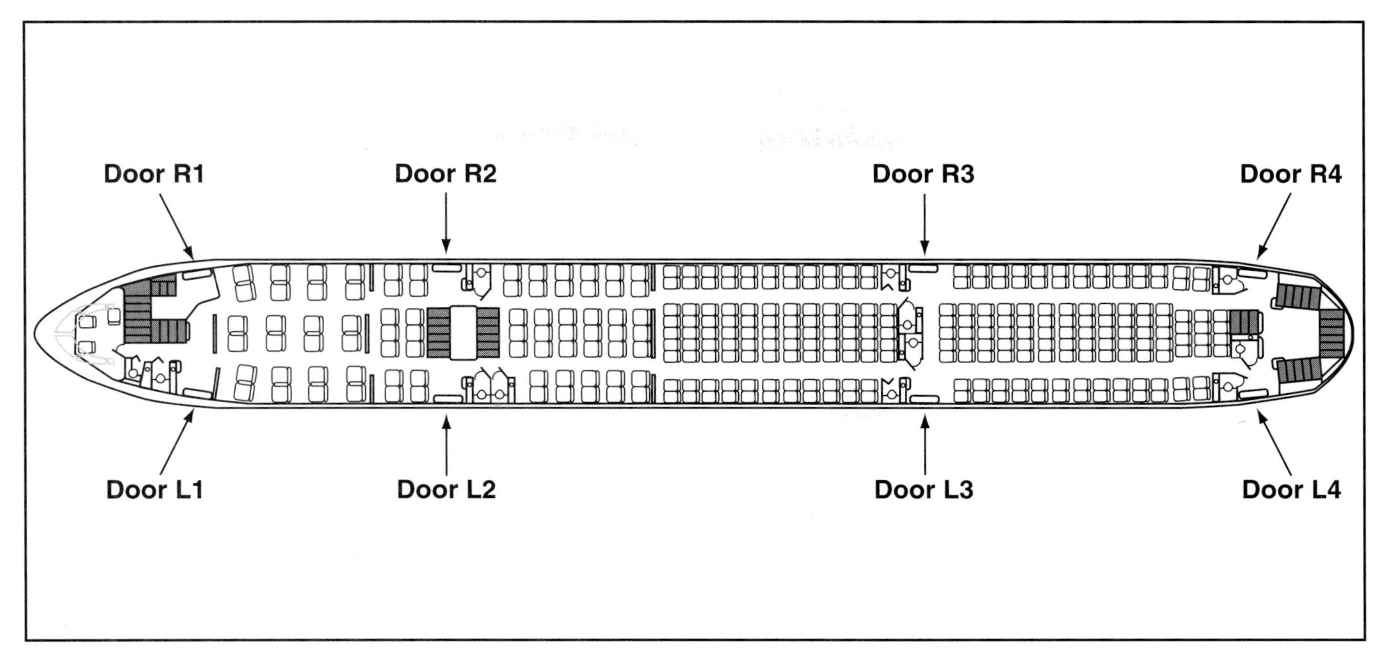

Figure 3.55 Door numbering is helpful when directing responding apparatus and personnel.

Figure 3.56 Cabin doors on general aviation aircraft may incorporate a two-piece door configuration.

Figure 3.57 Knowledge of door operation from the inside of the aircraft is critical when performing interior fire and rescue activities. *Courtesy of William D. Stewart.*

There are many different variations of cabin doors depending on the size and type of aircraft encountered. Opening and operating procedures can vary widely on doors found on the same aircraft, so spend time reviewing the aircraft that operate at your airport. Knowing how to operate the door from inside the aircraft also is vital should quick egress become necessary during interior operations (Figure 3.57).

Some wide-bodied commercial aircraft incorporate doors that open by moving upward into the fuselage. These doors also have an escape slide for egress but are designed to de-arm when opened from the outside. The doors move on tracks and are built with a pneumatic-assist feature that powers the door open when opened in the armed or emergency mode. The exterior operating handle can be found either forward or aft of the door on the side of the fuselage.

Commercial airliners are required to have floor lighting installed in the aisles to provide egress assistance for passengers. Track lighting is usually installed directly on the aisle floor or at the base of the seats. White or green lights lead to red lights to indicate the location of emergency exits. Additionally, exit signs are required to be installed at floor level (no more than 13 inches [325 mm] off the floor) at each emergency exit. If, while conducting an interior search, the need to evacuate arises, remember the lighting layout.

Small general aviation aircraft may have exit doors on both sides of the fuselage; others have them on one side only. Unlike most larger aircraft, some commuter aircraft have exit doors that are hinged on the bottom, open downward, and have steps built into them (Figure 3.58). These doors are assisted by compressed-air pistons or by heavy spring tension. Staying clear of the path of these doors is important to avoid injury.

Figure 3.58 Commuter aircraft have exit doors that are hinged at the bottom, swing out and down when opened, and are often equipped with steps. *Courtesy of William D. Stewart.*

If the bottom doorsill is greater than 6 feet (2 m) from the ground, the aircraft is required by the Federal Aviation Administration to be equipped with an inflatable emergency escape slide. Cabin doors are considered armed when the "girt bar" for the escape slide is secured to retention clips located at the bottom doorsill or when the interior door lever is moved to "armed mode." Slides and doors are armed by the cabin crew and remain armed from the time the aircraft leaves a gate to the time its flight ends at the next gate.

When cabin doors are in the armed mode, opening the door from outside the aircraft may be difficult due to resistance and may be dangerous due to the inflation and deployment of the escape slide. Because the slide deploys in a matter of a few seconds with explosive force, ARFF personnel must be very cautious when opening doors from the outside under emergency conditions (Figure 3.59). If ground ladders must be used to access these doors, they should be positioned beside the door on the side opposite the hinges. This is aft of the door on almost all commercial aircraft with hinged doors. As discussed earlier, some models of wide-body aircraft have doors which retract into the overhead area. On these aircraft, the slide can usually be de-armed by normal opening procedures from the outside of the aircraft. In this case, position the ladder on the side where the door controls are located, which may be found on either side of the

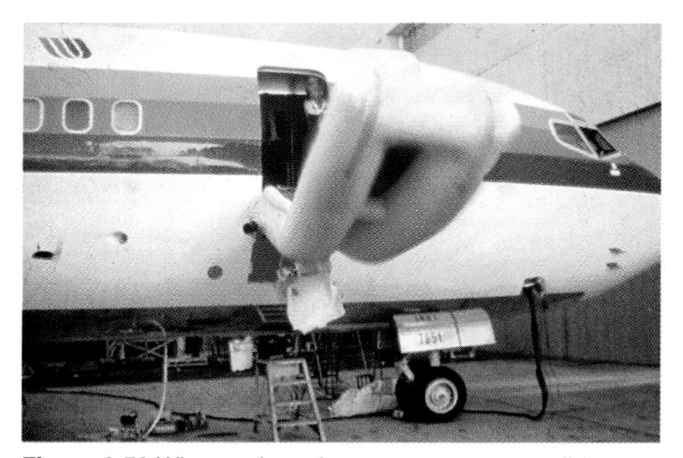

Figure 3.59 When activated, emergency escape slides can be fully inflated in a matter of seconds. *Courtesy of the National Audio Visual Center.*

door. Doors can be armed or de-armed from outside the aircraft on almost all wide-body aircraft and some newer narrow-body aircraft.

When the flight crew has initiated an evacuation, ARFF personnel should expect to see every usable exit open and slides deployed as soon as the aircraft comes to a stop (Figure 3.60). Once deployed, the escape slides must be protected from flame impingement and held steady in high-wind conditions. Flight attendants may have instructed several passengers to stay at the bottom of the slides to help evacuees. Available emergency responders should also assist when possible (Figure 3.61). Without assistance, people tend to pile up at the bottom of the slide, often causing additional injuries. Since the escape slides are extremely slippery due to a Teflon coating, as many as 10 to 15 percent of aircraft occupants suffer minor to moderate injuries going down the slides.

Most escape slides inflate automatically after the door is opened. A few older, narrow-body aircraft slides have a manual inflation handle. All escape slides have a pull-type, manual inflation handle somewhere at the top of the slide. It is usually marked and red in color. Sometimes escape slides fail to inflate for various reasons. Deflated slides still may be stretched out by evacuees and emergency responders and used for escape. They also can be disconnected from the aircraft and used as rafts. The top of the slide has a lanyard to detach the slide raft from the aircraft after passengers have boarded during a water evacuation.

Figure 3.60 If the areas outside the aircraft doors are clear of fire, all available exits will be used to evacuate passengers. *Courtesy of William D. Stewart.*

Aircraft Hatches

Commercial passenger-carrying aircraft may have escape hatches or doors over the wings. On high-wing models, these hatches are found under the wings. These exits, much like main cabin doors, all operate somewhat differently. One thing the hatches or windows that are on pressurized aircraft have in common is that they are plug-type in design. This plug-type design means these exits must be pushed inward from the outside or pulled inward from the inside to be opened. To aid escaping passengers in sliding off the wing to the ground, the pilot usually lowers the rear flaps (Figure 3.62). Many newer aircraft incorporate an inflatable escape slide that is activated when overwing exits are opened from inside the aircraft. Most overwing escape slides are designed to de-arm when the exit is opened from the outside. This escape slide is housed in the side of the fuselage and when activated expands outward and off the trailing edge of the wing. Again, aircraft familiarization training is necessary to

Figure 3.61 If possible, available emergency responders should assist in evacuation. *Courtesy of William D. Stewart.*

Figure 3.62 Flaps, when fully extended, offer an area for escaping passengers to slide down when exiting out overwing hatches. *Courtesy of William D. Stewart.*

ensure that all available means of evacuation are utilized during an emergency. On Boeing® 737 600-, 700-, 800-, and 900-series aircraft, the over-wing, plug-type hatch has been replaced with a spring-loaded over-wing hatch. When activated from either the inside or the outside, this hatch will spring upward and outboard of the fuselage (Figure 3.63). A firefighter should be aware of the rapid opening hatch and the hand-trapping hazards it may pose. Boeing calls this door an "automatic over-wing exit door."

Some commuter aircraft have over-wing or under-wing exits which are too small to accommodate a firefighter in full protective clothing with SCBA, but these exits offer a good opening for introducing handlines. Over-wing exits on larger aircraft are large enough to allow easy access to the interior, even for firefighters in full protective clothing and SCBA. If ground ladders are needed to access over-wing exits, they should be positioned at the leading edge of the wing.

Figure 3.63 Newer escape hatches are designed to open out and away from the aircraft. *Courtesy of American Association of Airport Executives.*

Figure 3.64 Some medium-frame commercial aircraft are equipped with a stair access at the rear of the aircraft. *Courtesy of William D. Stewart.*

Other Means of Egress

Rear stairs, emergency exit doors, overhead hatches and tail-cone jettison exits are some of the other devices that may be available to assist in evacuating an aircraft. A few medium-frame aircraft incorporate stairs built into the rear of the aircraft (Figure 3.64). Although not designed as a true emergency exit, these stairs provide an alternate means of accessing the main cabin if the aircraft is on its wheels. ARFF personnel should ensure that the aircraft is stabilized prior to entering so that if the aircraft shifts, the means of egress is not closed off. Smaller aircraft may be built with an exit specifically designed for emergency use only. These exits are commonly found on commuter aircraft and fall off the aircraft when opened. Due to the weight of the exit, personnel must stand clear and exercise extreme caution when opening. A few aircraft have overhead hatches that, when installed, provide another means of evacuating and ventilating an aircraft. Most are located over the flight deck and vary in method of operation.

The last type of alternate exit involves the tail-cone jettison system (Figure 3.65). The system is activated from inside or outside the aircraft by pulling on the activation handle, which is located at the left rear portion of the fuselage. Once pulled, the tail cone separates from the aircraft and falls to the ground. From the opening, an escape slide deploys and automatically inflates. Depending on the model, passengers exit the cabin through either a hatch or a standard-size cabin door located on the back wall of the cabin. If the tail cone has not been jettisoned, firefighters need to search that area for trapped occupants.

Figure 3.65 Many aircraft are equipped with a tail cone jettison system. *Courtesy of William D. Stewart.*

WARNING

Firefighters must use caution when walking under a tail cone as it may be jettisoned while they are underneath it.

Generally, all aircraft that operate above 14,000 feet (4,267 m) have the ability to pressurize the cabin. The pressurization system operates by controlling a motor-driven outflow valve that opens and closes to regulate the amount of cabin air that is exhausted outside the aircraft. The outflow valve (Figure 3.66) may be located on the left or right side of the rear fuselage or on some aircraft, on the left side of the fuselage just forward of the wing. During normal operation on the ground, the cabin should be depressurized, which is indicated by the outflow valve being fully open. Pressurization automatically commences just prior to takeoff and is usually maintained until just after landing unless a malfunction prevents the outflow valve from opening properly. It is impossible to open the main cabin doors or the over-/under-wing escape hatches if the cabin is pressurized. So, before attempting to open one of these doors, firefighters should find and force open the outflow valve. Some aircraft entry and cargo doors may also have pressure-releasing devices.

Emergency Cut-In Areas

Attempts to forcibly enter an aircraft should be made only after all other means of entry have failed.

Figure 3.66 Typical cabin out-flow valve that indicates when the cabin is pressurized. *Courtesy of William D. Stewart.*

Military aircraft have distinctive identification points for forcible entry that are bordered with contrasting markings and are stenciled with the words "CUT HERE FOR EMERGENCY RESCUE." Some civilian aircraft may have cut-in marks painted on the exterior of the fuselage to indicate those points where access to trapped occupants is possible. Cutting a fuselage is a time-consuming process that taxes the strength and endurance of personnel. Personnel should take time to become familiar with areas on an aircraft that are suitable for cut-in areas, but they should use this method as a last resort because it is one of the most hazardous and time-consuming means of forcible entry.

Data Recording Systems

Of critical importance to aircraft accident investigations are the so-called "black boxes." Identified as the flight data recorder (FDR) (Figure 3.67) and the cockpit voice recorder (CVR) (Figure 3.68), they are usually located in the tail section of the fuselage. Neither unit is black, but they are painted either international orange or bright red with a wide band of reflective material around them. As

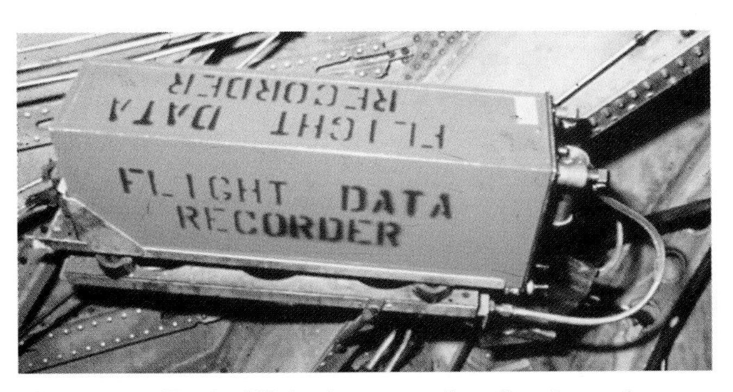

Figure 3.67 Typical flight data recorder. *Courtesy of National Audio Visual Center.*

Figure 3.68 Typical cockpit voice recorder.

with any other piece of evidence, these units should be protected in place and should only be removed by ARFF personnel if the units are in imminent danger of being damaged or destroyed. A chain of custody should be established and maintained until the proper authorities can remove the recorders from the area. If a unit is found submerged in water, it should be left in water because the recording device contains a metal ribbon that is conducive to rusting. If there is a chance of them being lost, remove them from the water and store them in fresh water until retrieved by the National Transportation Safety Board (NTSB).

Military Aircraft

The Armed Forces of the United States and Canada operate a wide variety of aircraft to support military objectives. Military aircraft fly over every part of the continent and often use or are based at civilian airports. Therefore, all firefighters should be familiar with the potential hazards associated with military aircraft.

This section describes the various types of military aircraft in use, the emergency systems incorporated with these aircraft and the procedures to follow if involved in a response to a military aircraft accident. Also covered are the various aircrew survival systems, actuating devices, aircraft emergency shutdown procedures, weapons and weapons systems, as well as appropriate fire fighting procedures.

Military Aircraft Types

Terminology unique to military aircraft is used to describe the particular aircraft types and to identify where weapons and ordnance may be found. Familiarity with these terms and with aircraft design variations allows ARFF personnel to better prepare for an emergency involving these military aircraft. It is important to remember that military operations are not limited to the areas around military bases but can extend to anywhere within the country. Military aircraft are often used to assist in civilian rescue operations, disaster relief, and other types of emergencies.

Military aircraft are assigned a designator letter which can be used by response workers to identify the aircraft type. These letters correspond to an aircraft and its assigned mission. Some aircraft may function in a dual role, but each carries the letter for which it originally was designed. These designator letters are the following:

A — Attack

B — Bomber

C— Cargo/passenger

E — Special electronic installation

F — Fighter

H — Helicopter

K — Tanker

O — Observation

P — Patrol

R — Reconnaissance

S — Antisubmarine

T — Trainer

U — Utility

X — Research

Some of the more common types are discussed in the following sections.

Fighter and Attack Aircraft

These aircraft are identified by an F or A (for example, F-22 [Figure 3.69] and A-10). Fighter and attack aircraft are designed to engage in air-to-air combat and/or to attack targets on the ground. Some attack aircraft may be as large as the four-engine AC-130 gunship, but most incorporate a one- or two-seat configuration. Weapons such as

Figure 3.69 The multiuse F-22 can accomplish a variety of missions for the military. *Courtesy of United States Air Force.*

internally mounted cannons, missiles, and bombs are carried beneath the wings and/or in the fuselage. Except for the AC-130 and similar large gunships, fighter and most attack aircraft are equipped with canopy-removal systems and ejection seats. The weapon racks and any externally mounted fuel tanks are designed to be jettisoned by using small explosive devices. The cargo-type aircraft that have been converted to attack aircraft are equipped with conventional weapons and carry a substantial amount of ammunition.

Bomber Aircraft

Identified with a B (B-1, B-2, and B-52, for example), these aircraft designed to carry and drop a large quantity of air-to-ground weapons are referred to as bombers (Figure 3.70). They can be four- to eight-engine aircraft and hold a crew of two to eight. They have explosive ejection seats and can carry weapons internally, externally, or both. A large fuel load and a significant quantity of high explosives are to be expected on a fully loaded aircraft of this type.

Cargo Aircraft

ARFF personnel may be most familiar with the C-5, C-17, C-130, and C-141, which are used to carry cargo. Military cargo aircraft may range from relatively small to quite large (Figure 3.71). Most military cargo aircraft are designed as dual-role aircraft in that they carry both cargo and/or personnel at the same time. These aircraft do not have ejection seats or canopy-removal systems; however, they may have jet-assisted takeoff (JATO) units attached to the sides of the fuselage. Cargo aircraft may carry a wide variety of cargo, which may include armored personnel vehicles, tanks, munitions, food, personnel, and supplies.

Figure 3.70 Long-range bombing is still a critical part of the military's mission. *Courtesy of United States Air Force.*

Tanker Aircraft

Tanker aircraft are cargo aircraft modified for inflight refueling of other aircraft and carry the letter K as a designator letter. Some examples include the KC-10 and the KC-130. They may be configured to perform the dual functions of cargo transport and fuel tanker at the same time (Figure 3.72). Obviously, their exceptionally large fuel load sets them apart from other cargo-type aircraft. Fuel load for this style of aircraft can be over 50,000 gallons (200 000 L), so responders may be faced with a very large volume of fire in the event of an accident.

Utility Aircraft

Identified with a U, utility aircraft are usually relatively small aircraft that perform a variety of support functions. They normally do not carry weapons or have ejection systems, and they are quite similar to

Figure 3.71 Military airlift command can support frontline operations through the use of large-capacity cargo aircraft. *Courtesy of United States Air Force.*

Figure 3.72 Inflight refueling can extend the reach of long-range missions. *Courtesy of United States Air Force.*

general aviation aircraft (Figure 3.73). Passenger load varies depending on the size and mission of the aircraft. An exception is the U-2 high-altitude reconnaissance aircraft. Although designated with a U, this is a specialized jet equipped with an ejection seat and sophisticated surveillance equipment.

Special-Purpose Aircraft

Special-purpose aircraft serve many functions for the military such as reconnaissance, command and control, testing, or electronic surveillance. Some special-purpose aircraft have a drastically different appearance, while others may be configured differently than other aircraft but from the exterior may not appear any different. Their designator letter varies depending on the mission of the aircraft.

The term *special-purpose aircraft* also describes the various uses of a common military aircraft. For example, the military version of the Boeing 707 is used as a C-135 cargo aircraft, KC-135 tanker, EC-135 electronics platform, E3A AWACS (airborne warning and control system) aircraft, and advanced range instrumentation aircraft (ARIA).

Helicopters

Helicopters play a major role in military operations and account for a large part of the aviation fleet. All branches of the armed services incorporate helicopters as part of operations (Figure 3.74). When ammunition and weapons are attached, they are usually carried inside the cabin or on pods attached to the fuselage. Auxiliary fuel tanks also may be carried internally or externally. Helicopters normally carry a crew of two to five but also may carry passengers and equipment. The AH-1 and AH-64 have hatches that can be jettisoned explosively.

Aircraft Emergency Systems

Military aircraft often operate in extremely hostile environments. Depending on the mission for which they have been designed, some aircraft have certain emergency systems to increase the chances of survival for the aircrew. While some of these systems are typical for any aircraft, some are quite unique to military aircraft. On a military aircraft with a hydrazine-powered emergency power unit (EPU), personnel must safety the EPU before de-energizing the electrical power. If this is not done and the power is shut off before the EPU is secured, the EPU will automatically start.

Fire Protection/Detection Systems

Almost all military aircraft have either halon or nitrogen fire extinguishing systems to protect the engines. Much like commercial aircraft, these systems can be activated by T-handles in the cockpit. T-handle activation usually shuts down the electrical power and fuel to the affected engine. Most of the fire systems are equipped to deliver "one shot" of extinguishing agent while others allow several shots to extinguish the fire. Some aircraft have extensive fire detection systems — all of which are monitored in the cockpit.

Emergency Doors/Hatches

As on any other type of aircraft, military aircraft have exit doors and hatches that may be used for

Figure 3.73 High-level reconnaissance can provide information to the combat mission. *Courtesy of United States Air Force.*

Figure 3.74 Low-level frontline strikes are often performed by fast-moving helicopters. *Courtesy of United States Air Force.*

emergency egress. The specific locations and operation of the systems vary with each model of aircraft, and some familiarity with the aircraft is necessary to effectively operate them under emergency conditions. Some military aircraft have hatches on the top of the fuselage; some aircraft, such as the C-5, KC-10, C-9, and T-43, have escape slides like those on commercial airliners.

Ejection Systems

One more method of flight crew egress involves the use of ejection systems to rapidly remove the aircrew when in danger. These systems, normally found in fighter, attack, bomber, and training aircraft, can be *extremely* dangerous to ARFF personnel and must be treated with extreme caution at all times.

Ejection seats may be rocket-powered or gas-powered. Some systems fire a single seat, some two seats, and some a complete module from the aircraft. Some systems, referred to as "zero-zero" systems, may be fired while the aircraft is on the ground and parked. Others require the removal of a hatch before the seat will fire while the aircraft is on the ground. Crew members fire the seats by pulling up on an arm rest, pulling a handle between their legs, or pulling a face curtain from behind their head (Figure 3.75).

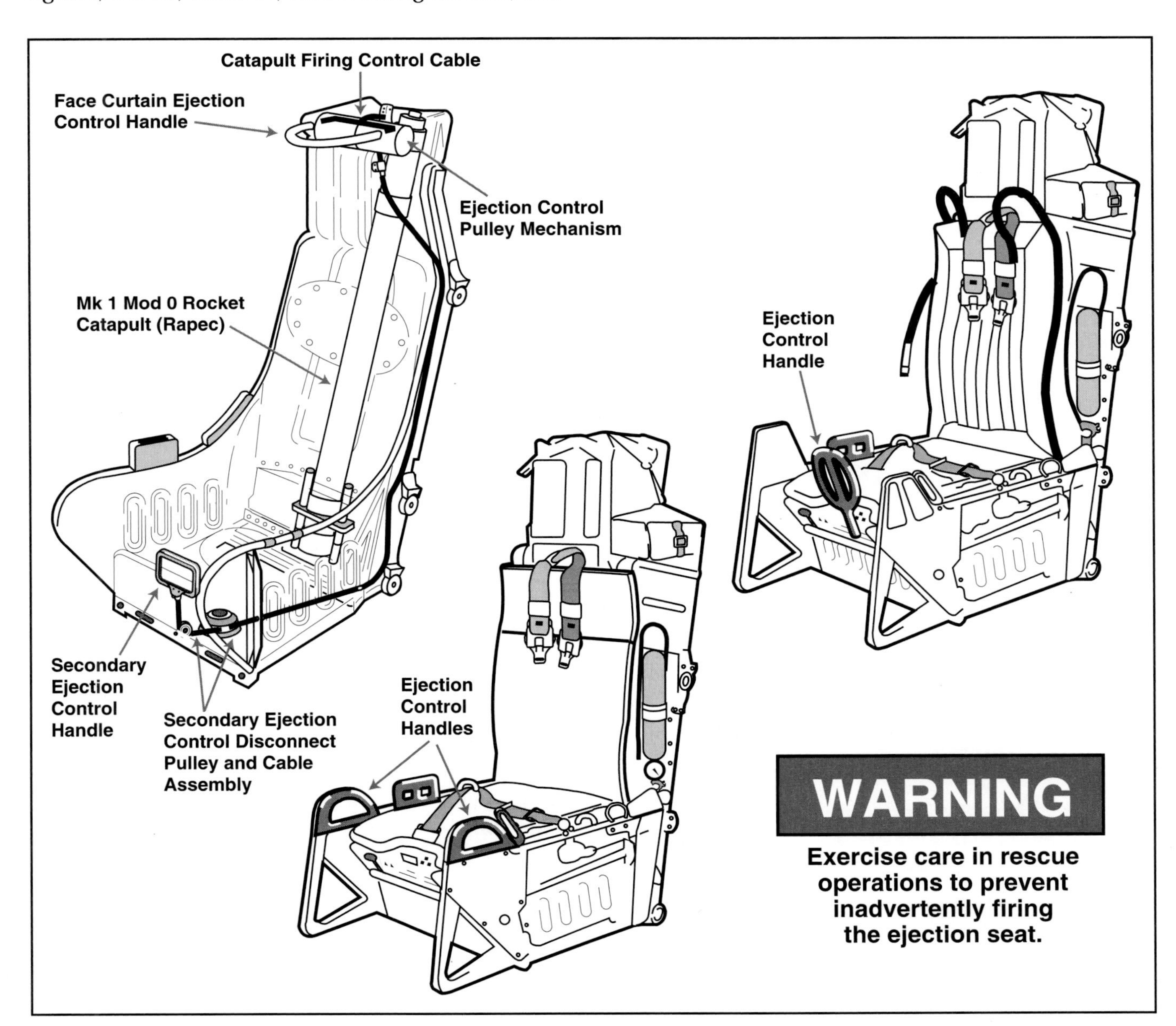

Figure 3.75 Ejection seat firing mechanisms vary with each type and model of aircraft, and ARFF personnel must use extreme caution when attempting to safety a seat.

Without proper training, ARFF personnel opening a hatch during an emergency could cause the seat to fire unless it is de-armed and placed in the safe mode. To safely de-arm an ejection system requires that safety pins be inserted into the correct positions on each seat, that catapult hoses be cut, or that the "head knocker" be pulled, depending on the type of seat (Figure 3.76). Because several pins may be required to safety a seat or because hoses may have to be cut in several places, hands-on training is the only way to become competent and confident in emergency procedures. For more specific information, see the "Securing Canopies and Seats" section later in this chapter.

Canopies

The canopy, which encloses the cockpit, consists of a metal framework with a transparent covering, usually of Lexan® or similar high-impact plastic. It is designed to protect the pilot or crew member while providing them with unobstructed visibility. Of the three types of canopies (Figure 3.77), the most common are the clamshell and the sliding types. The sliding canopy is easier to operate during rescues because it does not present as many restrictions as the clamshell canopy.

Canopies are actuated in various ways. Under normal conditions, they may be opened and closed pneumatically, electrically, hydraulically, or manually. In the event of malfunction or mechanical damage to the opening system, power-operated canopies may be opened manually. Once opened, they must be held or propped open with a canopy lock or strut, which prevents the canopy from slamming shut. Canopies weigh several hundred pounds (kilograms). Some canopies are disintegrated with explosives built into the shell or along the canopy frame, and the pilot is ejected through the debris. Only trained, qualified personnel should perform cutting operations to remove pilots.

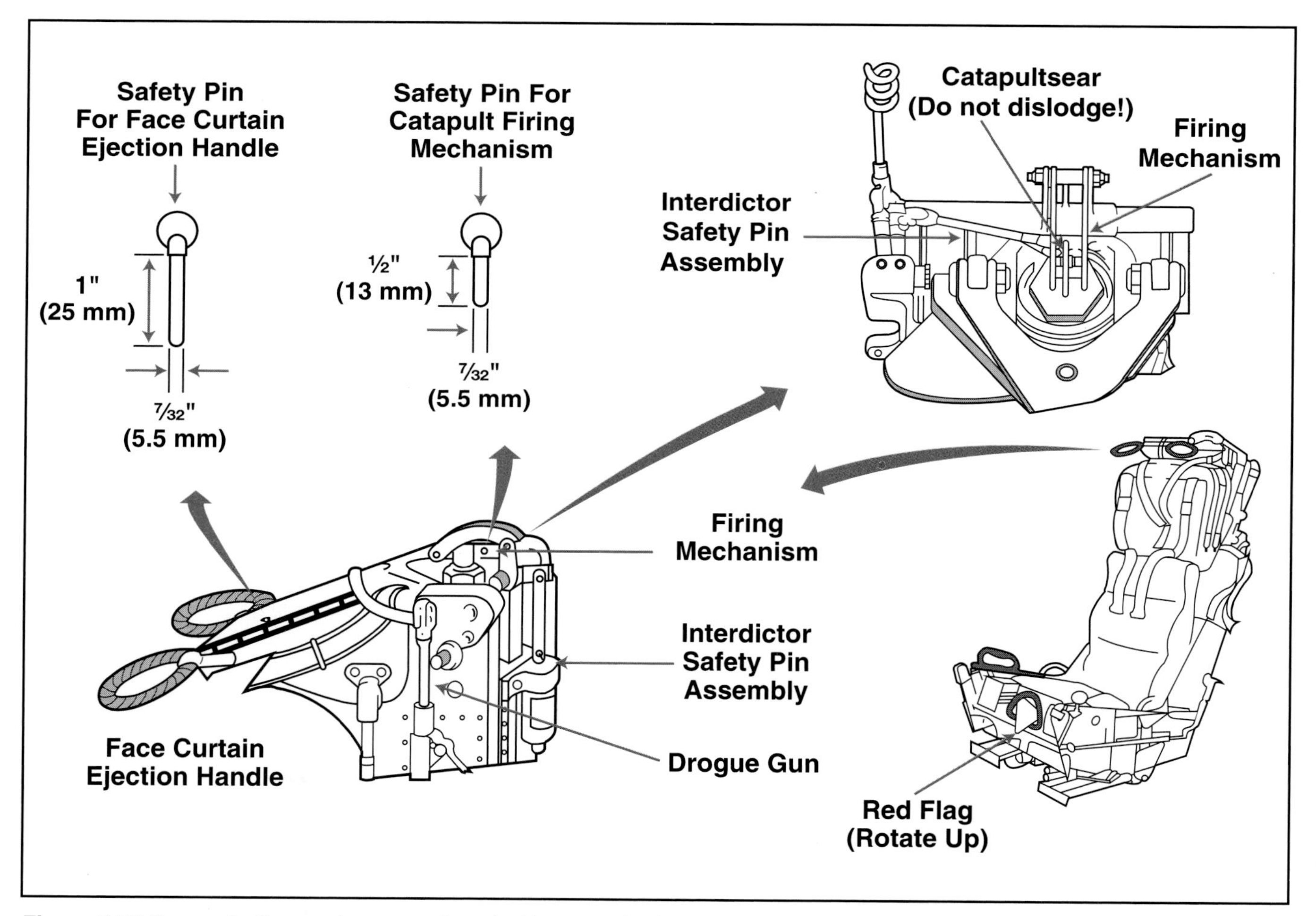

Figure 3.76 Some ejection seats are equipped with an arming/de-arming lever in the middle of the headrest, known as a "head knocker."

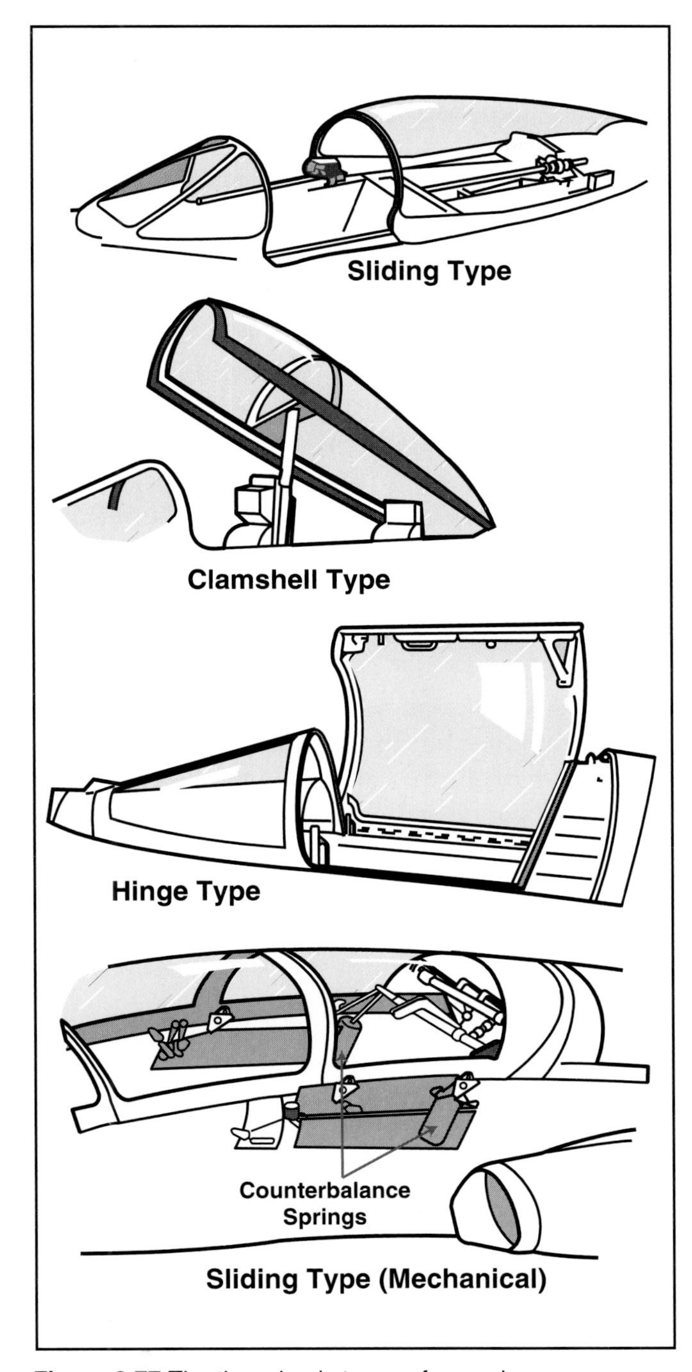

Figure 3.77 The three basic types of canopies.

Most military aircraft with ejection seats have an external means of jettisoning the canopy in an emergency. This system may be fired from either side of the aircraft. The canopy is fired by an explosive device that is intended to throw the canopy up and away from the aircraft. Caution should be used when firing this system as the canopy may fall back on the aircraft after it has been fired. Canopies or hatches should be jettisoned only if absolutely necessary.

CAUTION: Jettisoning the canopy may ignite fuel vapors. When jettisoning the canopy, follow the directions that are printed on the side of the aircraft.

Propellant Actuating Devices

Canopy-jettison and seat-ejection systems use explosive charges contained in propellant actuating devices. These devices include canopy removers, initiators, rotary actuators, explosive squibs, thrusters, and seat catapults. Each of these devices are components of a seat-ejection system. Automatic sequencing of these devices ejects crew members from the moving aircraft, although some systems have zero-zero capabilities that enable crew members to eject while the aircraft is at zero altitude and zero velocity.

Canopy Removers

Generally, canopy removers are gas-pressured telescoping devices that forcibly jettison the canopy in an emergency. When the actuating cartridge is fired, the rapidly expanding gases force the telescoping tubes to extend and jettison the canopy from the aircraft.

Initiators

Initiators are cylindrically shaped devices that provide the gas pressure required to start a sequence of events in the emergency ejection process (Figure 3.78). Some are fired by gas pressure and others by mechanical pressure. Some fire immediately upon actuation, while others have a time delay. When the

Figure 3.78 Initiators are part of larger explosive devices.

initiator pin is pulled, a firing pin strikes a cartridge, which in turn fires the initiator. The hot gases produced by the burning initiator propellant flow through a tube or hose causing other components to eject the canopy or seat.

Rotary Actuators

Rotary actuators perform various mechanical functions in aircraft or related equipment. They are activated by the gas pressure produced by other devices, such as initiators, or by an electrical current. Part of the "canopy remover pad release" system, they forcibly separate the crew member from the seat after ejection.

Thrusters

Thrusters are gas-operated devices that unlock or reposition various units in the escape system during the ejection sequence. For example, thrusters unlock the canopy latches just before canopy jettison, and they move seat and leg guards into position prior to ejection.

Explosive Squibs

Explosive squibs are small metal tubes closed at one end and plugged with a crimped-in rubber plug at the other end. The squib contains flammable mixtures that produce pressure or provide an ignition source when initiated. There are two types: the flash-vented type and the closed-end type. Flash-vented explosive squibs do not explode but emit a small flame, and they are often used to ignite rocket motors. Closed-end squibs are low-powered explosives that are normally used in explosive bolts, explosive release mechanisms, and fixed fire extinguishing systems.

Seat Catapults

Catapults are telescoping ejection devices used in the emergency ejection of the aircrew. They are designed either for upward or downward thrust depending on the type of aircraft. Two types of catapults are used for pilot ejection: cartridge type or rocket-motor type. Used in older ejection systems, the cartridge catapult propels the pilot and seat with sufficient force to clear the aircraft after the canopy has been jettisoned. The rocket catapult is more efficient and is employed in more advanced, high-speed aircraft. The rocket catapult provides increased thrust to make sure that the crew member successfully ejects, especially in low-altitude situations. During high-impact crashes and total breakup of military aircraft, all the dangerous components discussed may be scattered throughout the impact area or crash path.

Securing Canopies and Seats

Accidentally activating ejection seats and canopies may be *extremely* dangerous for ARFF personnel. The catapult containing the explosive charge for the ejection seat may hurl a 300-pound (136 kg) object at an initial rate of 60 feet (20 m) per second. Therefore, it is essential that ARFF personnel know how to safely secure canopies and ejection seats.

All seats have ground safety features that make the seats relatively safe for ARFF personnel to work around while removing crew members. The ejection system may be safetied by interrupting the firing sequence, cutting the initiator hose, or pinning the ejection handles. However, the specific method of safetying the ejection seat depends upon its manufacturer, the model of the seat, and how it may have been modified.

> **WARNING**
>
> **Safetying an ejection system is a hazardous operation. Rescue personnel should not attempt to safety an ejection system if they do not have the proper training and equipment necessary for the specific model of aircraft.**

Weapons and Weapon Systems

In an effort to ensure national security, military aircraft may carry a broad range of weapons and explosives at any time. These armaments may be carried in various forms such as ammunition for machine guns, pyrotechnics, rockets and missiles, and gravity bombs. Unless an aircraft is carrying external weapons, ARFF personnel may have no way of knowing if weapons are on board (Figure 3.79).

Figure 3.79 ARFF personnel should *always* expect the presence of weapons when dealing with military aircraft. *Courtesy of United States Air Force.*

Figure 3.80 Always approach military aircraft in an effort to avoid guns and missiles that may inadvertently activate. *Courtesy of Jeff Riechmann, Riechmann Safety Services.*

High Explosive (HE)

While HE itself is not a particular type of weapon, it is present to some degree in all weapons. Explosives must not be disturbed in any way by ARFF personnel. A military explosive ordnance disposal (EOD) team should be called to the scene. There are two distinct types of HE, *pressed* and *cast*; and they react somewhat differently when exposed to fire.

Pressed high explosive is pressed into an operational container such as a bomb case. In a fire situation where the ammunition case is not ruptured or broken open, radiant heat or direct-flame impingement will conduct heat through the case to the explosive. This buildup of heat in the explosive may eventually cause a detonation or deflagration of the ammunition. When dealing with confined high or low explosives in an area with fire and/or high heat, detonation most likely will occur.

If the ammunition case is ruptured or broken open, any excessive heat will cause the exposed explosive to burn. This burning explosive could result in a detonation or deflagration. Burning explosives produce flames of various colors. They may be red, greenish-white, yellow, or almost any other color. They normally burn with a very bright, flare-like luminance.

Cast high explosive is heated during the manufacturing process to become a thick liquid that is then poured into the ammunition case, where it slowly cools and re-solidifies. This type of explosive will react the same as pressed high explosive if the ammunition case is not ruptured or broken open. However, if there is any opening in the ammunition case when it is involved in fire, the HE will melt, run, and resolidify as it cools. Once the explosive resolidifies, it becomes *extremely* sensitive to shock or friction. Driving over or stepping on this resolidified HE may cause it to detonate. Regardless of the type involved, explosives should be expected to be scattered throughout the area in a military aircraft accident.

Ammunition

Another hazard to which ARFF personnel may be exposed is ammunition. Fighter and bomber aircraft normally carry internal guns with ammunition drums. This type of ammunition may react violently or discharge when involved in a fire. Guns may be located in the nose section or in the wing roots of fighter and attack aircraft. Personnel and apparatus should not be positioned to put themselves in line with the gun ports (Figure 3.80) They should take position at approximately a 45-degree angle off the nose or tail of the aircraft, provided this does not place them in front of or behind under-wing rockets or missiles.

Pyrotechnics

Another type of explosive found on military aircraft is pyrotechnics. All pyrotechnics present at least a minor explosive hazard. Photoflash cartridges, used for illumination during parachute drops, contain white phosphorous and produce a blinding white light when ignited. These units are *extremely* dangerous and may be found in different locations on many aircraft. Other pyrotechnics that may be found are units that dispense chaff and high-intensity flares. All these devices burn very hot and may ignite surrounding combustibles. Most pyrotechnics burn readily, and once ignited they are very difficult to extinguish because they contain their own oxidizers.

Rockets and Missiles

Rockets and missiles are self-powered weapons carried on various aircraft (Figure 3.81). The significant difference between rockets and missiles is that missiles have a guidance-and-control system, and rockets do not. Rockets must be aimed and fired in the direction of a target. However, there is no difference in the explosive potential of either type of weapon. Both types may be carried in internal bays, on wingtips, and on external pylons. Bomber aircraft may carry cruise missiles in both external and internal mounts.

Gravity Bombs

A significant destructive force can be found in the form of gravity bombs that come in many shapes and sizes. One of the largest bombs currently in the United States inventory is the 2,000-pound (907 kg) MK 84. Although many gravity bombs appear to be similar, they have various capabilities. Some have parachute-ejection devices (Figure 3.82), some separate and dispense smaller "bomblets," while some carry tear gas. When a bomb is involved in fire and cannot be cooled quickly, the area must be evacuated immediately to no less than 2,000 feet (600 m). If there is a detonation, other weapons in the area may also explode. Water is the agent of choice when attempting to cool a gravity bomb. Foam should not be used because it insulates the weapon and restricts dissipation of heat.

Nuclear Weapons

Nuclear weapons can be affixed to many different types of military aircraft. Nuclear weapons on aircraft are generally restricted to military installations, and aircraft do not usually fly with these weapons aboard unless they are directly involved in a war situation. Handling incidents involving nuclear weapons is the responsibility of military fire suppression personnel who have received specific training and guidance on these weapons systems. Because the chance of a nuclear detonation is extremely remote, the principal hazard of these weapons is the high explosives they contain.

Figure 3.81 Rockets and missiles may be found on a wide variety of military aircraft. *Courtesy of United States Air Force.*

Conventional Weapons/Munitions Fire Fighting Procedures

While the same fire fighting procedures used on civilian aircraft apply to unarmed military aircraft, there are distinct differences in fire fighting procedures when aircraft are carrying explosives. The primary effort must be directed toward accomplishing a quick knockdown of the fire and cooling of the munitions to maintain a survivable environment. When involved in fire, a weapon or explosive may be expected to detonate in 45 seconds to 4 or 5 minutes, depending on the type of weapon involved. Every effort should be made to extinguish and/or control the fire before the weapons become involved.

WARNING

Do not attempt to fight a fire in which a weapon is involved if it is not possible to extinguish the fire quickly. Because of the likelihood of detonation, all firefighters should withdraw at least 2,000 feet (600 m). If a rescue is in progress, continue to apply water (not foam) in copious amounts until the rescue is complete.

Figure 3.82 Gravity bombs with parachute-ejection devices. *Courtesy of United States Air Force.*

Chapter 4

ARFF Firefighter Safety

Job Performance Requirements

This chapter provides information that will assist the reader in meeting the following job performance requirements from NFPA 1003, *Standard for Airport Fire Fighter Professional Qualifications*, 2000 edition. Particular portions of the job performance requirements (JPRs) that are addressed in this chapter are noted in bold text.

3-1.1.1 General Knowledge Requirements. Fundamental aircraft fire-fighting techniques, including the approach, positioning, initial attack, and selection, application, and management of the extinguishing agents; limitations of various sized hand lines; **use of proximity protective personal equipment (PrPPE)**; fire behavior; fire-fighting techniques in oxygen-enriched atmospheres; reaction of aircraft materials to heat and flame; critical components and hazards of civil aircraft construction and systems related to ARFF operations; special defense area and limitations within that area; characteristics of different aircraft fuels; hazardous areas in and around aircraft; aircraft fueling systems (hydrant/vehicle); aircraft egress/ingress (hatches, doors, and evacuation chutes); hazards associated with aircraft cargo, including dangerous goods; hazardous areas, including entry control points, crash scene perimeters, and requirements for operations within the hot, warm, and cold zones; and **critical stress management policies and procedures.**

3-2.2 Communicate critical incident information regarding an incident or accident on or adjacent to an airport, given an assignment involving an incident or accident and an incident management (IMS) protocol, so that the information provided is accurate and sufficient for the incident commander to initiate an attack plan.

(a) *Requisite Knowledge:* **Incident management system protocol**, the airport emergency plan, airport and aircraft familiarization, communications equipment and procedures.

(b) *Requisite Skills:* Operate communications systems effectively, communicate an accurate situation report, implement IMS protocol and airport emergency plan, recognize aircraft types.

3-2.4 Perform an airport standby operation, given an assignment, a hazardous condition, and the airport standby policies and procedures, so that unsafe conditions are detected and mitigated in accordance with the airport policies and procedures.

(a) *Requisite Knowledge:* Airport and aircraft policies and procedures for hazardous conditions.

(b) *Requisite Skills:* **Recognize hazardous conditions and initiate corrective action.**

3-3.7 Attack a wheel assembly fire, given PrPPE, an assignment, an ARFF vehicle hand line and appropriate agent, so that the fire is controlled.

(a) *Requisite Knowledge:* Agent selection criteria, **special safety considerations, and the characteristics of combustible metals.**

(b) *Requisite Skills:* **Approach the fire in a safe and effective manner**, select and apply agent.

Every emergency to which a firefighter responds has the potential to be dangerous. Each firefighter not only is responsible for his or her own safety but also must look out for the safety of the entire team. Knowing some of the obvious safety concerns when responding to aircraft fires and other emergencies will help minimize the danger of being injured or killed.

Fire department standard operating procedures (SOPs) should cover health and safety programs. Each member of the department is responsible for reading and understanding the SOPs. NFPA 1500, *Standard on Fire Department Occupational Safety and Health Program,* is an excellent source for finding information on the aspects of health and safety for firefighters.

This chapter provides an overview of general safety issues that concern ARFF personnel in particular. The topics discussed are personal protective equipment including self-contained breathing apparatus and personal alert safety systems (PASS), the Incident Management System (IMS), personnel accountability, two-in/two-out rule, hazards associated with aircraft rescue and fire fighting, personnel decontamination, critical incident stress, and firefighter safety in the fire station.

Personal Protective Equipment

Self-Contained Breathing Apparatus

Due to the potential for respiratory injuries, SCBA must be worn at all aircraft fires. ARFF personnel operating in and around aircraft involved in fire face the same toxic atmospheres they would encounter in typical structural fires (Figure 4.1). There are many atmospheric hazards associated with burning aircraft. Combustion can produce carbon monoxide, hydrogen sulfide, hydrogen cyanide, hydrogen chloride, and phosgene. In addition to these dangerous gases, there may be other toxic and hazardous materials in the aircraft cargo. Many civilian and commercial aircraft use carbon and other graphite fibers in construction, creating a hazard similar to asbestos exposure. Other hazards include superheated air, oxygen deficiency, extinguishing agents, and combustible metals. A more detailed study of SCBA can be found in the IFSTA **Essentials of Fire Fighting** manual.

Personal Alert Safety Systems

NFPA 1982, *Standard on Personal Alert Safety Systems (PASS) for Fire Fighters,* established the standards for personal alert safety systems. The PASS device, which all ARFF personnel must wear when entering a hazardous atmosphere, sounds an alarm when a firefighter becomes incapacitated. The unit sounds automatically when the wearer is motionless for approximately 30 seconds, or it can be activated manually by the wearer. It should be capable of emitting an alarm of 95 decibels (dB) at a distance of 9.9 feet (3 m) for an uninterrupted period of at least one hour. Some PASS devices detect heat, some are integrated into the self-contained breathing apparatus (SCBA), and some even send a signal to a remote transmitter letting the Incident Commander (IC) know that a firefighter is in trouble.

The PASS device was designed to address a small part of the problem of accounting for firefighters who become incapacitated in a smoke-filled build-

Figure 4.1 Personal protective equipment for airport firefighters includes SCBA.

ing. Wearing a PASS device increases the chances of a firefighter being found in an emergency — but only if it is turned on and working properly.

As with any electronic device, the PASS device can develop problems, with the most common problem being dead batteries. It is a good idea to change the batteries of a PASS device at some regular interval. One idea for departments that do not activate their PASS devices often is to change the batteries when time changes in the spring and fall. Firefighters tell citizens to change the batteries in their smoke detectors in the spring and fall; this would make great sense for the firefighter's lifesaving device as well.

The most realistic operational problem with the PASS device is the firefighter not remembering to activate the PASS before entering the hazardous environment. Some manufactures have removed this problem by integrating the PASS into the SCBA (Figure 4.2). This PASS activates when the breathing-air system is activated. Then, the only way the PASS can be deactivated is by shutting off the breathing air. This serves two purposes: a reminder to turn on the PASS, and a reminder to turn off the air to the SCBA.

Hearing Protection

Personnel involved in routine and emergency operations around aircraft are exposed to noise that may exceed the accepted level of exposure. ARFF personnel are also exposed to high noise levels when operating in and around aircraft fire fighting vehicles. NFPA 1500, *Standard on Fire Department Occupational Safety and Health Program,* defines the maximum level of noise to which fire protection personnel are allowed to be exposed in the work environment (Figure 4.3).

Figure 4.2 SCBA with an integrated PASS device. *Courtesy of International Safety Instruments.*

Hearing protection should be made available for firefighters on all ARFF equipment. Earmuff-type protection provides excellent sound reduction for the wearer. Earplugs, which should be fitted for the individual, should be provided to each member of the department. The use of hearing protection is also important in and around the fire station where noise-producing equipment is operating. Generators, power saws, air compressors, and other equipment may produce noise levels from which the firefighter should be protected (Figure 4.4).

A hearing awareness program, supported by appropriate SOPs, should be implemented to

Figure 4.3 Hearing protection for ARFF personnel may be necessary during routine and emergency operation.

Figure 4.4 An airport firefighter wears hearing protection while checking an ARFF apparatus.

establish and maintain an awareness of hearing conservation. Periodic hearing tests should also be available to ARFF personnel. The bottom line with a hearing awareness program is to wear hearing protection. Airport firefighters are around more noises than most other types of firefighters. Hearing loss will affect firefighters the rest of their lives, so they should take the seconds needed to wear hearing protection.

Eye Protection

SOP-supported programs for awareness of vision hazards and protective measures are of critical importance to the safety of ARFF personnel. Supervisors are responsible for knowing and enforcing the rules for vision protection without exception.

ARFF personnel risk eye injury from many different sources during routine and emergency operations around aircraft or fire fighting apparatus and equipment. Numerous projections, such as pitot tubes on aircraft and outside mirrors and other appendages on fire fighting apparatus, are all potential sources of eye injury if personnel are not alert and are not wearing eye protection.

When in the down position, helmet-mounted face shields provide reasonable protection against many common eye injuries caused by walking into projections, by airborne debris, or by being splashed by fire-extinguishing agents or other fluids (Figure 4.5). Other activities, such as operating power tools that generate sparks or clouds of dust and debris, may require that the firefighter also wear goggles or self-contained breathing apparatus.

Personal Protective Clothing

Aircraft fires present serious problems for all ARFF personnel operating at the aircraft accident/incident. Firefighters must have adequate protection. Personnel must wear full protective clothing and SCBA during the initial approach and attack, while performing rescue, and during overhaul.

Firefighters assigned to an airport response should use proximity suits, which are aluminized ensembles that provide superior radiant and thermal heat protection. Every firefighter must know the shielding capabilities and limitations of their personal protective clothing. The flame resistance, strength, and weight of the material are critical to the clothing's usefulness at aircraft incidents.

Station/Work Uniform

Normal work uniforms should be made of flame-resistant material. However, these uniforms are intended for use under full protective equipment and are not intended to be used by themselves as protective clothing. The U.S. Environmental Protection Agency (EPA) classifies work uniforms as Level D protection — suitable only for routine support functions (Figure 4.6).

Structural Fire Fighting Protective Clothing

Firefighters must sometimes respond to aircraft emergencies with only structural fire fighting protective equipment. They also respond to non-ARFF calls on the airport property. A firefighter in structural personal protective clothing equipment (PPE), which consists of a turnout coat (with collar up), turnout pants, safety boots, leather gloves, flame-resistant hood, helmet (with earflaps down), and SCBA, is adequately protected from all but the most extreme conditions (Figure 4.7). Therefore, while limited in some applications, this type of protective equipment may still offer firefighters sufficient protection if they are aware of the hazard being encountered and of the limitations of the protective clothing.

Structural turnout gear is very resistant to cuts and abrasions resulting from contact with jagged metal edges common in damaged aircraft. This type of gear has a moisture barrier to protect firefighters from steam burns and a thermal barrier to provide some heat protection. However, structural clothing is susceptible to "wicking" hydrocarbon fuels, and it does not provide the reflective capabilities of proximity suits. The radiant heat produced by burning aircraft fuels can be extreme; therefore, it is recommended that proximity suites be used instead of structural clothing for aircraft fire fighting whenever there is a choice, as recommended by NFPA 1500.

Chemical Protective Clothing

While most aircraft accidents may contain hazardous materials, not all ARFF firefighters

Figure 4.5 Helmet-mounted face shields provide eye protection.

Figure 4.6 Firefighter wearing regular work uniform.

Figure 4.7 Firefighter wearing structural fire fighting protective clothing.

specialize in advanced haz mat operations. Refer to NFPA 471, *Recommended Practice for Responding to Hazardous Materials Incidents,* for information on proper haz mat protective clothing levels. It is the responsibility of every firefighter to understand what substances require them to wear specialized chemical protective clothing.

Proximity Suits

Proximity suits are designed for close proximity exposures to high radiant heat (Figure 4.8). Proximity clothing has a reflective outer covering designed to reflect radiant heat. With the addition of one or more layers of thermal barrier, they also can withstand exposure to steam, liquids, and some weaker chemicals.

Figure 4.8 Firefighters wearing proximity suits. *Courtesy of Robert Lindstrom.*

Firefighter Safety at the Scene

Incident Management System

Aircraft accidents require a well-organized and well-trained emergency crew using the Incident Management System (IMS). For example, consider the accident that occurred in Sioux City, Iowa, in July 1989. In this case, a United Airlines DC-10 made an emergency landing at an airport not designed for DC-10s. This small airport had to cope with a large-frame aircraft making a crash landing with little notice. Incidents like this demand a strong IMS.

Each department must use some type of incident management system for each emergency. This means everything from the small trash can fire to the aircraft crash. It is important to train using the IMS regularly. As an ARFF department, this means training with the outside agencies that will assist in case of a major accident.

The IMS was designed to make sense of the chaos a major aircraft disaster brings. The IFSTA **Essentials of Fire Fighting** manual gives a very good description of the different components of the IMS. For more information on implementing an IMS, see the **Model Procedures Guide for Emergency Medical Incidents**, published by Fire Protection Publications, and NFPA 1561, *Standard on Emergency Services Incident Management System.*

Personnel Accountability

When something tragic happens during the course of emergency operations, like a building collapse or flashover, it is imperative that the whereabouts of all personnel are identified immediately. This holds true in aircraft fire fighting as well as structural fire fighting. The ARFF firefighter has many of the same dangers as their structural fire fighting counterparts, with one exception — ARFF firefighters usually have to deal with a fuel-soaked crash site loaded with hidden surprises.

Every crash site is full of hidden dangers waiting to take the life of a would-be rescuer. Having a sound accountability system not only minimizes the potential for losing a fellow firefighter, but it also makes for a very well-organized emergency.

A good accountability system starts with an organized IMS. If a department follows good, standard IMS rules, it should have little problem accounting for its firefighters. An accountability system can be as simple as knowing where a single ARFF crew is located, to having good supervision, and to tracking dozens of resources in a long-term, complex incident.

So what does it take for an ARFF department to account for its personnel? The first step is to find a system that fits the needs of the department. The second is to write a comprehensive standard operating procedure that all members understand. The next element is to activate the personnel accountability system as written in the SOP. One of the main failures for personnel accountability systems is the lack of participation. Personnel should remember that it is everyone's responsibility to look out for one another. People get into trouble when they stray away from their assignments and when ICs fail to keep adequate record of firefighter location on the fire ground.

Two-In/Two-Out

Both OSHA and NFPA require the two-in/two-out policy for all interior fire fighting operations. In general, there must be at least four fully equipped and trained firefighters at the scene of an emergency before a team of two firefighters may begin interior fire fighting (Figure 4.9). One of the two outside standby firefighters can be the driver/operator or the incident commander. The only exceptions to the two-in/two-out rule are where a known life hazard situation exists and only immediate action could prevent the loss of life and in incipient fire situations. If a victim is known to be accessible to the rescuers, the two-in/two-out rule may be violated. Extreme caution is given to this statement to note that firefighter personal safety must be considered before making this decision. It should also be noted that a department could not use this statement as an excuse for not following the two-in/two-out rule. (**NOTE:** Canadian regulations call for fire service personnel to follow the two-in/two-out rule *with no exceptions.*)

As with any interior fire fighting operation, two-in/two-out is very important to ARFF firefighters. Fighting interior aircraft fires is a dangerous task. A burning aircraft interior has been described as an aluminum inferno filled with burning, dripping plastics and toxic-gas-producing upholstery, hav-

Figure 4.9 ARFF crew members follow the two-in/two-out rule during a training evolution. *Courtesy of Robert Lindstrom.*

ing tight enclosed quarters, and filled with 100 or more occupants at times. The need for two-in/two-out is obvious.

The concept of two-in/two-out is no different from what is used for structural fires. The team on the outside has to maintain contact with the interior crew and be ready to respond to assist a downed firefighter.

Crews who are operating inside the aircraft must maintain visual or physical contact so that they can help each other if something does go wrong. PASS devices can be manually activated in case of an emergency to help assist the exterior firefighters in locating the downed personnel. One of the firefighter's best defenses in finding his or her way around the cabin is becoming familiar with the aircraft before they actually ever encounter a burning aircraft. Aircraft familiarization is a life-saving tool for any firefighter. ARFF personnel should learn their department's two-in/two-out policy and practice often enough to have the procedure become second nature.

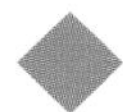

Hazards Associated with ARFF

All aircraft accidents pose several serious hazards. Each firefighter must understand the most obvious hazards associated with ARFF. The following list names only a few. Each accident is different and has its own unique problems.

- ***Aircraft jet engines.*** Aircraft jet engines may continue to run after the crash. These engines can ingest firefighters and overturn vehicles with the jet wash.
- ***Aircraft reciprocating engines.*** A reciprocating engine that is not properly shut down can restart if the prop is moved.
- ***Military aircraft.*** Weapons, ejection seats, flares, large loads of fuel, and hazardous cargo present problems on these aircraft.
- ***Landing gear.*** Because of the metals used in its construction, landing gear burns at high temperatures and reacts violently when water or foam is applied. There also could be the risk of the tires exploding when reaching extremely high temperatures. Caution must be used always when approaching burning or overheated landing gear; approaching from the front and rear is recommended. Unstable landing gear should be pinned by authorized personnel.
- ***Fuel.*** Jet fuel is a known carcinogen, and vapors and smoke can cause chemical pneumonia (in addition to the problems associated with the fires caused by jet fuel).
- ***Wreckage.*** Sharp, jagged edges can tear personal protective clothing and cause injuries.
- ***Energized electrical lines.*** Aircraft have very large electrical systems. Transport aircraft usually have 115-volt AC electrical systems and 24- or 28-volt DC electrical systems. Energized electrical lines may injure or electrocute personnel.
- ***Hydraulic and pneumatic lines.*** These lines contain flammable and toxic fluids and gases under very high pressures.
- ***Oxygen systems.*** Aircraft use pressurized oxygen systems, chemically generated oxygen, and liquid oxygen systems. Each of these systems poses a significant risk of explosion when engulfed in flames; extreme caution must be used at all times when approaching.
- ***Composite fibers.*** As discussed in Chapter 3, modern aircraft are constructed of composite materials. The dust, smoke, and very small fibers resulting from the cutting or the combustion of the aircraft skin presents a respiratory hazard to firefighters.
- ***Biohazards and chemical hazards.*** Most, if not all, aircraft crash sites contain significant amounts of biohazards and chemical hazards. The fuels alone pose a significant threat to the firefighter. Biohazards come mainly from the bodily fluids of occupants of the aircraft. They may also be present on debris contaminated with lavatory waste or may come from items such as donor blood or organs being shipped onboard the aircraft. All crash sites should be considered biohazard sites until properly decontaminated or determined otherwise.
- ***Other aircraft accident/incident hazards.*** Dense vegetation and uneven, soft, or wet terrain can make an emergency scene a difficult and dangerous place to work. Adverse weather may pose further complications. Extinguishing

agent foam blankets can make aircraft surfaces slippery and can hide trip hazards. Large, unstable fuselage sections may collapse, roll, shift, or slide. There are fall hazards from the significant heights encountered with large-frame aircraft. Depleted uranium used for counterweights and energized radar systems can be threats to emergency responders. Heat stress can be a serious problem when working in full PPE in warm and humid climates. Safety Officers and supervisors should continually monitor the scene for hazards and take steps to protect emergency personnel. Incident Commanders should make sure that a rehab area is established, that personnel are rotated out of the incident scene in a timely manner, and that their personal needs are attended to.

It is important for firefighters to understand the different types of hazards associated when responding to all aircraft emergencies. Each situation brings different hazards and requires the firefighter to stay alert at all times. The list above is small in comparison to what could be encountered at any one aircraft incident.

Personnel Decontamination

As stated in the Hazards Associated with ARFF section, the aircraft accident site is full of hazardous materials and biohazards. Every crash site must be evaluated for hazards and crews properly protected. There always will be the need for personnel decontamination at a crash site. Each firefighter, EMS responder, or victim may need decontamination prior to departing from the hot zone.

The procedure for decontamination should be established according to the contaminant that is present. Refer to NFPA 471, *Recommended Practice for Responding to Hazardous Materials Incidents,* for guidance in determining what type of decontamination procedure is appropriate for a specific contamination.

Critical Incident Stress

Stress

One of the greatest silent killers in any emergency responder's life is stress. We go from zero to warp speed in a moment's notice. Being an ARFF firefighter comes with many different types of stress. There is the stress of responding to a serious-sounding aircraft in-flight emergency and not knowing how the situation will end. There is the stress of responding to a major aircraft crash with no survivors and the stress of being away from your loved ones for a 24-hour shift.

Coping with Stress

So how do ARFF personnel deal with all this? Not everyone needs to seek professional help for everyday events, but everyone does need to know when to ask for help. The main reason why stress affects physical and emotional health is that the mind and body can work against each other. Each person has a way of coping with stress. Some ways of coping with stress are good and some are bad. An example of a bad way of coping would be to start consuming alcohol to "ease the pain." This will affect a person not only physically but also mentally. Because alcohol is a depressant, consuming it causes further depression. An example of coping with stress in a good way would be to take some time to run, jog, or walk. Physical exertion creates a stimulus reaction that reduces negative stressors and makes us feel better. Taking the time to talk to coworkers about a troubling incident also can help. Responders sometimes blame themselves for not being able to do more for someone in need, even though they know that they have, in fact, done all they can do to prevent death and/or a serious outcome.

Critical Incident Stress Debriefing (CISD)

Critical incident stress debriefing is peer-group or professional interaction immediately after a major incident. The recommended incidents to debrief include mass-casualty situations, loss of a child, and serious injury or loss of a coworker. The question comes to everyone: How soon should personnel seek CISD? The answer is plain — it depends on the situation.

If there has been a large passenger aircraft crash at the airport and there are no survivors, personnel should begin CISD as soon as the fires have been put out and while waiting for the investigative teams to arrive. It is very traumatic to the responder to see the death and destruction. It is especially hard when they have all these skills and specialized equipment and still everyone was lost. It is not by any means the fault of the rescuer; it lies in the events that caused the crash.

Because the injuries suffered by victims sometimes can be extremely gruesome and horrific, firefighters and any others who had to deal directly with the victims should participate in a CISD process. Because individuals react to and deal with extreme stress in different ways — some more successfully than others — and because the effects of unresolved stresses tend to accumulate, participation in this type of process should not be optional. The process should actually start *before* firefighters enter the scene if it is known that conditions exist there that are likely to produce psychological or emotional stress for the firefighters involved. This is accomplished through a prebriefing process wherein the firefighters who are about to enter the scene are told what to expect so that they can prepare themselves.

If firefighters are required to work more than one shift in these conditions, they should go through a minor debriefing, sometimes called *defusing*, at the end of each shift. They should also participate in the full debriefing process within 72 hours of completing their work at the incident.

Firefighter Safety at the Fire Station

Safety is the concern of every firefighter. It is not the responsibility of the department safety officer to look out for all the safety hazards in the fire station. Personnel should take a common-sense approach to firefighter safety at the fire station:

- Practice good housekeeping. Keep floors and all walking surfaces clean, dry, and clear of clutter. Ensure that exit areas are lighted and free of obstructions.
- Store hazardous materials, including flammable liquids, properly. Keep material safety data sheets (MSDS) for all hazardous materials kept (this includes aqueous film forming foam [AFFF] concentrate), and keep them where they can be easily retrieved.
- Use proper lifting and carrying techniques when moving equipment or heavy objects (Figure 4.10).
- Follow tool and equipment safety rules.
- Place portable heaters used in stations so that they are out of travel routes and away from combustibles. Use only the type that if turned over, it will turn itself off.

Personnel that observe any situations that warrant a safety concern should bring them to the attention of the health and safety officer.

For further information, please refer to the IFSTA **Fire Department Occupational Safety** manual.

Figure 4.10 Using proper lifting techniques is a part of practicing firefighter safety. *Courtesy of Robert Lindstrom.*

Chapter 5

Aircraft Rescue and Fire Fighting Communications

Job Performance Requirements

This chapter provides information that will assist the reader in meeting the following job performance requirements from NFPA 1003, *Standard for Airport Fire Fighter Professional Qualifications,* 2000 edition. Particular portions of the job performance requirements (JPRs) that are addressed in this chapter are noted in bold text.

3-2.2 Communicate critical incident information regarding an incident or accident on or adjacent to an airport, given an assignment involving an incident or accident and an incident management system (IMS) protocol, so that the information provided is accurate and sufficient for the incident commander to initiate an attack plan.

(a) *Requisite Knowledge:* Incident management system protocol, the airport emergency plan, airport and aircraft familiarization, **communications equipment and procedures**.

(b) *Requisite Skills:* **Operate communications systems effectively, communicate an accurate situation report,** implement IMS protocol and airport emergency plan, recognize aircraft types.

3-2.3 Communicate with applicable air traffic control facilities, given a response destination on or adjacent to an airport and radio equipment, so that all required clearances are obtained.

(a) *Requisite Knowledge:* **Communications equipment and frequencies, tower light signals, fire department and airport terminology.**

(b) *Requisite Skills:* **Operate communications equipment effectively.**

Communication, both direct and indirect, plays a vital role in operating an airport. Aircraft movements, vehicular movements, and ramp operations are just a few items controlled by constant communications. When an emergency arises, ARFF personnel should be able to communicate clearly with emergency dispatchers and air traffic controllers in order to locate and respond to the incident site. In some instances, the ARFF Incident Commander may be provided the luxury of talking directly to the pilot of the aircraft involved in the emergency. The success of incident management depends on clear and understandable communication at all levels. Clearly communicated orders reduce confusion and help to maximize the use of available resources. Clear communication promotes teamwork and reduces the likelihood of freelancing by individual units. It also provides the Incident Commander (IC) with a clearer picture of the incident as various operations are being conducted. Because other fire and law enforcement agencies and the local media monitor public safety frequencies, the manner in which communications are handled projects an image of the department.

Each jurisdiction should establish standard operating procedures (SOPs) for dispatch and emergency scene communications. Ideally, these communications should be coordinated with other agencies within the area. These procedures should include clearly defined lines of communication, specified frequencies, and guidelines for their use. To be most effective, all participating agencies and their individual units should use the established procedures in their day-to-day operations and should exercise the procedures through regular training.

The suggested methodology for planning and implementing ARFF communications can be found in the Federal Aviation Administration (FAA) Advisory Circular 150/5210 – 7C, July 1, 1999, *Aircraft Rescue and Firefighting Communications*. Recommendations regarding this advisory circular along with all other aspects of the airport fire and rescue communications system are covered in this chapter.

Notice to Airmen (NOTAM)

Notice to airmen, commonly called a NOTAM, is information that is issued by the airport operator or air traffic control personnel; a NOTAM addresses important information about airport operations involving runways, taxiways, and essential services. An example may be a NOTAM involving runway construction that may be for a specific date, for an amount of time, or until further notice. Fire personnel should be made aware of, and post, NOTAMs at the time the airport operator issues them. If fire protection apparatus and services fall below the requirements identified in FAR Part 139.319 for longer than a 24-hour period, fire department personnel must notify the airport operator so that a NOTAM can be issued.

Airport Communication Systems

Depending on the size of the airport, ARFF communications may be handled by a local fire department located off the airport or by a dedicated ARFF dispatch center located on the airport (Figure 5.1). Dispatchers, mutual aid personnel, and all ARFF personnel should be familiar with the terminology common to the airport community and with communication procedures of control tower personnel. Mutual aid and support organizations should be made aware of airport communication procedures in an effort to eliminate confusion when responding to an incident. A communication plan should be established, tested, and implemented if multiple types of radio systems are being used and are not compatible. At many airports, the activating authority (control tower, flight-service station, airport manager, fixed-based operator, or airline office) communicates directly with the airport fire department and should be able to communicate directly with support agencies such as emergency medical services (EMS), airport maintenance, and police. Airport communication systems for ARFF operations include a means for sounding audible alarms, using direct-line telephones, and using radios.

Air traffic control personnel usually provide the following basic information regardless of the method used to notify firefighters of an aircraft incident/accident. This may vary from airport to airport or at military installations. The firefighter taking the information should ask additional questions as necessary to fully understand the situation.

- Make and model of aircraft
- Name of air carrier
- Response category such as Alert 1 (local standby), 2 (full emergency), or 3 (aircraft accident). Fire

Figure 5.1 An airport communications and dispatch center. *Courtesy of William D. Stewart.*

service personnel may elect to upgrade or modify the response based on the information received. (NOTE: Response categories are not standardized.)

- Emergency situation
- Number of persons on board
- Amount of fuel on board, usually in pounds, but sometimes in hours of flying time remaining
- Any other pertinent information the reporting party may have knowledge of, such as hazardous cargo on board, non-ambulatory persons aboard, etc.

Audible Alarms

When an actual or potential emergency is discovered, the activating authority will activate audible alarms to alert any or all of the following:

- Airport or facility occupants
- Regular ARFF personnel
- Auxiliary ARFF personnel
- Essential support services, such as airport security, local law enforcement, EMS providers, and others located on or off the airport.

Commonly, a direct-line telephone, bell, Klaxon® or similar device, or combinations of devices are used to alert ARFF personnel in the airport fire station (Figure 5.2). When airport auxiliary firefighters are used, they may be notified by pagers, tone-activated radio receivers, or a siren or horn that is easily heard above normal noise level.

Figure 5.2 Emergency alarms for alerting ARFF personnel. *Courtesy of Michael T. Defina, Jr., Metro Washington Airports Authority Fire Department.*

Direct-Line Telephones

In earlier times, direct-line communication was limited to that between the control tower and the ARFF station. As time evolved and experience showed the importance of notifying additional resources quickly, direct-line telephone conference circuits were established between the control tower and multiple emergency agencies. These agencies may include airline station managers, medical transport organizations, area hospitals, and mutual aid fire departments (Figure 5.3). Such a telephone circuit provides a primary means of aircraft accident/incident notification. To enhance their reliability, these lines should be tested regularly and monitored continuously. A means should be provided for their immediate repair when necessary. This type of system can be used to notify and request resources from multiple organizations at the same time. Some organizations may be provided with a one-way monitor that allows notification of an incident or accident but not two-way conversation.

Radio Systems

The most efficient means for communicating with personnel on the emergency scene operations is the two-way radio. Radios should have a sufficient number of channels to allow the necessary com-

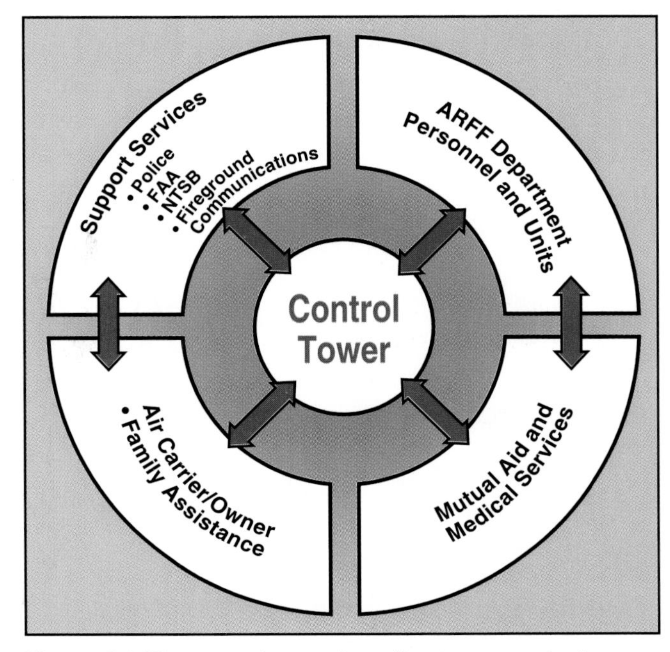

Figure 5.3 The control tower has direct communications with several emergency service agencies.

mand, tactical, and support functions to operate on separate channels, and the IC should have the ability to communicate with agencies operating on other frequencies (Figure 5.4).

Although each agency or group of agencies is assigned a specific radio channel that it may use for routine and emergency messages, all agencies concerned must have one or more common channels for mutual aid operations. Strict radio discipline must also be exercised to facilitate the proper and efficient use of shared radio channels. In addition, these agencies should have multichannel scanning capability in order to monitor local radio channels for critical emergency information.

The use of radio codes or esoteric terminology must be suspended during multiagency operations. A code "10-10" may mean one thing to a fire department, something entirely different to the police, and absolutely nothing to another agency. The use of Clear Text, as specified in the Incident Management System (IMS), helps eliminate confusion.

Airport and ARFF management should ensure that fire service radio systems comply with Federal Communications Commission (FCC) Part 89, "Rules Governing Public Safety Radio Service," or the regulations of the authority having jurisdiction. Only a technician licensed by the FCC is authorized to adjust transmitters, including base station, mobiles, and portables. Radios and communications networks that are not in daily use should be tested regularly to ensure satisfactory operation. Defective units should be repaired or replaced immediately. Personnel are prohibited from transmitting false or misleading information; unassigned call signals; and indecent, obscene, or profane language. Current computer-monitored systems record and time-stamp verbal communications to help ensure that procedures are followed.

Figure 5.4 An incident commander communicates with different agencies using a two-way radio with multichannel capabilities. *Courtesy of Michael T. Defina, Jr., Metro Washington Airports Authority Fire Department.*

The command or communications/dispatch center is responsible for ensuring the proper operation of the radio system. Some of the more important functions performed include:

- Clearing the air as soon as possible
- Maintaining discipline on the air
- Determining the order of priority for simultaneous transmissions

Aviation Radio Frequencies

The firefighter may use or monitor several radio frequencies that are unique to the aviation environment. The exact frequency number for the following radio channels varies from airport to airport. One of the most important is the ground control frequency. At controlled airports, which have an active air traffic control tower, a ground control frequency will be used to obtain clearance for driving on the aircraft movement area during routine and emergency situations.

Another important frequency is local control or the tower frequency. Each runway may have its own frequency. An airport with several runways may have several tower frequencies (Figure 5.5). Aircraft are on this frequency from the time they are transferred to the airport air traffic control tower (ATCT) from approach control, which is usually

Figure 5.5 Typical air traffic control tower radio. *Courtesy of Michael T. Defina, Jr., Metro Washington Airports Authority Fire Department.*

approximately ten miles out, to the point they turn off the runway onto a taxiway. If not interfering with monitoring ground control instructions, firefighters should monitor this frequency during in-flight emergencies and listen to the conversation between the pilot(s) and the tower personnel. Often, responding firefighters are able to hear information of importance and interest to them. Tower personnel usually rebroadcast or convey an edited version of this information on the ground control frequency.

Flight Service Stations (FSS) may operate a radio frequency at airports without an active tower or airport advisory. This is an air traffic control (ATC) facility that can provide in-route communications, provide VFR (visual flight rules) search and rescue services, assist lost aircraft and aircraft in emergency situations, relay ATC clearances, and broadcast information and airport advisories. This frequency may be used only during normal working hours.

UNICOM, or unified communications, is a private, nongovernmental frequency that may provide information or access to services and is usually found at general aviation airports. It is sometimes monitored by airport personnel or various airport tenants. It could be used by pilots to declare an emergency.

The common traffic advisory frequency (CTAF) is used on airports without an operating ATCT or when the tower is closed. The frequency used for this purpose may be UNICOM, FSS, or one of the tower frequencies. Pilots broadcast their positions, intended flight activity (takeoff or land), or ground operation (taxi route) on this frequency. Vehicle operators, such as ARFF firefighters, announce their intended ground operation, such as where they are going or what they are doing.

The Automated Terminal Information Service (ATIS) is a frequency that has a continuous radio broadcast on weather and airfield information. It identifies the runway(s) in use, taxiway closures, and NOTAM information.

Firefighters should continually monitor the appropriate aviation frequencies when responding or operating on aircraft movement areas. If firefighters must communicate on the ground control frequency, they should follow the correct order of information:

- Name of facility being called — "Airport name, Ground"
- Vehicle identity, such as "ARFF 1"
- Firefighter location
- Request of clearance to desired area
- Preferred route to take (optional)

After giving this information, the firefighter should end communication by saying "Over."

Ground control specifies a route if one is not requested. A requested route also may be changed due to aircraft movements or other reasons. ARFF personnel should repeat the tower instructions before acting and should not hesitate to ask for clarification if uncertain of the tower's instructions. If ground control advises of aircraft traffic or a hazard in the area of travel, the firefighter should acknowledge that the aircraft or hazard is in sight. The ARFF vehicle should proceed only after receiving the appropriate clearance, and personnel should inform ground control when they are clear of the aircraft movement area.

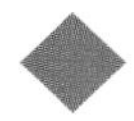

Pilot/ARFF Incident Commander Communications

Advancements in communication technology have provided the pilot of an inbound emergency aircraft the ability to talk directly to the ARFF incident commander. This allows the ARFF incident commander to provide the pilot with information regarding the visible condition of the aircraft, status of ARFF equipment, and specifics relating to the emergency. The ARFF incident commander can advise the pilot of conditions outside the aircraft so that the pilot can make key decisions regarding passenger evacuation (Figure 5.6). The flight crew may provide information to the ARFF IC such as the number of people on board, amount of fuel remaining, and any hazardous materials that may be carried on the aircraft. Because of the workload by the flight crew during the emergency, the pilot should initiate this communication. It is important to remember that the pilot is ultimately responsible for the aircraft and its occupants. ARFF

Figure 5.6 The ARFF incident commander talks directly with the pilot. *Courtesy of Michael T. Defina, Jr., Metro Washington Airports Authority Fire Department.*

Incident Commanders must remember only to advise the crew of the conditions of the aircraft and not to convey evacuation instructions unless specifically requested. It is important that all parties discuss the guidelines that govern the procedure during preincident planning. Guidance for initiating the pilot/ARFF communications procedure can be found in FAA Advisory Circular 150/5210-7C *Aircraft Rescue and Firefighting Communications.*

An alternate system available to ARFF personnel and others to communicate with aircraft crew members is the interphone system. Air carrier maintenance, mechanics, ramp, and pushback personnel use this system to communicate to various areas of the aircraft during routine operations. Some fire departments use it during emergencies to talk with the pilots. A special headset can be plugged in to an interphone jack location, which is usually found near the flight deck, ground power connection, or nose gear. There are two systems. The flight connection allows communication only with the flight deck and pilots. The service connection allows communication with the flight deck as well as various compartments (air conditioning, accessory, cargo), wheel wells, rear empennage access areas, fueling and APU panels, and other areas on the aircraft.

Proper Radio/Telephone Use

ARFF personnel should follow departmental procedures when calling another unit. To aid in clear communication, personnel should observe the following guidelines for proper radio/telephone use:

- Speak directly into the microphone, holding it at a 45-degree angle to and no more than one and one-half inches (40 mm) from the mouth.
- Speak distinctly, calmly, and clearly.
- Pronounce each word carefully, but convey messages in natural phrases — not word by word.
- Use a conversational tone and a moderate speed.
- Speak only as loudly as you would in ordinary conversation. If surrounding noise interferes, speak louder, but do not shout.
- Try to speak in a low-pitched voice because low-pitched tones transmit better than high-pitched tones.

It is important to maintain a calm, clear tone when issuing orders or making reports over the radio. This prevents the need to repeat messages numerous times.

International Civil Aviation Organization (ICAO) Phonetic Alphabet

When atmospheric or other conditions make radio transmissions difficult to hear, it is standard practice to spell out critical information, substituting certain standard words for individual letters of the alphabet. These are often used to indicate an aircraft identification number or a building number or location. This practice reduces the confusion created by certain letters of the alphabet that sound alike. The ICAO phonetic alphabet is used exclusively for this purpose. Also, a specialized vocabulary of words and phrases has been developed to simplify and clarify radio messages as well as keep them brief. Personnel operating radios should use this vocabulary and the phonetic alphabet, when necessary, to help ensure that messages are understood correctly. Listed here are the letters of the alphabet and their corresponding phonetic names.

A — Alpha (al-fah)	S — Sierra (see-air-rah)
B — Bravo (brah-voh)	T — Tango (tang-go)
C — Charlie (char-lee or shar-lee)	U — Uniform (you-nee-form or oo-nee-form)
D — Delta (dell-tah)	V — Victor (vik-tah)
E — Echo (eck-oh)	W — Whiskey (wiss-key)
F — Foxtrot (foks-trot)	X — X-ray (ecks-ray)
G — Golf (golf)	Y — Yankee (yang-key)
H — Hotel (hoh-tel)	Z — Zulu (zoo-loo)
I — India (in-dee-ah)	1 — Wun
J — Juliett (jew-lee-ett)	2 — Too
K — Kilo (key-loh)	3 — Tree
L — Lima (lee-mah)	4 — Fow-er
M — Mike (mike)	5 — Five
N — November (no-vem-ber)	6 — Sicks
O — Oscar (oss-cah)	7 — Sev-en
P — Papa (pah-pah)	8 — Ait
Q — Quebec (kwee-beck)	9 — Nin-er
R — Romeo (rom-me-oh)	0 — Zero

Sample Vocabulary

The following sample vocabulary is not all-inclusive but is representative of communications words and phrases in most common usage in the airport environment. To be effective in radio communication, personnel must be thoroughly familiar with these terms and their meanings.

A

Acknowledge — "Confirm that you have received and have understood the message."

Advise intentions — "Explain what you plan to do."

Affirmative — "Yes," "permission is granted," or "that is correct."

Air Traffic Control (ATC) — Service operated by appropriate authority to promote the safe, orderly, and expeditious flow of air traffic.

B

Base leg — The flight path at a right angle to the landing runway off the approach end.

Base to final — Turning into final approach position.

Blind (dead) spot — An area from which radio transmissions cannot be received. May also be used to describe portions of the airport not visible from the control tower.

Broadcast — Transmission of information for which an acknowledgment is not expected.

C

Confirm — "Verify" or "recheck."

Correction — "An error has been made in the transmission, and the corrected version follows."

D

Downwind leg — A flight path parallel to the landing runway in the direction opposite to landing.

E

ETA — Estimated time of arrival.

Expedite — "Prompt compliance is required."

F

Final approach — That portion of the landing pattern in which the aircraft is lined up with the runway and is heading straight in to land.

Flameout — Unintended loss of combustion in turbojet engines resulting in the loss of engine power.

Fuel on board — Amount in pounds (6 to 7 lb per gallon [0.7 kg to 0.8 kg per liter]) on aircraft remaining.

G

Gear down — Landing gear in down and locked position (have green light in the cockpit).

Go ahead — "Proceed with your transmission."

Go around — Maneuver conducted by a pilot whenever a visual approach to a landing cannot be completed.

H

Hold your position — "Do not proceed! Remain where you are."

How do you hear (read/copy) me? — A question relating to the quality of the transmission or to determine how well the transmission is being received.

Hung gear — One or more of the aircraft landing gear not down and locked (no green light indication in the cockpit).

I

Immediately — "Action is required without delay."

I say again — "The message will be repeated."

J

Jet blast — Wind and/or heat blast created behind an aircraft with engines running.

L

Low approach — An approach over a runway or heliport where the pilot intentionally does not make contact with the runway.

M

Make a 90, 180, or 360 (degree turn) — Instructions normally given by the control tower to the aircraft to indicate the degree of turn the pilot is to execute; also frequently used by the control tower to direct vehicles on the ground.

Make your best time — "Expedite."

Mayday — The international radio distress signal.

Minimum fuel — Indicates that an aircraft's fuel supply has reached a state where it can accept little or no delay before landing.

Missed approach — A maneuver conducted by a pilot whenever an instrument approach cannot be completed into a landing.

N

Negative — "No"; "permission not granted"; "that is not correct."

O

Out — "The conversation is ended, and no response is expected."

Over — "My transmission is ended; I expect a response."

Overhead approach (360 overhead) — A series of standard maneuvers conducted by military aircraft (often in formation) for entry into the airfield traffic pattern prior to landing.

P

Prop or rotor wash — Windblast created behind or around an aircraft with engines running.

Proceed — "Go" or "go to."

R

Read back — Repeat the message back to the sender to ensure accuracy.

Received (copied) — "Message has been received and understood."

Repeat — Request operator to say again.

Roger — "Message received and understood."

NOTE: "Roger" *should not* be used to answer a question requiring a "yes" or "no" answer. Use *affirmative* or *negative*.

S

Say again — Request a repeat of last transmission.

Speak slower — Request to reduce rate of speech.

Stand by — The person transmitting will pause,

and those receiving transmission should await further transmission.

Stand by to copy — "Prepare to receive detailed information that should be written down."

T

That is correct — Indicates agreement with how message is understood.

U

Unable to — Indicates inability to comply with a specific instruction, request, or clearance.

V

Verify — Request for confirmation of information.

W

Wilco —"Received message, understand, and will comply."

Wind direction and velocity — Wind direction is given to the nearest 10 degrees, and velocity is given in knots. A report of "wind at 330 at 10" would mean the wind was blowing from 330 degrees (30 degrees from north) at 10 knots (12 mph).

Words twice — Indicates that communication is difficult; request that every phrase be said twice.

Computers

As computers have evolved, so has their use in ARFF. The types of computers vary from laptops to mobile data terminals (MDTs) to global positioning systems (GPS). The computers can provide the following:

- Data on airport layouts
- Prefire plans of airport buildings
- Diagrams and information on various aircraft
- Information on dealing with hazards associated with dangerous goods
- Ability for personnel to give the dispatch center the status and location of ARFF apparatus
- On-screen messaging between apparatus and dispatch

As computer technology continues to develop both in hardware and software, the use of computers will continue to expand as an information tool, a communications system, and a fire scene management system.

Light Signals

At uncontrolled airports, aircraft and vehicle operators follow specific regulations, procedures, and airport markings to allow for orderly aircraft and vehicle movement. The routes and patterns are established to illustrate the desired flow of ground traffic for the different runways or airport areas. The routes are designed to keep aircraft away from high-hazard and traffic areas if at all possible.

At controlled airports, control tower personnel issue clearances, instructions, and information to vehicles operating in aircraft movement areas. Vehicle driver/operators must be in radio or visual contact with tower ground control. Control tower personnel direct aircraft and vehicular traffic by two-way radio on the ground-control frequency or with light signals if radio communication is lost. Because radio communication is much more reliable and efficient, it is highly desirable that all ARFF apparatus have multichannel radio equipment. Responding mutual aid companies not familiar with the airport, and/or not equipped with the necessary radio frequencies, will have to be escorted to the incident scene by airport personnel or by fire department personnel familiar with the airport layout.

Using radios is only one method of communication used for traffic control. The other means of traffic control in aircraft movement areas is through light signals. The tower controller uses a light gun to direct a colored light beam at a vehicle or aircraft (Figure 5.7). Before being allowed to operate a vehicle in aircraft movement areas, operators should memorize the light gun signals and their meanings as described in Figure 5.8 (and also in Appendix A, "Ground Vehicle Guide to Airport Signs and Markings"):

- *A steady green light* means that it is clear to cross, proceed, or go.
- *A steady red light* means to stop!
- *A flashing red light* means to clear the taxiway/ runway.
- *A flashing white light* means to return to the starting point on the airport.
- *Alternating red and green lights* mean to exercise extreme caution.

Figure 5.7 An air traffic controller signals aircraft with a light gun. *Courtesy of Michael T. Defina, Jr., Metro Washington Airports Authority Fire Department.*

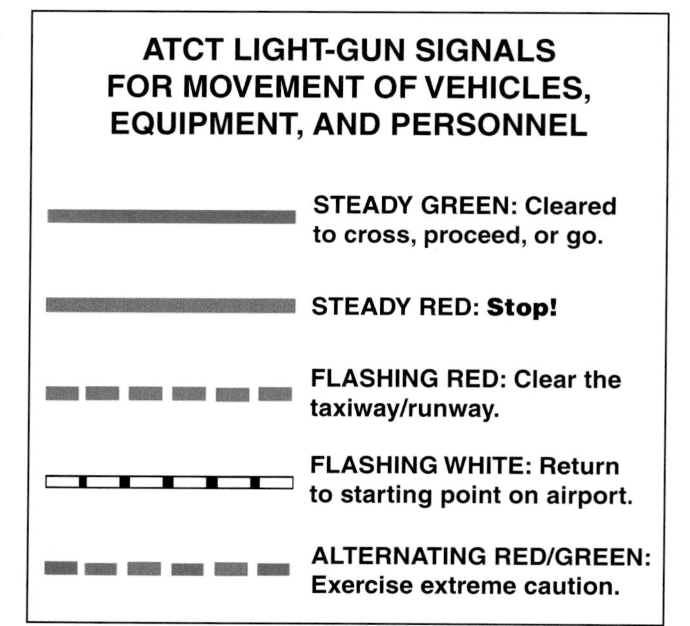

Figure 5.8 Air traffic control tower light gun signals.

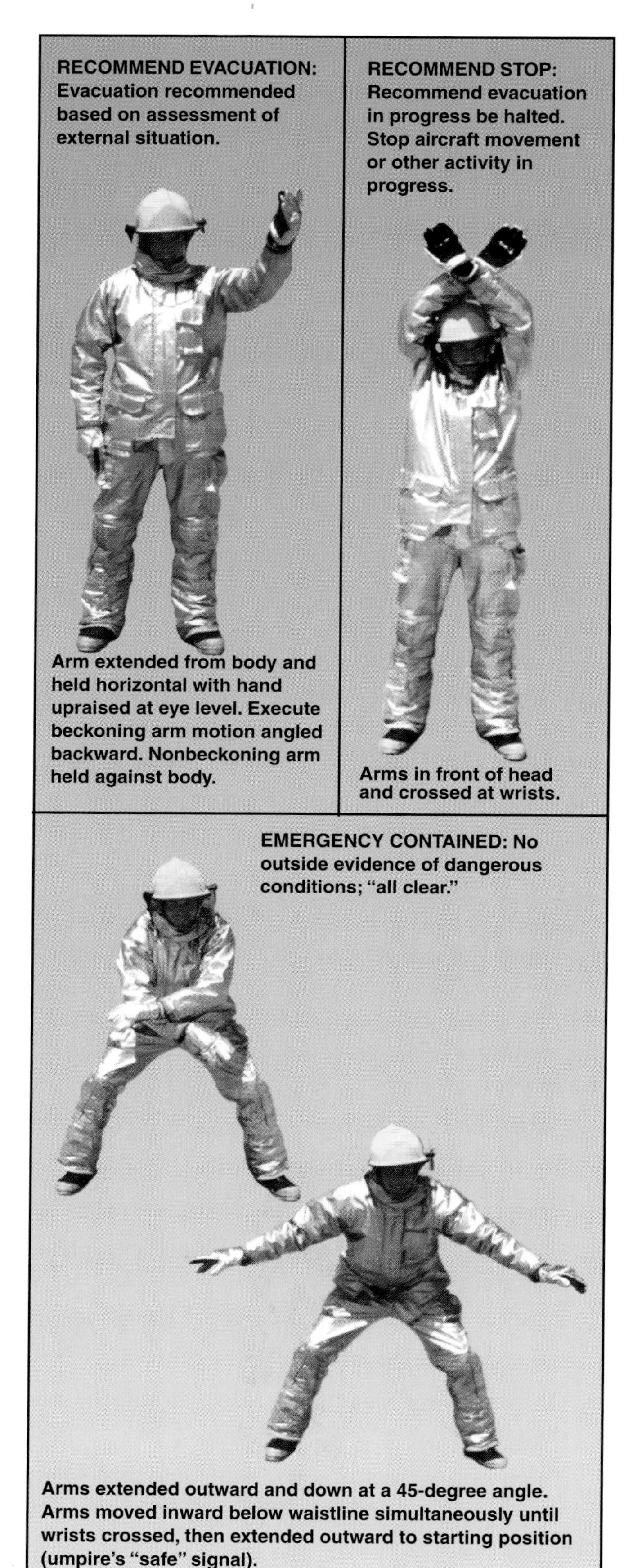

Figure 5.9 Emergency hand signals.

Hand Signals

Because of the high noise levels common to aircraft crash scenes, airport fire departments developed a system of hand signals with which an officer could communicate with a vehicle operator. These signals have been used extensively as a method of communicating when conducting fire-fighting operations. New signals have been introduced that allow ARFF personnel to communicate with the flight crew of an emergency aircraft (Figure 5.9). These signals are designed to advise the crew of recommendations regarding evacuation operations. Even though advances in portable voice-activated transceivers allow ARFF personnel to incorporate radio communications from under their hoods and helmets, personnel should have a basic knowledge of hand signals in the event that radio communications break down.

It may be necessary for an individual airport fire department to devise additional hand signals to fit its particular procedures and/or apparatus. For example, one department uses one up-raised finger to request water only to be discharged from the turret and two fingers to request that foam be discharged. It is most important that all ARFF personnel within the department know and understand whatever signals are adopted for use by the department, and this knowledge can only be accomplished with thorough training and frequent use.

Other Signals for Aircraft Accident Operations

- ***Back out or retreat*** — Sound all audible devices (horns, sirens, etc.) for obviously extended time (1 to 2 minutes).
- ***Apparatus is running out of agent*** — Flash headlights and sound the siren.
- ***Open or close handline*** — Tap hand firmly on the desired nozzle barrel.
- ***Change handline nozzle/stream pattern*** — Place wrists together and clap hands.
- ***Advance with handline*** — Pat shoulder.
- ***Back out with handline*** — Tug coattail sharply, or with the hands in front of the chest, give series of pushing motions.

Chapter 6

Aircraft Rescue and Fire Fighting Apparatus

Job Performance Requirements

This chapter provides information that will assist the reader in meeting the following job performance requirements from NFPA 1003, *Standard for Airport Fire Fighter Professional Qualifications*, 2000 edition. Particular portions of the job performance requirements (JPRs) that are addressed in this chapter are noted in bold text.

3-3.3 Extinguish an aircraft fuel spill fire, given PrPPE, an ARFF vehicle turret, and a fire sized to the AFFF flow rate of 0.13 gpm (0.492 L/min) divided by the square feet of fire area, so that the agent is applied using the proper technique and the fire is extinguished in 90 seconds.

(a) *Requisite Knowledge*: **Operation of ARFF vehicle agent delivery systems**, the fire behavior of aircraft fuels in pools, physical properties and characteristics of aircraft fuel, agent application rates and densities.

(b) *Requisite Skills*: **Apply fire-fighting agents and streams using ARFF vehicle turrets**.

3-3.9 Replenish extinguishing agents while operating as a member of a team, given an assignment, an ARFF vehicle, a fixed or mobile water source, a supply of agent, and supply lines and fittings, so that agents are available for application by the ARFF vehicle within the time established by the authority having jurisdiction (AHJ).

(a) *Requisite Knowledge*: **Resupply procedures, operation procedures for ARFF vehicle replenishment**.

(b) *Requisite Skills*: Connect hose lines, operate valves.

Aircraft rescue and fire fighting vehicles are the backbone of any ARFF fire department. The airports they serve are as diverse as the vehicles themselves. Aviation industry standards require that ARFF apparatus be able to reach the scene of an aircraft emergency in far shorter time periods than those normally associated with a municipal fire department's response to a structure fire. With the heavy fuel loads and the large numbers of passengers carried by aircraft today, ARFF vehicles must be ready for all conditions and problems. Vehicles used for ARFF operations should be designed as self-contained units with the ability to discharge adequate quantities of extinguishing agents within a short period.

This chapter discusses apparatus requirements; types of aircraft rescue and fire fighting apparatus, vehicle options, turret types, handlines, resupply methods, and apparatus maintenance (Figure 6.1).

Figure 6.1 ARFF apparatus are manufactured to include various options and equipment. *Courtesy of Robert Lindstrom.*

ARFF Apparatus Requirements

Levels of Protection

The minimum types of ARFF apparatus that are required at an airport are determined by the index rating of that airport. Airports are categorized or indexed based on the type of aircraft using the facility and the daily average number of departures. The FAA, NFPA, and IACO each have their own rating systems (Table 6.1). Based on its assigned index or category, an airport must maintain certain minimum levels of ARFF apparatus and equipment at all times. Should any of the required apparatus or equipment become inoperative at certificated airports in the U.S. and an equal replacement is not available within 48 hours, the airport management must notify the Federal Aviation Administration and the affected carriers of the reduction in operational readiness.

Although the outcome differs among the organizations, NFPA, FAA, and ICAO use similar methods to determine the number of vehicles and the amount of extinguishing agent required. Each uses the length of aircraft landing at an airport to determine the ARFF apparatus requirements (Figure 6.2). However, there are some differences in the actual formulas they use. While FAA limits its requirements to certificated airports that serve aircraft having a certain passenger capacity, NFPA and ICAO use the length of all aircraft landing at an airport. The Department of Defense also has specific requirements as it pertains to aircraft sizes and configurations. Military aircraft operations are governed by NFPA 403, and military airports do not have categories. It is important that the appropriate agency requirements be reviewed to ensure compliance.

Figure 6.2 The length of the aircraft routinely using an airport is a factor in determining the airport's index or category. *Courtesy of Robert Lindstrom.*

NFPA 403 also has requirements for the amount of foam and auxiliary extinguishing agents and number of ARFF apparatus that should be stationed at various sizes of airports. These requirements are based on the NFPA system of categorizing airports by size.

Apparatus Design

There are a number of agencies and standards that cover ARFF apparatus design and types of apparatus required at any given airport. The following requirements pertain to airport fire fighting vehicles and should be considered when preparing specifications for these vehicles or determining the apparatus necessary to protect a particular airport:

- Federal Aviation Regulations (FAR) Part 139.317 and as outlined in FAA Advisory Circular (AC) 150/5220-10B
- International Civil Aviation Organization (ICAO) *Airport Services Manual Part I, Rescue and Fire Fighting*
- NFPA 414, *Standard for Aircraft Rescue and Fire Fighting Vehicles*
- NFPA 403, *Standard for Aircraft Rescue and Fire Fighting Services at Airports*
- NFPA 412, *Standard for Evaluating Aircraft Rescue and Fire Fighting Foam Equipment*

Table 6.1
Airport Categories

Airport Category U.S. NFPA	FAA	ICAO	Overall Length of Aircraft Up to but Not Including ft	(m)	Maximum Exterior Width Up to but Not Including ft	(m)
1	GA-1	1	30	(9)	6.6	(2)
2	GA-1	2	39	(12)	6.6	(2)
3	GA-2	3	59	(18)	9.8	(3)
4	A	4	78	(24)	13.0	(4)
5	A	5	90	(28)	13.0	(4)
6	B	6	126	(39)	16.4	(5)
7	C	7	160	(49)	16.4	(5)
8	D	8	200	(61)	23.0	(7)
9	E	9	250	(76)	23.0	(7)
10			300	(91)	25.0	(8)

Aircraft Rescue and Fire Fighting Apparatus

Aircraft rescue and fire fighting apparatus must operate effectively in both paved and unpaved areas (Figure 6.3). In responding to aircraft accidents, ARFF apparatus may have to be driven across terrain that might be difficult to traverse in typical structural fire apparatus. This terrain also may be littered with aircraft wreckage, victims, and both ambulatory and nonambulatory survivors. These vehicles may have to discharge extinguishing agents while moving into or out of fire fighting positions.

Because of the large volumes of fuel involved in aircraft fires, mass application of extinguishing agents may be required very quickly in order to protect the occupants of the aircraft. ARFF personnel use specialized aircraft fire fighting vehicles equipped with turrets, handlines, ground sweeps, undertruck nozzles, and extendable turrets to apply the extinguishing agents. Additionally, ARFF vehicles today carry more medical supplies, ladders, and rescue tools and equipment.

Figure 6.3 ARFF apparatus capabilities include off-road maneuvering.

ARFF Apparatus Classifications

For specifying performance criteria of ARFF apparatus and agent systems, NFPA 414 divides apparatus into three different groups based on vehicle water-tank capacities:

- 60 to 528 gal (227 L to 1 998 L)
- >528 and ≤ 1585 gal (>1 998 L and ≤ 5 999 L)
- >1585 gal (>5 999 L)

The FAA provides a class rating system for ARFF apparatus based on the amount of water carried on board. These classes are covered in Table 6.2.

Table 6.2 FAA Classes of ARFF Vehicles	
Class	**Minimum Usable Water (Rated Capacity) gal (L)**
1	1,000 (3 785)
2	1,500 (5 678)
3	2,500 (9 463)
4	3,000 (11 355) and over

Source: FAA AC 150/5220-10B, *Guide Specification for Water/Foam Aircraft Rescue and Firefighting Vehicles.*

Structural Apparatus

In addition to the ARFF vehicles previously discussed, some structural apparatus may be equipped for aircraft rescue and fire fighting (Figure 6.4). These vehicles are specially adapted to be used for ARFF applications. The modern trend in structural fire fighting apparatus has been to install a fixed foam proportioning system, which can be used with either attack lines or piped turrets for foam application. These systems may be limited in their capacity of sustaining foam fire fighting operations as they typically have smaller foam concentrate storage tanks than do ARFF apparatus. Nevertheless, fixed foam proportioning systems give firefighters more capability than external inline foam eductors because of the friction loss, lack of mobility, and other problems associated with these devices. Structural apparatus can also be ordered with all-wheel drive, turrets, aircraft rescue equipment, secondary agent systems, (dry chemical, clean agents, Class D agents) and mass casualty supplies and equipment. Firefighters should be

Figure 6.4 This structural apparatus is equipped for ARFF operations. *Courtesy of Ron Jeffers.*

knowledgeable in the capabilities of the apparatus and foam systems in the event of an aircraft accident/incident.

WARNING

Attempting to extinguish a large fire that requires more than the capability of the apparatus may place firefighters and equipment in danger.

It may be possible to use a structural fire apparatus that is not equipped with a fixed foam proportioning system to produce foam by pouring foam concentrate directly into the apparatus water tank. This method is commonly referred to as *batch mixing*. Table 6.3 shows the proper amounts of concentrate that are needed for various water-tank sizes. Once the foam concentrate has been distributed throughout the tank, the apparatus fire pump is operated in the standard manner to produce a foam fire stream. This method may be used only with regular AFFF foam concentrates. It is not suit-

Table 6.3
Amount of Concentrate Needed for Various Sizes of Water Tanks

	Foam Concentrate Proportioning %	Water Tank Size in Gallons (Liters) 500 (2 000)	750 (3 000)	1,000 (4 000)	1,500 (6 000)	2,000 (8 000)
Concentrate to be added in gallons (liters)	1%	5 (20)	7.5 (30)	10 (40)	15 (60)	20 (80)
	3%	15 (60)	22.5 (90)	30 (120)	45 (180)	60 (240)
	6%	30 (120)	45 (180)	60 (240)	90 (360)	120 (480)

able for use with alcohol-resistant AFFF concentrates. Caution should be used when batch mixing to make sure all the foam has been flushed from the system after the operations are completed. This prevents the vehicle's fire fighting system from becoming damaged by the long-term effects of foam.

Support Vehicles and Equipment

Most ARFF departments operate a host of support equipment, which may include command post vehicles, mobile water supply vehicles (Figure 6.5), foam supply trailers, hazardous materials vehicles and trailers, ambulances, mass-casualty trailers, heavy rescue apparatus, and even buses. Some departments have adapted mobile air stairs, food service vehicles, and other such equipment for use as elevated platforms. Modifications such as adding preconnected hoselines, hose reels, ventilation fans, and other innovations make these useful devices. These apparatus are intended to meet specific needs for particular airports; however, they are generally not considered as credit towards the minimum index requirements for fire protection.

Apparatus Features and Options

With today's changing technology and the many features available on ARFF apparatus, airport firefighters have many advantages that the firefighters of yesterday did not have. ARFF vehicle features and options include the following:

- Antilock brake systems
- Central inflation/deflation tire systems
- Driver's enhanced vision systems (DEVS)
- High-mobility suspension systems (independent suspension systems)
- Extendable turrets

Antilock Brake Systems

Antilock brake systems provide the driver with greater control when operating under poor road conditions such as surfaces slick from ice and rain. Antilock brakes keep the vehicle wheels from skidding, which can result in a loss of control. However, this brake system can also give a driver a false sense of security if the driver completely relies on the braking system to prevent the vehicle from losing control. The driver must continue to drive with due caution at all times whether or not the vehicle has antilock brakes.

Central Inflation/Deflation Tire System

This technological advance allows the driver to deflate the vehicle tires while the vehicle is moving or stationary in order to improve the vehicle traction (Figure 6.6). These systems can be operated at predetermined speeds without interrupting the fire fighting capability of the vehicle. Deflating the tire gives the tire a larger surface area and aids in removing mud and debris from the tread, making the aggressive tread of the modern vehicle tires more effective. As with any advanced system, the tire deflation system can increase downtime for the vehicle when requiring maintenance. The minor

Figure 6.5 Mobile water supply vehicle. *Courtesy of Charlevoix (MI) Fire Department.*

Figure 6.6 Parts of a central inflation/deflation tire system can be seen on this wheel rim. *Courtesy of Robert Lindstrom.*

problem of having the vehicle out-of-service for longer times is greatly outweighed by the increase of ability to respond to incidents in more adverse driving conditions.

Driver's Enhanced Vision System

The driver's enhanced vision system (DEVS) brings the ARFF vehicle into the 21st century when it comes to modern technology. Most airports experience weather conditions that can delay ARFF response times. The DEVS allows the ARFF driver to use modern technology to make a safer, quicker response under adverse conditions.

The DEVS is composed of three subsystems:

- *Night vision* — An infrared camera and monitor that enhances vision in smoke, fog, adverse weather, and darkness (Figure 6.7)
- *Navigation* — A differential global positioning system (DGPS) receiver and moving map display inside the cab
- *Tracking* — A digital radio datalink between the command center and vehicles over which accident information, vehicle position reports, and other messages are sent.

This combination gives ARFF vehicle driver/operators a distinct advantage when responding under adverse driving and vision conditions. The advantage is that the driver/operator may be able to more easily locate a crash site that would otherwise be obscured by the conditions. As with any other advanced systems, the DEVS requires the driver to train often under the conditions for which the system was designed.

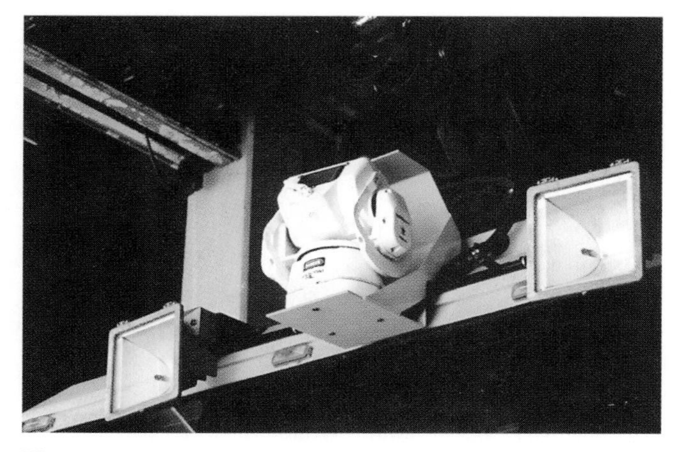

Figure 6.7 The night vision monitor is a part of the driver's enhanced vision system. *Courtesy of Robert Lindstrom.*

High-Mobility Suspension Systems

The high-mobility suspension system gives the ARFF vehicle greater mobility when responding both on and off paved surfaces. The idea behind this system is to keep the wheels in contact with the surface as much as possible. Standard straight-axle vehicles have a tendency to lose contact with the surface when encountering extremely uneven terrain. With the high-mobility suspension, each wheel and axle is independent from the others and will articulate over the terrain allowing each wheel to maintain greater contact with the surface. An example would be a vehicle crossing a ditch at an angle. With a straight axle, this vehicle has the possibility of one wheel losing contact with the surface because the axle is rigid and does not allow for much flexibility. The high-mobility suspension uses independent drive suspension that allows the drivelines to maintain contact with the road surfaces.

Apparatus Fire Suppression Equipment

ARFF vehicles are capable of applying various extinguishing agents from turrets, handlines, ground-sweep and undertruck nozzles, an extendable turret, or a combination of all or any of these devices.

Fire Pumps

Every major ARFF vehicle has a fire pump rated for that specific vehicle. All ARFF vehicles are capable of delivering large quantities of water to the fire fighting systems. Also, the fire pumps in ARFF vehicles can operate while the vehicle is in motion. This capability allows the operator to attack the fire on initial approach. Because the method of transferring power between the pump and engine during pump-and-roll operations may vary according to manufacturer, ARFF vehicle operators should practice and become familiar with the pump-and-roll characteristics of the vehicle.

Some ARFF apparatus have structural fire fighting capability (Figure 6.8). This capability allows the vehicle to be operated as a structural pumper to draft from a water source, work from a fire hydrant, and operate from the vehicle water tank. The advantages to structural systems on an

Figure 6.8 An ARFF apparatus with structural fire fighting capabilities.

ARFF vehicle include the ability to better control water pressure when making aircraft interior attacks and the expanded capability of the vehicle when it is used at airports that do not have a dedicated structural pumper. A majority of the manufacturers who provide structural capability usually provide foam to all discharges. The foam delivery systems also vary from one manufacturer to another. Because these systems vary, driver/operators must become familiar with the specific vehicles at their department.

Turrets

For mass application of agents during rescue and fire extinguishment, vehicles should have one or more turrets. The turrets may be mounted on the tops of the cabs (Figure 6.9) or on the front bumpers of the vehicles. They may be operated either manually or by remote control and should be capable of discharging extinguishing agents in various patterns ranging from straight-stream to fog patterns. Some turrets may also have automatic oscillating

Figure 6.9 Roof and bumper turrets and undertruck nozzles in operation. *Courtesy of Christchurch International Airport, New Zealand.*

features. Vehicles having remote electrical or hydraulic controls for turret operation should also have manual override controls.

There are three types of turret nozzles used presently — aspirating, nonaspirating, and dry-chemical injection. The advantages and limitations of each type are discussed in Chapter 9, "Extinguishing Agents."

Handlines

Handlines are needed to extinguish interior fires in the fuselage that cannot be reached by turrets, to provide protection for rescue personnel, and to extinguish peripheral fires after rescue operations are completed. Most ARFF vehicles are equipped with one or more of the following handlines: preconnected noncollapsible booster hose stored on a reel or standard collapsible hose stored in a hose bed (Figure 6.10). Both types of hoses must be equipped with variable pattern, shut-off nozzles. Nozzles may be aspirating or nonaspirating.

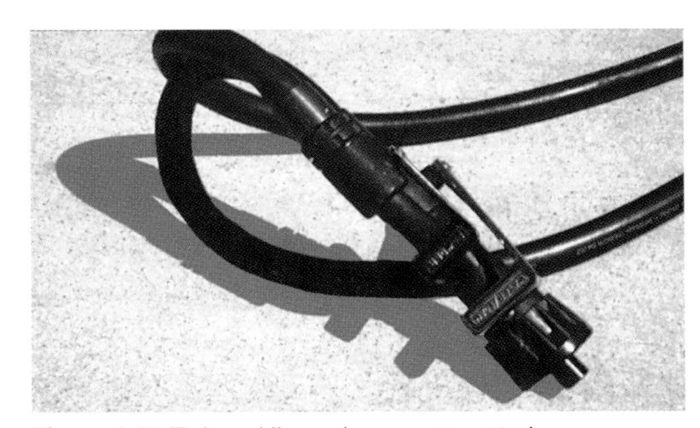

Figure 6.10 Twinned line using preconnected, noncollapsible booster hose.

Ground-Sweep and Undertruck Nozzles

Ground-sweep nozzles are used to lay a blanket or path of foam in front of the vehicle so that it can move into extinguishment and/or rescue positions without endangering the apparatus. Operation of ground-sweep nozzles should be controlled from within the cab of the vehicle.

Undertruck nozzles discharge extinguishing agents directly beneath the vehicle chassis. They are designed to protect the ARFF vehicle and equipment from the possibility of fuel and flames floating back and igniting beneath the vehicle itself. Controls for the undertruck nozzles are located inside the vehicle cab.

Extendable Turrets

This innovative fire fighting equipment has become increasingly popular on major ARFF vehicles in recent years. The extendable turret gives the firefighter an advantage when fighting aircraft fires. It is designed to attack the fire at the base of the flames – commonly called the low-sweep method. This places the agent stream where the foam can best attack the fire.

Some extendable turrets also are equipped with a piercing nozzle that is designed to penetrate the aircraft skin and apply agent to the aircraft interior without placing firefighters in danger. Testing of extendable turret piercing nozzles has shown their impressive ability to contain and control interior fire spread and flashover conditions. The piercing nozzles on extendable turrets can flow in excess of 250 gpm through the piercing device. The extendable turret with Forward Looking Infared (FLIR) can be used to locate hidden hot spots in cargo holds or to assist in locating the seat of a fire in a cabin interior fire. Because the extendable turret is mounted to the top of the ARFF vehicle, the driver/operator of the vehicle must realize that the vehicle may be more top-heavy than other ARFF vehicles. It may be necessary to compensate for this factor when driving the vehicle. Because these devices are complicated to operate, driver/operators will require continual training in their operation and in tactics and strategies.

Auxiliary Agent Delivery Systems

Most major ARFF vehicles carry some type of auxiliary agent. These agents include dry chemical, halon, and other clean agents and may be delivered through separate hose reels, twinned with a foam handline, or piped directly to a dry-chemical injection nozzle.

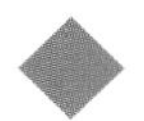

Agent Resupply Methods

When it comes to fighting large fuel-spill fires, rapid resupply can be as important as fighting the fire. All ARFF vehicles should have the ability to quickly resupply with both water and foam concentrate (Figure 6.11). Resupply can be critical when operating under large-scale emergency conditions. Each department should spend time training for rapid resupply using whatever method it plans to use in real situations.

Water-Fill Methods

As with structural fire apparatus, the primary method of refilling the ARFF apparatus water tank is through hose intake connections on the side or rear of the apparatus. These intakes may route water either through the pump or directly into the water tank. According to NFPA 414, the piping must be designed so that the tank may be filled in two minutes at an inlet pressure of 80 psi (552 kPa).

There are also several methods of supplying water to the ARFF apparatus tank inlets. Water can come directly from the fire hydrant, a mobile water supply,

Figure 6.11 Rapid resupply is necessary for efficient ARFF operation. *Courtesy of Charles F. Dusha, Logan Rogersville Fire District, Springfield, MO.*

a supply line extended from a pumper, or a fixed, overhead-fill hose found in most fire station apparatus rooms. The most effective of these usually is the supply line extended from the pumper.

CAUTION: When filling from the pumper, use caution so as to not over-pressurize the piping or water tank of the ARFF apparatus, as this could cause it to rupture.

All ARFF vehicles also have overhead-fill method capability. The overhead method is not as fast or safe as filling from the side of the vehicle. Persons using the top-filling methods should use caution when walking on top of the vehicle, as it may be slippery.

Figure 6.12 Three sizes of foam concentrate containers. *Courtesy of Chemguard.*

Foam Resupply

Foam concentrates are available in different sizes of containers (Figure 6.12). Foam can be resupplied several different ways, including:

- Direct filling from 5-gallon (20 L) containers
- Overhead gravity filling in the fire station
- Using a mechanical or hand foam concentrate pump to transfer foam concentrate from large storage containers or a foam tender

The least desired method is direct filling with 5-gallon (20 L) containers. This method is slow and requires a large number of personnel. Filling from foam tenders is a convenient method, allowing the vehicle to be serviced closer to the incident site (Figure 6.13). Whatever method a department uses must be flexible and work efficiently when it is needed.

Figure 6.13 A foam resupply trailer. *Courtesy of Robert Lindstrom.*

Auxiliary Agent System Servicing

Vehicles equipped with auxiliary agent systems typically store the agent in pressurized containers. The most important thing for ARFF personnel to know when refilling is what type of agent their department uses. It is not acceptable to mix different types of agents or expellants. The system should only be filled according to the system manufacturer's directions. At no time should there be deviations from the manufacturer's recommended servicing instructions. Dry-chemical systems should be serviced in a well-ventilated area, and persons filling them should use respiratory protection. It is extremely rare for these systems to be serviced during the course of an incident.

Although all agent systems should be thoroughly flushed after each use, it is of utmost importance that dry-chemical piping and hoselines be completely flushed, or blown down, after any discharge. Dry chemical left in the piping or hoselines tends to attract moisture and cake. Without proper blow down, the line may become packed and unusable.

Apparatus Maintenance

According to FAR Part 139.319 and ICAO Annex 14, all ARFF apparatus must be maintained so that they are always in operational condition. Therefore, ARFF apparatus and its equipment should be inspected immediately after shift change and after each use and serviced as necessary. Ideally, the fire apparatus mechanic should occasionally monitor routine preventive maintenance inspections as a means of improving the quality of the inspections by explaining mechanical functions and service requirements to ARFF personnel during the inspection. It is also good practice to use a detailed inspection checklist so that each item can be checked off as it is inspected or serviced.

Personnel should keep a complete record for each vehicle. These records should include mileage, engine hours, fuel and oil consumption, tire-replacement information, and parts information (when required, ordered, and installed), as well as total expenditures and out-of-service time. For each pumper, a record of annual pump performance tests also should be included in these permanent files.

Chapter 7

Aircraft Rescue Tools and Equipment

Job Performance Requirements

This chapter provides information that will assist the reader in meeting the following job performance requirements from NFPA 1003, *Standard for Airport Fire Fighter Professional Qualifications*, 2000 edition. Particular portions of the job performance requirements (JPRs) that are addressed in this chapter are noted in bold text.

3-3.5 Attack a fire on the interior of an aircraft while operating as a member of a team, given PrPPE, an assignment, an ARFF vehicle hand line, and appropriate agent, so that team integrity is maintained, the attack line is deployed for advancement, **ladders are correctly placed when used, access is gained into the fire area,** effective water application practices are used, the fire is approached, attack techniques facilitate suppression given the level of the fire, hidden fires are located and controlled, correct body posture is maintained, hazards are avoided or managed, and the fire is brought under control.

(a) *Requisite Knowledge:* Techniques for accessing the aircraft interior according to the aircraft type, methods for advancing hand lines from an ARFF vehicle, precautions to be followed when advancing hose lines to a fire, observable results that a fire stream has been applied, dangerous structural conditions created by fire, principles of exposure protection, potential long-term consequences of exposure to products of combustion, physical states of matter in which fuels are found, common types of accidents or injuries and their causes, the role of the backup team in fire attack situations, attack and control techniques, techniques for exposing hidden fires.

(b) *Requisite Skills:* Deploy ARFF hand line on an interior aircraft fire; **gain access to aircraft interior;** open, close, and adjust nozzle flow and patterns; apply agent using direct, indirect, and combination attacks; advance charged and uncharged hose lines up ladders and up and down interior and exterior stairways; locate and suppress interior fires.

3-3.8 Ventilate an aircraft through available doors and hatches while operating as a member of a team, given PrPPE, an assignment, tools, and mechanical ventilation devices, so that a sufficient opening is created, all ventilation barriers are removed, the heat and other products of combustion are released.

(a) *Requisite Knowledge:* Aircraft access points; principles, advantages, limitations, and effects of mechanical ventilation; the methods of heat transfer; the principles of thermal layering within an aircraft on fire; the techniques and safety precautions for venting aircraft.

(b) *Requisite Skills:* **Operate doors, hatches, and forcible entry tools**; operate mechanical ventilation devices.

3-4.1 Gain access into and out of an aircraft through normal entry points and emergency hatches and assist in the evacuation process while operating as a member of a team, given PrPPE and an assignment, so that passenger evacuation and rescue can be accomplished.

(a) *Requisite Knowledge:* Aircraft familiarization, including materials used in construction, aircraft terminology, automatic explosive devices, hazardous areas in and around aircraft, aircraft egress/ingress (hatches, doors, and evacuation chutes), military aircraft systems and associated hazards; **capabilities and limitations of manual and power rescue tools** and specialized high-reach devices.

(b) *Requisite Skills:* **Operate power saws and cutting tools, hydraulic devices, pneumatic devices, and pulling devices; operate specialized ladders** and high-reach devices.

3-4.2 Disentangle a trapped victim from an aircraft, given PrPPE, an assignment, and rescue tools, so that the victim is freed from entrapment without undue further injury and hazards are managed.

(a) *Requisite Knowledge:* **Capabilities and limitations of rescue tools.**

(b) *Requisite Skills:* **Operate rescue tools.**

The tools and equipment used for aircraft accidents/incidents vary to some degree from the tools and equipment used in structural fire fighting. However, conventional tools are adaptable for aircraft rescue and forcible entry in most cases. In addition to the tools and equipment covered in this chapter, there are many other tools (such as bolt cutters, wrecking bars, hacksaws, shovels, and door openers) used similarly for both structural and airport rescue and fire fighting. This chapter discusses general tools along with the equipment and the tools unique to aircraft rescue and fire fighting procedures.

Both the National Fire Protection Association (NFPA) and the International Civil Aviation Organization (ICAO) have published lists of recommended tools to be carried on ARFF apparatus. Aircraft rescue firefighters should be familiar with the tool requirements of the authority having jurisdiction over their airport operations.

Considerations in Using Rescue Tools and Equipment

A number of tools and equipment have been designed specifically for aircraft rescue. Many of these tools are appropriate for a variety of rescue situations and are, therefore, well-known and widely distributed. Other tools are very specialized and are designed for specific applications.

Aircraft rescue tools include both hand tools and power tools (Figure 7.1). These can be divided into four groups, based on the manner in which they are used: cutting, prying, pushing/pulling, striking. A rescue tool may belong to more than one of these groups, and many hand tools can be used for prying and spreading as well as for cutting, striking, or even lock-entry work. These tools are often referred to as "multipurpose" or "utility" tools, and some tools may be used in combination with other tools. Sometimes, using power tools is much easier than using hand tools when doing aircraft rescue work because of the tremendous mechanical advantage these tools provide. Some power tools generate over 20,000 psi (140 000 kPa) of mechanical energy. In other cases, due to restricted access and mobility, small hand tools may be needed.

Figure 7.1 Rescue tools and equipment stored on an ARFF apparatus.

Flammable Atmosphere

Aircraft carry large amounts of highly volatile flammable liquids. During the dynamics of a crash, often the fuel system is compromised, which can create a flammable atmosphere. ARFF personnel should carefully consider this and take proper precautions to avoid a "flash" or ignition of the spilled flammable materials. During cutting operations for example, a saw blade that contacts a steel cable, rivet, or other similar material can cause sparking that would lead to a fire. The aircraft incident scene needs to be rendered safe to use rescue tools and equipment. Fuel leaks should be identified, stopped, or controlled. Leaks can be plugged or patched using a wide variety of materials and techniques. Leaking fuel can be contained by diking or can be captured in portable containers. Spilled fuel can be covered with a foam blanket — followed by frequent reapplications — as well as dirt or other absorbent materials. ARFF crew members should eliminate obvious ignition sources by shutting down the aircraft power on the flight deck and disconnecting the battery.

Stability of Aircraft

Firefighters always should consider the stability of the aircraft before making entry. Unless stabilized, the aircraft may move, shift, roll, etc., trapping and possibly injuring occupants and rescue crew members as well as causing more fuel to be released. Structural conditions of the aircraft must be constantly monitored. Any undercarriage fire creates a potential for aircraft collapse or the explosive disintegration of affected components. Personnel also must consider the structural integrity of the aircraft when positioning apparatus. Should the aircraft shift, it may hit improperly placed apparatus.

Many tools, equipment, and materials can be used to stabilize the aircraft. Cribbing, airbags, heavy timber, and jacks can be used to prevent the aircraft from rolling, sliding, twisting, or shifting (Figure 7.2). Dirt can be pushed up against the fuselage, or heavy equipment can be parked against it. ARFF personnel can use ropes, cables, and chains to help secure large aircraft wreckage. In addition, shoring, such as wood timbers, hydraulic speed shores, and ladders, can be used to support sections of the aircraft and prevent collapse.

Figure 7.2 Firefighters position and inflate an air lifting bag to stabilize an aircraft.

Training of ARFF Personnel

ARFF personnel must have hands-on training using rescue equipment. When practical, training should involve an actual aircraft so that firefighters can learn the capabilities and limitations of these tools on aircraft. Many techniques that work in auto extrication, for example, do not work as effectively in the aircraft environment and can only be learned through training and practical experience.

Safety

Operating rescue tools and equipment during aircraft rescue operations can be very dangerous. The number of personnel in the operational area should be limited to the minimum number necessary to complete the task. All personnel in the hazard zone should be in full protective gear, including especially eye and hand protection. Because rescue tool operations are usually very noisy, rescue personnel should wear hearing protection. New advanced aerospace materials (composites) and other respiratory hazards encountered at aircraft incidents necessitate the use of breathing apparatus. Personnel should maintain a natural body position and solid footing when using tools and equipment. A low worker-to-supervisor/safety officer span of control should be maintained, and rescue teams should coordinate their efforts to avoid adversely affecting each other's efforts.

Assorted Rescue Tools and Equipment

Many general tools and equipment can be used in ARFF operations. Most conventional tools and equipment commonly used in structural rescue and fire fighting can be adapted to an aircraft rescue and fire fighting situation. An assortment of conventional and specialized tools and equipment should be available to carry out the functions of ARFF, including rescue and a variety of other purposes. For example, after entry into an aircraft, firefighters may have to cut seat belts or harnesses in order to free occupants. Pads or salvage covers may be needed to cover jagged egress openings that could injure people or damage equipment. Tools and equipment may be needed to plug leaking fuel or oil lines. ARFF personnel should plan ahead so that they have the appropriate tools and equipment available to perform forcible entry and rescue operations in a timely and effective manner. NFPA 403, *Aircraft Rescue and Fire Fighting Services at Airports*, and *ICAO Airport Services Manual, Part 1, Rescue and Fire Fighting*, specify the types and amounts of tools to be carried on ARFF vehicles. The following sections discuss ordinary and commonly used tools and the primary or most common use for each.

Hand Tools

Hand tools generally can be defined as tools that rely on human force to transmit power directly to the working end of the tool. A standard fire apparatus toolbox commonly holds various hand tools and accessories that can be used for rescue-related tasks, including screwdrivers, Allen wrenches, socket sets, open-end and boxed wrenches, ratchets, drivers, assorted pliers, hammers, handsaws, hammers, chisels, punches, and cutters. The following are examples of some of these tools and how they are often used in aircraft rescue and fire fighting:

- ***Screwdrivers.*** Standard screwdrivers may be used to open access panels secured with Dzus fasteners. It is recommended that personnel carry screwdrivers, which also may be very helpful in pulling out the handle on a flush-mounted operating lever or for other tasks.
- ***Pike poles.*** Pike poles, either conventional or specialized crash poles, are useful for pushing and pulling operations such as holding open doors on cargo aircraft.
- ***Rescue tool assembly.*** A rescue tool assembly consists of the equipment typically carried on a rescue belt or in a tool roll: a V-blade harness-cutting knife, pliers, flashlight, hatchet, and other small tools (Figure 7.3).
- ***Axes.*** A wide variety of axes can be used on aircraft. It may be necessary to make holes for inserting the tips of hydraulic spreaders during forcible entry, and on some commercial aircraft, holes may be made with an axe.
- ***Assorted prying tools.*** Standard crowbars, wrecking bars, and other types of pry bars can be used for leverage and for bending and prying objects. They can also be used to force open doors and hatches on light-construction, nonpressurized aircraft.
- ***Harness-cutting knife.*** A harness-cutting knife, usually with a V-blade, is used to cut seat belts, parachute straps, and webbing (Figure 7.4).

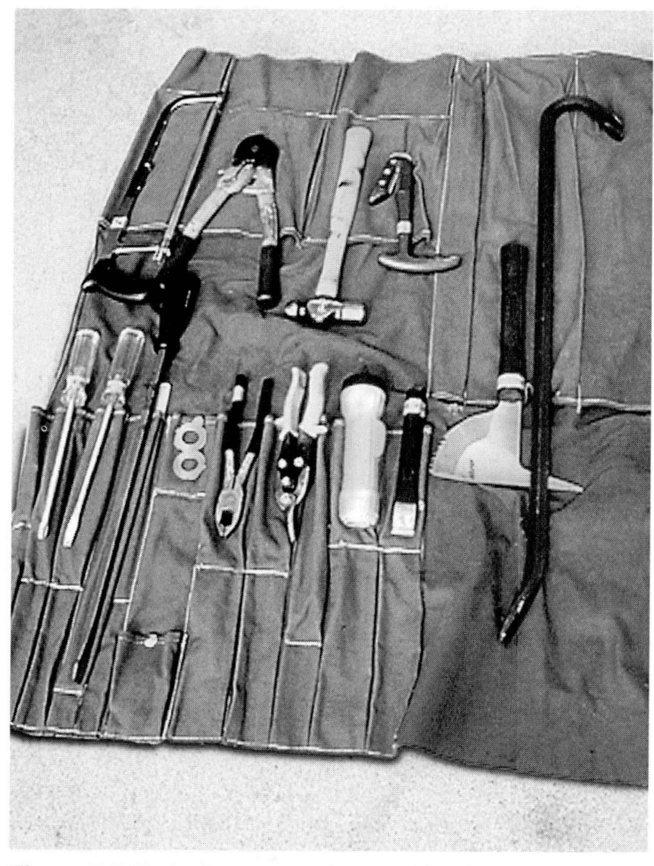

Figure 7.3 Typical rescue tool assembly. *Courtesy of Aviation Emergency Training Consultants.*

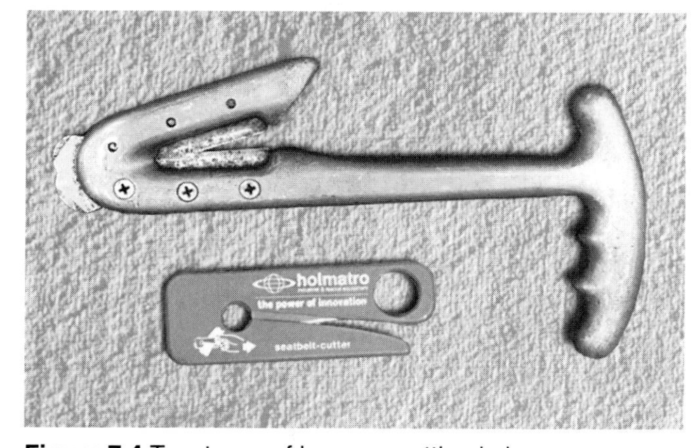

Figure 7.4 Two types of harness-cutting knives.

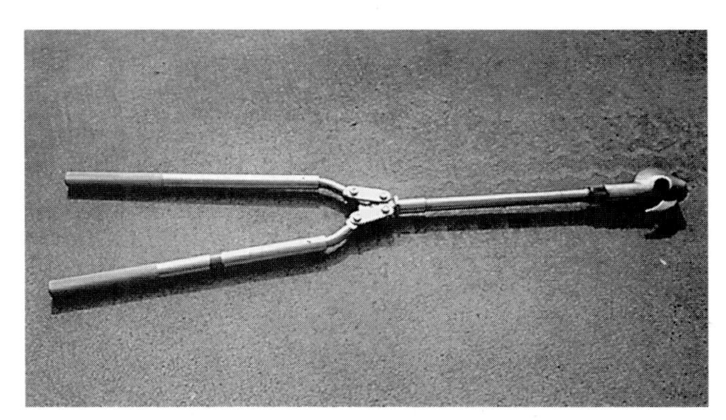

Figure 7.5 De-arming tool. *Courtesy of Aviation Emergency Training Consultants.*

- ***Cutters.*** A cable cutter is used primarily to sever cables, small hoses, and metal tubing. A de-arming tool is used to sever gas-initiator lines and safety some ejection seats, making it safer for personnel to work near them (Figure 7.5). Also, many other types of wire and bolt cutters can be used during aircraft rescue.

CAUTION: ARFF personnel must be aware of the aircraft construction in order not to cut in areas of the aircraft that may create safety hazards for the rescuers or passengers.

Power Tools

There are four major types of power tools, classified by the medium used to generate force:

- Electric
- Hydraulic
- Pneumatic
- Pneumo-hydraulic

Electrically powered tools use stored energy from a battery or convert electrical energy into mechanical energy through an electric motor. Hydraulic tools have pumps that produce and transmit pressure through a liquid (hydraulic fluid) to the working end of the tool. Hydraulic pumps are either hand-operated or powered by gasoline-driven or electric motors. Pneumatic tools use either an air compressor or stored pressurized air to transmit energy to the working end of the tool. Pneumo-hydraulic devices combine both air and liquid force in an air-driven hydraulic pump that generates power for operating the tool. While these power units are safer to use in flammable atmospheres, they are not widely used.

Electric

Power saws used for aircraft forcible entry operations should be equipped with blades that are capable of cutting metal. Circular saws, either electrical or gasoline-powered, should be rated heavy-duty and be capable of cutting a variety of materials. On large-frame aircraft, personnel may need to use large blades in order to penetrate the fuselage. In other situations, chain saws and reciprocating saws may also be useful. Many different types of reciprocating saw blades are available. ARFF crew members can use these types of saws and blades in tight quarters or in areas that are difficult to access.

A full assortment of rotary saw blades should be carried for the various types of cutting that may be required. Common types of cutting blades are multipurpose or composite, carbide- and diamond-tipped, and serrated. Composite blades can wear out quickly, chip, or become pinched because on rounded aircraft fuselages the angle of the cut is constantly changing. Blades should be color-coded with the legend clearly posted on the compartment or carrying case. Power saws are the tool of choice for rapid, clean cuts. Drawbacks to their use include excessive tool and cutting noise, as well as the possibility of sparks when the blade contacts ferrous metals (steel and magnesium).

CAUTION: If using a gasoline-motor-driven saw, do not store the fuel container in the same compartment or box with composite-type saw blades because fuel fumes have been known to damage these saw blades, causing them to fail.

Another electric tool used by ARFF personnel is the battery-powered/electric drill/driver. When used with a socket drive, they may be used to open a variety of compartments. Rescue workers should use caution not to exceed the recommended revolutions per minute (rpm) of the tool when accessing these compartments, or they may damage the opening mechanisms.

Hydraulic

Hydraulically operated tools may be used for spreading or forcing apart structural members of an aircraft during extrication operations. The hydraulic pressure can be produced either manually through a hand pump or through a power unit (usually either a gas-driven engine or a pneumatic or electrical pump) (Figure 7.6). Because these tools are so versatile in their application, they are highly recommended for airport fire departments.

On apparatus that remain at the scene, the tools and equipment should be stored so that they are readily available during rescue operations. For example, if ARFF personnel place the hydraulic rescue tools and equipment on a major crash apparatus that may return to the station to refill with water or agent, then personnel must take the rescue equipment off the apparatus so that it can be left at the scene.

Electric, pneumatic, or gasoline-powered hydraulic spreaders and cutters commonly used in auto extrication also have some application in aircraft incidents. Personnel may need to make a hole in the aircraft skin between the structural members for inserting the tips of the spreader. The spreaders tear the aircraft skin and bend structural members more than they cut. Sometimes the aircraft skin will roll up or orange-peel ahead of the spreading tips, or the aircraft fuselage metal may be brittle. A technique for quickly opening a large hole in the side of an aircraft is to use a hydraulic spreader in tandem with a hydraulic cutter. As the spreader is used to tear a progressive cut in the skin, hydraulic cutters are used to reach through the cut and sever the structural members underneath.

Figure 7.6 A portable hydraulic power unit.

CAUTION: Hydraulic spreaders may project metal fragments in all directions during rescue operations. In addition, some hydraulic tools are heavy and may need to be operated by two rescue personnel.

In flammable areas rescue workers should consider using these spreaders and cutters because they do not produce sparks (as opposed to an electric power unit or a gas-driven unit). They also do not produce the type of noise that is associated with a gas-driven unit.

WARNING

Be aware of the potentially flammable atmosphere of the aircraft accident/incident. When using power tools around aircraft, always wear complete protective gear, including SCBA, and have a charged handline in place.

Pneumatic

ARFF personnel may perform numerous cutting tasks with an air chisel (or air hammer) during aircraft rescue operations. An air chisel may be powered by compressed air from a breathing apparatus cylinder (Figure 7.7), from a compressor or cascade system, or from an air system on ARFF vehicles. The air chisel cuts by applying thousands of short-distance impacts per minute against a metal object. Different tips are available, and the lightness and size of an air chisel allow the firefighter to use it from a ladder. Air chisels can also be used to chip out some rigid plastic windows.

WARNING

Never use compressed *oxygen* to power pneumatic tools. Mixing pure oxygen and grease or oils found on the tools will result in a fire or violent explosion.

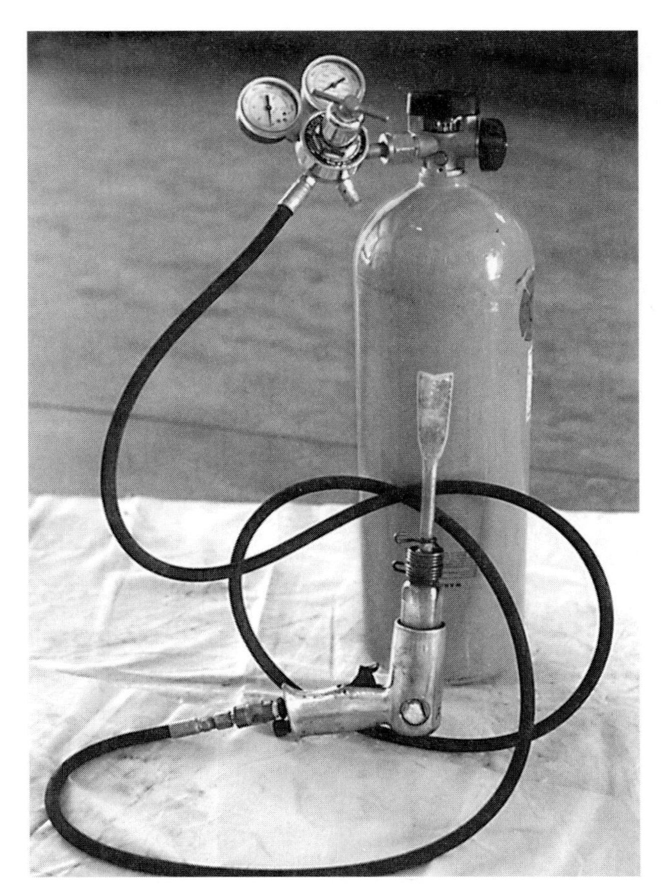

Figure 7.7 Compressed air from an SCBA cylinder powers this air chisel. *Courtesy of Vespra (ONT) Fire Department.*

Lifting and Pulling Tools and Equipment

Often it is necessary to perform lifting and pulling tasks to free trapped victims or to gain access to the interior. Rescue and fire fighting personnel should be familiar with all types of tools that can perform these functions. The following paragraphs discuss the more common lifting and pulling tools and equipment.

Truck-Mounted Winch

The use of a truck-mounted winch in the aircraft rescue and fire fighting environment is somewhat limited. There may be occasions, however, where the use of such devices could prove critical; these uses include:

- To quickly move a piece of wreckage in order to gain access to an area
- To stabilize an aircraft or component
- To assist with forcible entry (such as pulling a door open)

WARNING

Operate any truck-mounted winch in accordance with the manufacturer's recommendations. Failure to follow such specifications as the maximum weight limits may cause the winch to fail, injuring or killing the operators or personnel in the immediate area.

Come-Along

The come-along provides much the same application as the truck-mounted winch, only this device is more portable. This tool increases the pulling ability of a lever to its maximum through a ratchet/pulley action. The come-along is anchored to a secure object, and the cable or chain is run out to the object to be moved. Once both ends are attached, the lever is operated to pull the movable object toward the anchor point. The most common sizes or ratings of come-alongs are 1 to 10 tons (907 kg to 9 072 kg).

Rope

The use of ropes in the fire service is widespread, and its applications are well-known. In the aircraft rescue and fire fighting environment, the applications are the same. The primary uses include pulling, stabilization, movement of tools and equipment, and creating barriers. (**NOTE:** Consult the IFSTA **Essentials of Fire Fighting** manual for more information on the use, application, and care of ropes.)

Chains

Chains are used primarily in conjunction with other devices or tools. They are often used to extend the distance for pulling operations (for example, they can be attached to the spreaders of a hydraulic rescue tool). Chains are stronger than rope and are more suitable for some applications.

Webbing

Webbing is easily carried by individual ARFF personnel and can be used in confined environments. Commonly used for personnel applications, such as seats or slings, this versatile tool should be carried by all ARFF personnel. An 8- to 10-foot (2.5 m to 3 m) length can be carried in the pocket of turnout gear for easy accessibility.

Rope, chain, and webbing also can be used to suspend tools and lights when working inside the aircraft.

Other Equipment

There are many other tools, equipment, and devices that can be applied to the aircraft rescue and fire fighting function. The use and applications are limited only by the need and the initiative of ARFF personnel. Some of the more common types of other equipment needed to perform ARFF operations are listed here.

Repair Plugs

Repair plugs made of wood or rubber are used to plug leaking lines such as fuel lines and hydraulic fluid lines. Rescue crews in most jurisdictions carry a wide variety of shapes and sizes of plugs to ensure that a suitable one is available when needed. Adjustable plugs are available also.

Pins and Other Locking Devices

Some fire departments carry landing gear pins or other devices that can be used to lock and prevent the movement of landing gear assemblies. Some military aircraft, especially fighters, have a wide variety of pins for securing guns, canopy jettison systems, seat ejection systems, emergency power units, and other hazardous systems. Munitions safety pins act as a way of securing the aircraft's weapons from accidentally activating while on the ground.

Salvage Covers

Salvage covers may be used to cover jagged openings in order to prevent injury to personnel and passengers. Different-colored salvage covers may be used to designate collection points for equipment and personnel.

Ladders

All types of ladders (ground and aerial) are used for the passage of personnel to and from elevated sections of an aircraft (Figures 7.8 a and b). Personnel should consider laddering the leading edge of the wing, all doors, and other access points.

Figure 7.8a Aircraft firefighters use a ground ladder at the leading edge of the wing to access the aircraft.

Figure 7.8b Mobile stairs can be used to access aircraft doors.

Air Lifting Bags

The air lifting bag is a versatile device that is easily applied to rescue and aircraft stabilization work. It simply transmits the force of compressed air, usually supplied by compressed-air cylinders, throughout the surface of the bag. Although working pressure does not exceed 200 psi (1 400 kPa), the pressure is multiplied over every square inch (centimeter) of the bag, producing enough force to lift or displace enormous objects. There are two types of air lifting bags: low pressure and high pressure, both of which can be stacked (Figures 7.9 a and b).

Figure 7.9a A low-pressure air lifting bag.

Figure 7.9b A high-pressure air lifting bag. *Courtesy of Joel Woods.*

Lighting Equipment

During night operations it can become critical to provide additional lighting to carry out tasks safely. The importance of readily available lighting cannot be overstated (as is required equipment on crash apparatus, for example). ARFF personnel should know how to quickly set up portable lighting as well as how to operate all lighting sources on the apparatus.

Electric Generators

Portable electric generators with floodlights may be used to illuminate forcible entry and rescue points on an aircraft. These generators may also be used to operate electric chain saws, circular saws, reciprocating saws, and various other power tools. Generators that are mounted on apparatus should be removable to allow the generator to be transported to limited-access areas.

Portable Lights

Portable lights should be readily available to the airport fire department (Figure 7.10). Using lights on apparatus does not always provide the direct

Figure 7.10 One type of portable lighting unit.

lighting necessary for interior operations or other areas remote (and perhaps inaccessible) for apparatus. Portable lights and associated equipment, such as cords and stands, should be carried, along with a generator having enough capacity to provide power to lights and to any other electrically operated tools and equipment.

Vehicle-Mounted Lights

Most modern crash apparatus are equipped with a variety of lighting installed on the apparatus (Figure 7.11). It is common to see high-powered floodlights mounted in the front of apparatus to light up large areas during approach and subsequent operations. Most apparatus also have side- and rear-mounted lights (generally of less power). Elevating or extending lights can provide additional lighting.

Lighting Equipment Safety

Care should be taken to avoid using portable lighting equipment in a flammable atmosphere as it could provide a source of ignition. Personnel should also use caution to recognize the dangers associated with electricity and practice the necessary safeguards because this equipment often is used in a wet environment. Safety tips that ARFF personnel should follow when working around electricity include:

- Maintain a safety zone around the generator, electrical cords, and lighting.
- Guard against electrical shock.
- Treat all wires as "hot" and being of high voltage.

Figure 7.11 One of the lights mounted on an ARFF vehicle.

- Only use approved devices in good working condition.
- Wear full protective clothing, and use only insulated tools.
- Exercise care using ladders, hoselines, or equipment near electrical lines and appliances.
- Ensure all cords and devices have the proper ground wire.
- Do not touch any tool, equipment, or apparatus that is in contact with electrical wires because body contact will complete the circuit to ground resulting in electrical shock.
- Do not drape cords across fences, metal guardrails, or through water or foam.

Gaining Access into an Aircraft

Forcible entry should be considered only as a last resort. The conditions of the wreckage, flammable atmospheres, and many other factors (such as weather and nighttime operations) can create an environment that is hazardous not only to victims but also to rescue personnel.

The easiest and quickest way for rescue personnel to gain access to an aircraft is through normal doors and hatches. These openings usually have external releases (Figure 7.12). The same rule of forcible entry in structural fire fighting also applies to aircraft: ***try before you pry*** — try the normal means of opening the door before attempting to force it open (Figure 7.13). If occupants are attempting to exit the aircraft, do not impede their egress. In this case, ARFF personnel may need to force entry at another location, such as an emergency exit.

Aircraft windows may be used for ventilation, and personnel may use windows during rescue. Some windows are modified for use as emergency exits (Figure 7.14). On most aircraft, these exits are identified and have latch releases both outside and inside the cabin. Unlike egress doors, most window exits open toward the inside.

Many modern aircraft rely on electrical or other systems to operate doors (under normal conditions). For example, if the electrical system on an

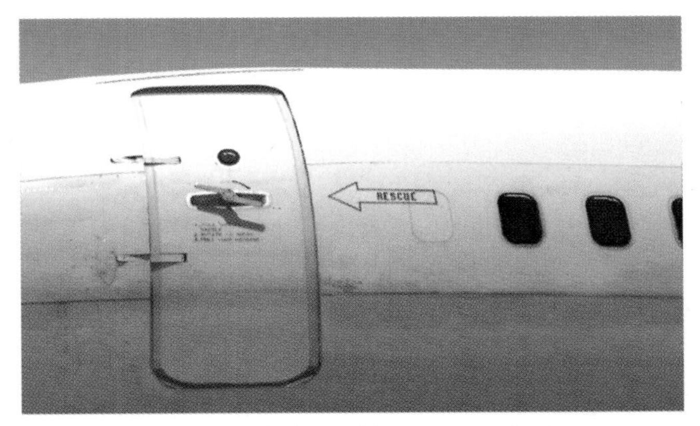

Figure 7.12 An aircraft door with an external release.

Figure 7.13 Firefighters may find that the cabin door can be opened from the outside without forcing entry.

Figure 7.14 A window emergency exit on a business/corporate type of aircraft.

Figure 7.15 Aircraft exterior cut-in areas. *Courtesy of Williams Training Associates.*

MD-11 is inoperative, there is a backup pneumatic system that can operate the door. This is automatic and activates from inside or outside when the door is opened in the emergency mode. It uses a small pressurized nitrogen cylinder that, when activated, assists in operating the mechanism that lifts the 500-pound door without electrical power.

Because of the impact forces on the aircraft, doors and hatches can become jammed and inoperable. It may be possible to force open jammed doors by using a pry bar around the frame or at the hinges on only the lightest and smallest of aircraft. When forcible entry is necessary for rescue, ARFF personnel must be *extremely* careful to prevent injury to survivors who may be just inside the fuselage. Just as in buildings, a forced opening should be cleared of as much broken material as time permits whenever a victim must be removed through it. A blanket or felt pad can be laid on jagged edges to prevent injury and to protect hoselines that extend through the opening.

Emergency escape slides can also be dangerous to ARFF personnel and can be obstacles to gaining access. These slides inflate with enough force and pressure to cause severe injury or death to anyone in the immediate area. The slides are generally hooked to the fuselage with a bar attached to the bottom of the slide. Some of these require manual operation by pulling an inflation ring or ball, but some may release automatically when the door opens.

If all other methods of gaining access fail, personnel must attempt to cut through the wall of the fuselage. Military aircraft have areas of the fuselage designed for cutting. These areas are outlined with yellow or black dashed lines and are labeled "CUT HERE FOR EMERGENCY RESCUE"; there are usually only one or two marked places, and they tend to be small in area (Figure 7.15). Civilian aircraft generally do not have these markings, so ARFF personnel need to know the areas to cut. Before cutting, the ARFF crewmember must envision the interior, including the arrangement of the bulkheads, partitions, decks, and armor plates (in military aircraft) and the locations of fixed equipment. If it is absolutely necessary to cut, the cutting should be done around or near the windows. Except for electrical wires, these are areas where rescue crewmembers have the least chance of cutting into aircraft systems. If the power has been shut down on the flight deck and the battery has been disconnected, contacting electrical lines is not a problem.

Cuts should sever the fewest possible reinforcing channels, stiffeners, ribs or longerons. Structural reinforcements of the skin are almost always parallel or perpendicular to the length of the fuselage. This creates rectangular sections of skin surface between horizontal stringers and longerons, and between curved vertical ribs or bulkheads. (**NOTE:** See Chapter 3, "Aircraft Familiarization," for more details on aircraft construction.)

The area selected for cutting should consist of one or more of these rectangular skin surfaces, where cutting involves the least number of heavier sections supporting it. The area should also be of a size sufficient for access and egress of personnel and equipment. Usually the rivet pattern indicates the presence of structural members and heavier sections located underneath. If there are a lot of rivets, the firefighter should not make a cut at that area. Another area to avoid is the main deck and the area immediately below. The deck, fuselage, and diagonal stiffeners form a highly reinforced area where most of the systems are found.

WARNING

Beware of the potential hazards when it is necessary to penetrate the skin of any aircraft in areas not marked as cut-in areas. Hazards include high-pressure hydraulic lines, compressed-gas cylinders, pneumatic lines, advanced aerospace materials, lavatory waste, all aircraft systems, and on military aircraft, unexploded ordnance.

Chapter 8

Aircraft Rescue and Fire Fighting Apparatus Driver/Operator

Job Performance Requirements

This chapter provides information that will assist the reader in meeting the following job performance requirements from NFPA 1003, *Standard for Airport Fire Fighter Professional Qualifications*, 2000 edition. (**NOTE:** This chapter also provides an overview of the job performance requirements found in Chapter 7, "Aircraft Rescue and Fire-Fighting Apparatus," of NFPA 1002, *Standard on Fire Apparatus Driver/Operator Professional Qualifications*, 1998 edition.) Particular portions of the job performance requirements (JPRs) that are addressed in this chapter are noted in bold text.

3-3.3 Extinguish an aircraft fuel spill fire, given PrPPE, an ARFF vehicle turret, and a fire sized to the AFFF flow rate of .13 gpm (0.492 L/min) divided by the square feet of fire area, so that the agent is applied using the proper technique and the fire is extinguished in 90 seconds.

(a) *Requisite Knowledge*: Operation of ARFF vehicle agent delivery systems, the fire behavior of aircraft fuels in pools, physical properties and characteristics of aircraft fuel, agent application rates and densities.

(b) *Requisite Skills*: **Apply fire-fighting agents and streams using ARFF vehicle turrets.**

3-3.6 Attack an engine or auxiliary power unit/emergency power unit (APU/EPU) fire on an aircraft while operating as a member of a team, given PrPPE, an assignment, ARFF vehicle hand line or turret, and appropriate agent, so that the fire is extinguished and the engine or APU/EPU is secured.

(a) *Requisite Knowledge*: Techniques for accessing the aircraft engines and APU/EPUs, methods for advancing hand line from an ARFF vehicle, **methods for operating turrets**, methods for securing engine and APU/EPU operation.

(b) *Requisite Skills*: Deploy and operate ARFF hand line, **operate turrets**, gain access to aircraft engine and APU/EPU, secure engine and APU.

The ARFF apparatus driver/operator is responsible for safely transporting firefighters and apparatus to and from the scene of an emergency or other call for service. Once on the scene, the driver/operator must be capable of operating the apparatus properly, swiftly, and safely (Figure 8.1). The driver/operator must also ensure that the apparatus and the equipment it carries are ready at all times. The driver/operators of ARFF vehicles have the monumental tasks of locating downed aircraft quickly and of fire suppression and agent conservation. The ARFF apparatus is designed

Figure 8.1 ARFF apparatus. *Courtesy of Oklahoma Air National Guard, Will Rogers World Airport, Oklahoma City, OK.*

to deliver large quantities of extinguishing agent in a short period of time, which requires a skilled and alert driver/operator. Many ARFF apparatus have only one firefighter assigned to them — the driver/operator. This places a large responsibility on these individuals.

This chapter discusses duties and responsibilities of the ARFF driver/operator, including inspection and maintenance, safe vehicle operation, agent discharge, and familiarization with auxiliary systems and equipment. In general, driver/operators must be mature, responsible, and safety conscious. Because of their wide array of responsibilities, driver/operators must be able to maintain a calm, can-do attitude during stressful emergency situations. Psychological profiles, drug and sobriety testing, and background investigations may be necessary to ensure that the driver/operator is ready to accept the high level of responsibility that comes with the job.

To perform their duties properly, all driver/operators must possess certain mental and physical skills. The required levels of these skills are usually determined by each jurisdiction. In addition, NFPA 1002, *Standard for Fire Apparatus Driver/Operator Professional Qualifications,* sets minimum qualifications for driver/operators. In particular, Chapter 7 of this standard references ARFF apparatus driver/operators. It requires any driver/operator who will be responsible for operating a fire pump to also meet the requirements of NFPA 1001, *Standard for Fire Fighter Professional Qualifications,* for Fire Fighter II. For more information on the general qualifications for driver/operator, refer to the IFSTA **Pumping Apparatus Driver/Operator Handbook.**

Apparatus Inspection and Maintenance

The reasons for keeping vehicles in top operating condition goes without saying. ARFF vehicles are expected to respond to accidents/incidents at a moment's notice without delay. A complete apparatus inspection and maintenance program helps keep vehicles in top operating condition.

Few vehicles are the same, and each department should have an SOP for the inspection and maintenance of its vehicles. The purpose of this section is not to teach the ARFF firefighter how to inspect the vehicle but rather to teach the importance of a systematic approach to the department's inspection and maintenance program (Figure 8.2).

General Inspection Procedures

Each department must set a plan for inspecting its vehicles according to the manufacturers' requirements and the needs of the department. For more details of a model inspection program, see Chapter 3 of the IFSTA **Pumping Apparatus Driver/Operator Handbook**.

The main reason for vehicle inspection is, of course, to make sure that the vehicles operate properly when they are needed. Another good reason for daily vehicle inspection is to keep the driver/operators skilled at operating the vehicle. The more practice they have operating their vehicle, the more skilled they are when operating at the scene of an emergency.

Inspection programs greatly differ among departments. A department that rarely uses its vehicles does not need a program requiring inspections as frequently as those of active departments. The obvious things accomplished in a daily inspection include checking all engine fluid levels, checking fire fighting agent levels, setting the mirrors and safety equipment appropriately for the person operating the vehicle, and performing an operational test to make sure the vehicle performs normally. All problems should be passed on to the proper authorities to ensure any problems are corrected before the vehicle is placed back into service.

Figure 8.2 A firefighter performs a daily inspection on one of the department's apparatus.

For airports having only the vehicles required to maintain their operational status, a vehicle out-of-service due to maintenance problems is a serious concern. A program should be in place to ensure that vehicles are operational and ready at all times so that aircraft will not have to be diverted from the airport because of insufficient protection levels.

Agent Dispensing Systems

ARFF apparatus have different types of extinguishing systems. Many modern ARFF vehicles carry and dispense at least two different types of extinguishing agents. The following sections cover the different types of extinguishing systems common on apparatus and briefly discuss how to inspect them.

Foam Systems

Each major ARFF apparatus has a foam system, and just as the vehicles become diverse, so do the foam systems. They differ in capacity, application methods, and proportioning systems. Inspection of these systems on a daily basis usually consists of making sure that the agent tank is full (Figure 8.3). On a somewhat less-frequent basis, the inspection includes testing the foam proportioning system to ensure that the appropriate ratio of foam concentrate and water is being discharged from the vehicle. Improper foam concentrate proportioning may result in either one of two problems:

- The foam-to-water mixture may be too lean (not enough foam concentrate in the foam/water solution). This will cause the foam bubbles to drain sooner than desired.
- The foam concentrate may be too rich (too much foam concentrate in the foam/water solution). This will cause foam concentrate to be wasted.

Figure 8.3 ARFF personnel test the foam proportioning system. *Courtesy of Robert Lindstrom.*

NFPA standards allow any one of four methods to be used for testing a foam proportioning system for calibration accuracy:

- Foam concentrate displacement method
- Foam concentrate pump discharge volume method
- Foam solution refractivity testing
- Foam solution conductivity testing

These methods are explained in detail in the IFSTA **Pumping Apparatus Driver/Operator Handbook** and **Principles of Foam Fire Fighting** manual. Personnel should follow the apparatus manufacturer's recommended procedures for conducting this testing.

One other inspection concern is the handline delivery method used for the department's vehicles (Figure 8.4a). Some types of handline delivery are

Figure 8.4a Handlines must be inspected regularly.

1½-inch or 1¾-inch (38 mm or 45 mm) hoselines, booster reels, and multiagent handlines. It is important to follow normal maintenance, care, and reloading procedures for these hoses (Figure 8.4b).

Dry-Chemical Systems

Many ARFF vehicles carry dry chemical, usually Purple K® (PPK), as an auxiliary extinguishing agent. The reason PPK is the dry-chemical agent of choice for multiagent applications is its compatibility with AFFF (Figure 8.5). The size of dry-chemical systems on ARFF apparatus usually starts at 500 lb (227 kg) and can be much larger on some apparatus. The dry chemical is stored on the ARFF apparatus in a manner similar to that of the typical portable fire extinguisher. A large vessel carries the agent, and compressed nitrogen stored in a separate cylinder is used to expel the agent. It is very important to be familiar with the type of dry-chemical system on the department's vehicles. These systems require regular attention so that they perform properly when needed.

Dry chemical is dispensed three different ways. The most common way of application is to dispense using a handline stored at some position on the vehicle. Other popular ways include piggybacking and water stream injection. On piggyback systems, the manufacturer has mounted an independent dry-chemical nozzle directly over the water/foam nozzle on the roof or bumper turret. Water stream injection systems work by injecting the dry chemical directly into the water/foam stream of the main turret. This allows the dry chemical to stay inside the water stream of the turret, greatly extending the effectiveness of the agent.

Dry chemical is often used for three-dimensional fires on aircraft engine nacelles or for running fuel fires. This makes water stream injection effective not only by applying the dry chemical to fight the fire but also by supplying the water/foam solution to stop the spill fires associated with three-dimensional fires.

Testing the dry-chemical system is expensive and time-consuming. The only way to ensure that the system functions correctly is to activate the system on some sort of time schedule. System manufacturers recommend set safety procedures to follow when returning dry-chemical systems to service. Do not deviate from these procedures. Older dry-chemical systems may require agent fluffing on a regular basis to loosen the dry chemical. This is necessary because after a vehicle is driven over a period of time, the dry chemical can begin to settle and pack tightly inside the container. This can cause the agent to cake and not work properly. Fluffing involves inserting a device (usually made of metal) into the storage cylinder and agitating the chemical. Wooden devices are not recommended as they could splinter and block the agent delivery system. Stirring should be done carefully to avoid damaging the agent piping and the internal tank parts.

Figure 8.4b Hose should be loaded properly in the apparatus.

Figure 8.5 Dry-chemical system. *Courtesy of Robert Lindstrom.*

Checking the dry-chemical handline is also important. Make sure after each use that the handline is completely blown out to keep the agent from blocking the hose, rendering the system useless (Figure 8.6). Personnel should follow the manufacturer's recommendations when blowing out the hose and returning the hose to the hose reel.

Figure 8.6 Firefighters blow out a handline after using dry chemical. *Courtesy of Robert Lindstrom.*

Clean Agent Extinguishing Systems

Some ARFF apparatus are equipped with clean agent extinguishing systems. Clean agents extinguish fire by interrupting the chemical chain reaction of combustion without leaving any residue after they have been used. Clean agents are useful because they save damage on expensive jet engines and sensitive electrical components.

In the past, the primary clean agent used on ARFF apparatus was Halon 1211. However, halon extinguishing agents harm the ozone layer of the earth's atmosphere, and their future production has been globally banned. Some companies have developed new clean agents that may be used in place of halon agents and are not harmful to the environment. These agents are usually stored in pressurized vessels on the ARFF apparatus and, like dry chemical, may be dispensed using other gases such as argon. These also may be dispensed using handlines, which are stored on the vehicle.

These systems require very little inspection other than checking the agent level and pressure of expellant gas. When servicing these systems, personnel should follow the manufacturer's procedures to the letter.

WARNING

Firefighters have been killed from not following directions and improperly servicing these types of systems. Always follow the manufacturer's instructions whenever servicing any pressurized system.

Principles of Safe Vehicle Operation

After learning a vehicle's capabilities, the driver/operator must take the time to learn how to safely operate the vehicle. ARFF vehicles are larger and heavier than the personal vehicles driven to and from work and can be dangerous if operated by untrained personnel. These vehicles typically are even larger than structural fire apparatus, and previously-trained drivers may require additional training (Figure 8.7). This section stresses the im-

Figure 8.7 In a large, open space a driver/operator practices maneuvering an ARFF vehicle. *Courtesy of Robert Lindstrom.*

portance of safe vehicle operations under a number of situations. NFPA 1002, *Standard for Fire Apparatus Driver/Operator Professional Qualifications*, Chapter 7, describes required driving skills for the ARFF driver. It is suggested that the operators review these skills and those that apply to the ARFF driver/operator in the IFSTA **Pumping Apparatus Driver/Operator Handbook**.

NFPA 1002 specifies a number of practical driving exercises that the driver/operator candidate should be able to successfully complete before being certified to drive the apparatus. The standard requires that driver/operators be able to perform these exercises with each type of apparatus they are expected to drive. Some jurisdictions prefer to have driver/operators complete these evolutions before allowing them to complete the road test. This ensures that driver/operators are competent in controlling the vehicle before they are allowed to drive it in public.

Before Leaving the Station

Safe vehicle operation begins before even leaving the fire station. The driver/operator and occupants must wear seat belts! Seat belts do save lives. The next concern is to see that all heavy objects in the vehicle cab are secured. A loose piece of equipment such as a portable radio can become a missile in a vehicle collision. Personnel should take the time to find a place to safely store these items in the vehicle.

Making sure vehicle safety equipment is set for the driver/operator is important also. Taking the time to adjust the mirrors, seats, steering column, and radio volume can help avoid a collision.

Vehicle Control Variables

When operating the vehicle, the driver/operator must understand what affects the way the vehicle operates in different situations. Because the ARFF vehicle driver/operator is operating a specialized vehicle under other-than-normal driving conditions, he or she may be faced with variables rarely encountered when, for example, driving a car to the grocery store.

Braking Reaction Time

Braking reaction time can best be described as the time it takes the driver/operator to react to a situation. Knowing the braking characteristics of a vehicle can prevent a serious collision. When added to other factors like ice, snow, and water, the friction is reduced under the tires, which makes the vehicle braking even less effective.

Load Factors

In recent years, there has been a serious problem with ARFF vehicles rolling over. This has been attributed to several factors, one of which is the reaction a vehicle with a higher center of gravity has when making turns. Remember that when making a turn onto a taxiway, a 1,500-gallon (6 000 L) four-wheel-drive ARFF vehicle reacts differently than a pickup truck. Take a moment to think where all the weight lies on these vehicles. The water tank is typically on or above the main vehicle frame rails. This already creates a disadvantage: If the weight is quickly thrown in one direction and the driver/operator oversteers a corner, the vehicle is at risk for rolling over. One way to prevent a rollover is to understand the characteristics of the vehicle — know where the weight is and how the vehicle reacts when making a turn. The more things placed on ARFF vehicles, the heavier they become. This added weight can contribute to the reaction the vehicle has when it makes a turn at high speeds. The driver/operator should know the vehicle's load factors and drive according to what is safe for that vehicle.

General Steering Reactions

An object in motion tends to stay in motion at a constant velocity unless acted upon by an unbalanced force. This principle holds true when making sharp turns with ARFF vehicles at high rates of speed. If the weight of the vehicle is thrust quickly in one direction, the momentum of that weight will continue in that direction unless an equal or greater force stops it. If the momentum is greater than the resistance posed by the vehicle suspension system, then the vehicle is likely to overturn. When most driver/operators are asked about the last thing they remembered when their vehicle rolled over, they state that they remember making a turn. They never seem to know at what point the vehicle started to roll over.

All new vehicles purchased with FAA funds require a device that indicates to the driver/operator the attitude of the vehicle when making a turn. The driver/operator should take the time to reference this device to better understand the forces the vehicle receives when making a turn.

Speed and Centrifugal Force

Speed affects several important factors in driving the ARFF vehicle. When a vehicle is driven in a straight line, its speed has a direct impact on the required stopping distance. Simply put, the faster the vehicle is driven, the longer it will take for it to be brought to a safe stop.

Speed also impacts a vehicle's ability to be turned. The faster a vehicle moves, the more centrifugal force is exerted when the vehicle is turned. As centrifugal force increases, so do the chances of the vehicle overturning if the force exceeds the capabilities of the vehicle suspension system to resist the force.

Those airports that rely on emergency access roads for responding to aircraft emergencies must understand what speed and centrifugal force mean to ARFF vehicles. Hopefully, these roads will be engineered to reduce the hazards created by speed and centrifugal force.

Skid Avoidance

ARFF vehicles not only are heavy but also travel at high speeds, requiring a longer distance to stop. This makes avoiding a skid sometimes hard to accomplish. With the addition of poor weather conditions, there is a greater potential for problems. The power needed by the vehicle's drivetrain to move the vehicle at high speeds can make braking difficult. To avoid a skid, the driver/operator must know the vehicle stopping distances in normal driving conditions. Understanding vehicle's braking reactions on dry surfaces can help the driver/operator perceive how the vehicle will react when braking on wet or frozen surfaces. Another way is for personnel to take the time to train under the types of adverse conditions commonly encountered in their jurisdiction. This does not mean finding an icy patch on the airfield, racing toward it, and attempting to control the vehicle; this means starting really slowly and then testing the brakes and steering. Understanding how the vehicle reacts under training conditions can pay off in emergency driving conditions.

Acceleration/Deceleration

In order to maintain proper control of the vehicle, the driver/operator must first understand the effects of acceleration and deceleration on vehicle control. When responding to an emergency, the driver/operator's main concern is to arrive safely and as quickly as possible. However, the driver/operator must remember that he or she is driving a heavy vehicle moving at speeds of up to 60 miles per hour (97 km/h); these large ARFF vehicles require longer stopping distances than do passenger vehicles driven at the same speed. The driver/operator should anticipate the actions of the vehicle and accelerate accordingly when responding to emergencies.

Another concern of over-acceleration and excessive deceleration is the burden this places on the vehicle engine, transmission, and braking systems. The cost associated with maintaining these systems can be high if the drivers do not properly operate the vehicles. As usual, when in doubt about the safe operation of ARFF vehicles, personnel should follow the manufacturer's recommendations.

Shifting and Gear Patterns

Most modern ARFF vehicles are equipped with automatic transmissions. Automatic transmissions are far more efficient than manual transmissions, allowing the operators to concentrate on driving and not on shifting while responding to emergencies. However, some of the other vehicles used to deliver ARFF services may be equipped with manual transmissions. It is important to learn the proper techniques for operating the vehicles with manual transmissions. For more information on driving vehicles with manual transmissions, see the IFSTA **Pumping Apparatus Driver/Operator Handbook**.

Safe ARFF Vehicle Operation

The majority of the driving is under normal driving conditions. Driver/operators should begin to learn safe vehicle operations while in a normal driving mode, so when faced with adverse surface conditions, such as ice and snow, they are better prepared to deal with these operations (Figure 8.8).

Figure 8.8 ARFF driver/operators must be thoroughly familiar with the rules for operating apparatus on runways and taxiways. *Courtesy of Robert Lindstrom.*

On the Airport

The majority of driving performed by driver/operators of ARFF vehicles is done while on the airport property. The ARFF driver/operator must be very familiar with all aspects of the airport layout, including taxiways, runways, ramp areas, and service roads. Aircraft parking areas present special challenges to the driver/operator because of the varied activities and obstacles that are found there. These include taxiing aircraft, fuel trucks servicing aircraft, and ground handling equipment that clutters the area.

Safe operation of the ARFF vehicle is paramount in these situations. Driver/operators must move through these areas with a watchful eye. Many of these dangers may seem to jump out in front of a driver/operator who is not paying attention. Driver/operators should spend enough time driving around the aircraft parking ramps to have a comfortable feel of what to expect while responding to emergencies.

Probably one of the most important driving situations is the route taken when responding to emergencies. Unlike structural firefighters, ARFF crew members do get advanced warning of emergencies from time to time. Predesignated response routes are given for reaching standby parking positions along the runways. The driver/operator must know the route without having to look at a map or without thinking about the best route to take when responding. While situations may change, the majority of responses to the runway standby positions often may not change. Driver/operators should know how to safely enter these areas while working with the air traffic ground controllers, and they should remember that these routes are not designed for making high-speed turns and movements like on highways. ARFF driver/operators should allow ample time to brake and safely corner when responding to emergencies.

Loose or Wet Soil

Operating on soil in poor condition is one factor each driver/operator may face when responding to emergencies. Each driver/operator must know the off-road capabilities of his or her vehicle. Advances have been made in the last few years in dealing with vehicle operations in off-road conditions. One of these advances is the Central Inflation/Deflation System (CIDS). The CIDS is a system that allows the vehicle operator to deflate the tires to increase the traction surface of the tire. This also helps keep the tire tread from filling with mud/dirt and decreasing traction. These systems use an onboard air compressor and are controlled in the vehicle cab. Many airports have chosen this system to help deal with poor off-road conditions.

Steep Grades

As everyone is familiar with the terrain of their airports, they may understand that there are places having steep grades. All major ARFF vehicles are designed by manufacturers to function on steep grades (Figure 8.9). This does not mean the vehicle can be driven the same as on flat ground. The vehicle's center of gravity is definitely changed when

Figure 8.9 ARFF apparatus must be able to maneuver on all different types of terrain. *Courtesy of Robert Lindstrom.*

it is climbing up and down steep grades. If forced to operate the vehicle on a steep grade, the driver/operator should be extremely cautious and not make sudden changes in direction. Other factors such as mud, snow, or generally poor surface conditions may present an additional safety issue, adding to the danger of operating on these grades.

Vehicle Clearance of Obstacles

One of the major contributing factors to a vehicle getting stuck while operating off-road is it catching on something underneath it. ARFF vehicles are designed to allow for maximum ground clearance whenever possible. This, however, does not prevent the vehicle from getting hung up. The operator must understand the ground clearance of the vehicle and learn to visualize objects to demonstrate this clearance. One way to practice this skill is to set up an object that will not harm the undercarriage of the vehicle, and then cross over the top to gain a better understanding of the vehicle's capabilities. Knowing this will help the driver/operator decide whether or not the vehicle may clear an object.

Limited Space for Turnaround

Many taxiways and aircraft parking areas provide limited space for turning a vehicle, and ARFF vehicles are, by design, very large vehicles. One major obstacle to overcome when training is the space it takes to turn the vehicle around. The driver/operator must be capable of visualizing the vehicle's turning radius. This is best accomplished in a training environment such as by setting out traffic cones or some other nondestructible objects and driving around them to learn the vehicle's steering pattern. The turning-around exercise required by NFPA 1002 also is excellent for giving the driver/operator experience in making these maneuvers.

Side Slopes

A serious concern when driving the ARFF vehicle off-road is approaching different types of terrain. One problem is the angle of attack taken when approaching a steep side slope. The vehicle should be designed to traverse a side slope when operating off-road. The driver/operator not only should know the vehicle's capability of approaching a side slope but also should be able to visualize the approach to a ditch or ravine. The center of gravity of the vehicle changes while attempting to traverse a side slope. How the driver/operator operates on the side slope affects the center of gravity. As a general rule of thumb, the driver/operator should not try to travel along the side slope at a right angle. Rather, he or she should approach side slopes at angles to reduce the effect on the vehicle's center of gravity.

Night Driving

After dark, the airfield appears far different from how it does during the daylight hours. While taxiways are marked with lighted signs and the taxiways are lined with blue lights, the airfield still takes on a whole new appearance. Each driver/operator should access the airfield at night and practice finding his or her way around. As pilots practice instrument landings and flying while using only instrumentation, the ARFF driver/operator should practice driving with the driver's enhanced vision system (DEVS) on a regular basis. This practice should be accomplished under driving conditions similar to those they may encounter in adverse weather and in darkness (Figure 8.10).

Adverse Environmental or Driving-Surface Conditions

Adverse environmental conditions are commonly contributing factors in delayed ARFF responses to aircraft accidents/incidents. Heavy rainfall, blinding snowfall, fog, smoke, and darkness of night cause visibility to be diminished and slow the response to an emergency. In such conditions, it can be difficult to see an aircraft that has no lights and

Figure 8.10 The driver's enhanced vision system is a feature that helps driver/operators access an emergency scene in darkness. *Courtesy of Robert Lindstrom.*

that is not burning. Driver/operators must use extreme caution when approaching the suspected accident/incident site to avoid accidentally striking the downed aircraft or people who are fleeing or who have been thrown from it.

Knowing the possible location of the accident/incident helps the driver/operator take the best route, which keeps the responding vehicle out of harm's way. Driver/operators should use all their senses when responding in poor visibility, including slowing down and opening windows to listen for any strange noises.

One more very important part of navigating the route across the airfield is to be very familiar with the layout. Airfield familiarization training during daylight hours will help driver/operators find their way in low-visibility conditions. In addition, knowing the runway and taxiway marking and lighting systems will help driver/operators find their way in poor lighting conditions.

Maneuvering and Positioning ARFF Vehicles on the Accident/Incident Scene

Personnel must consider what they may encounter and what to do when they arrive on scene. Each accident/incident is different. As discussed in Chapter 4, "ARFF Firefighter Safety," the aircraft accident/incident site is full of problems. The driver/operator must be alert to the conditions found when arriving on site so as not to further harm any passengers or crew members (Figure 8.11).

Wreckage Patterns

The factors that determine the wreckage pattern include direction and speed on impact, weather conditions, size of aircraft, type of crash (high-impact or low-impact), and location of the crash site. Different types of terrain have different effects on the wreckage pattern also.

In addition, the wreckage itself causes problems for personnel approaching the accident/incident site. They should be prepared to find large debris, victims, fire, hazardous materials, and anything else imaginable. Knowing a little bit about the wreckage patterns themselves can provide some insight into the dangers to expect.

Survivors

When responding to the accident/incident, the driver/operator must always consider where survivors may be found. If the passengers and crew are capable of escaping the aircraft on their own, many of them will be out of the aircraft before the first ARFF units/emergency vehicles arrive. This in itself poses a significant problem for the initial setup and fire attack as often victims come toward the emergency vehicles. Some departments have taken on the practice of shutting off the emergency lights so that the survivors do not move towards their vehicles. When operating powerful turrets, the driver/operator must remember to not directly hit any victims with the stream. At night it may be

Figure 8.11 Reaching an accident scene may be complicated because of the nearby terrain and the aircraft wreckage. *Courtesy of George Freeman, Dallas (TX) Fire Department.*

difficult to spot victims. Personnel should use whatever means possible to light the scene when working at night.

Terrain

The type of terrain where an accident/incident occurs definitely affects the driver's approach to the aircraft. Many airports are situated away from metropolitan areas, near mountain regions, at the edge of a body of water, or even in the middle of a city that has grown around the airport. These terrain factors add to the difficult task of positioning the ARFF apparatus. It is important for personnel to be familiar with the terrain of the airport and nearby areas.

A basic rule of thumb to remember when positioning at a downed aircraft is to stay uphill, upwind, and upstream of the accident/incident site whenever possible. This basic rule often must be violated due to the terrain.

Some other factors that cause problems with terrain are mud, steep inclines, lack of access roads to accident/incident site, water crossings, poor bridges, and rocky/hilly areas (Figure 8.12). All these elements can delay response and make operations very difficult. The key once again is anticipating problems and being thoroughly familiar with the airport and surrounding areas. Airport topographical maps provide driver/operators with an excellent overview of response area terrain.

Agent Discharge

Effective agent application has a definite effect on the outcome of the accident/incident fire. The driver/operator must be familiar with the extinguishing agents and their characteristics and how to apply them effectively to extinguish fires. Chapter 9, "Extinguishing Agents," provides greater details of agent application and fire extinguishment.

Agent Management

Agent management is one very important aspect to controlling a large fuel-spill fire surrounding an aircraft accident/incident site. The days of hurrying to the aircraft, dumping everything, and driving like crazy back to resupply are long gone. The driver/operator must have a thorough familiarization of the agents being used and know how to manage

Figure 8.12 This small bridge may not withstand the weight of an ARFF apparatus.

their use. These basics are no different from the information in the IFSTA **Essentials of Fire Fighting** or **Principles of Foam Fire Fighting** manuals.

The most important aspect associated with agent management is knowing how much of each agent is carried on the apparatus and what the realistic fire fighting capabilities of that amount of agent are. By training on lifelike scenarios, the ARFF driver/operators should be able to develop a sense of the capabilities their vehicles possess. This will allow the driver/operators to make good judgments on how and where to use the agent during true emergency conditions.

Effects of Terrain and Wind

As mentioned earlier, the terrain has a definite effect on agent application. The main issue is the flow of agent when applied on a hillside or wet area. Planes never seem to crash in perfect areas, so personnel must be prepared to deal with whatever situations occur.

Because airports are wide-open spaces, they tend to be windier than more built-up locations (Figure 8.13 a). Wind has detrimental effects on all types of fire streams (Figure 8.13 b). When discharged into or across the prevailing winds, the agent stream will be broken up and its reach considerably shortened. This is particularly a problem with dry chemical and clean agents. These agents are so light in weight that wind easily disperses them and carries them off. The problem with dry chemicals can be partially lessened by systems that inject the dry chemical into the water/foam stream discharged from a handline or turret. This somewhat decreases the effect that wind has on the dry chemical.

Figure 8.13a If the situation allows, it is best to discharge extinguishing agent with the direction of the wind.

Figure 8.13b The agent stream is adversely affected when directed into the wind.

Reach, Penetration, and Application

Each major ARFF vehicle has certain characteristics that allow that particular vehicle to achieve a certain agent-application rate, pressure, and distance. It is very important that the driver/operator understand agent-application capabilities of the vehicle (Figure 8.14). Knowing the vehicle's turret reach allows the driver/operator to judge the distance from the fire and keep the vehicle at a safe distance. If the vehicle gets too close, its windows could be damaged, and possibly the vehicle could catch fire.

Each vehicle turret has specific capabilities. The specific gallons (liters) per minute, pressure, and turret location drive these capabilities on the vehicle. Some vehicles are equipped with an extendable turret; this turret has a much farther reach than the conventional roof turret. The extendable turret also has the ability to attack the fire from lower positions that places the agent closer to the base of the fire, reducing the effects that fire has on breaking up the foam as it is applied to the fire.

Figure 8.14 An extendable turret on an apparatus allows agent to be discharged from a point closer to the fire. *Courtesy of Clark County Fire Department, Las Vegas, NV.*

Application Techniques

Each driver/operator must develop a technique for operating the vehicle's fire fighting systems. One of the most difficult systems to operate is the roof/bumper turret. Reaching the fire with the agent is very important for the obvious reasons, but doing so is a skill that requires practice. With the environmental concerns of today, it can be difficult to get the much-needed real fire experience. One way to learn the vehicle's reach is by setting up a course using a traffic cone and a softball. Place the softball on the traffic cone, and begin making attacks on the softball attempting to knock it off the cone without knocking over the cone (Figure 8.15). This exercise may sound easy, but it is not. It takes a skilled driver/operator knowing the vehicle's turret reach and control to even knock the ball off the cone. By practicing drills such as these, driver/operators can develop their techniques for operating the turrets.

Effectiveness

The driver/operator must be able to determine how the agent will affect the fire and, as discussed earlier, will have to make sure the agent is actually reaching the fire and extinguishing it. Discharging the agent a few seconds at a time can best do this. This allows the foam to work and spread across the burning fuel and conserves agent.

Figure 8.15 The driver/operator aims a turret stream at a ball set on a traffic cone. *Courtesy of Robert Lindstrom.*

Pump-and-Roll Capability

One feature of all major ARFF vehicles is the ability to pump and roll — driving towards a fire and after engaging the fire fighting system, begin fire attack. Pump and roll takes practice to gain the skills needed to effectively fight fire (Figure 8.16). The movement of the vehicle can make turret control difficult. Once again, practice makes perfect. Each day during morning checkout is a good time to practice pumping and rolling by picking a target and attempting to hit it while on the move.

Figure 8.16 Off-road pump-and-roll practice.

Resupply

One very important ingredient to a successful fire attack operation is a well-planned method for agent resupply. The most common agent resupply need during ARFF operations is the need to refill major ARFF vehicles with water. This is because all ARFF apparatus have foam concentrate tanks large enough to produce foam for several tanks full of water. For example, an ARFF vehicle with a 200-gallon (800 L) foam concentrate tank and a 1500-gallon (6 000 L) water tank and using 3% foam concentrate uses four tanks of water for each tank of foam concentrate. This makes water resupply one very important factor in keeping the ARFF vehicle fighting fire.

Foam concentrate resupply can be as easy as filling a few gallons (liters) of foam from 5-gallon (20 L) containers to using foam tenders to fill the ARFF vehicles. Each department needs to devise a resupply system that fits its needs. Some departments have central supply points around the airport in order to cut back on the travel time for refilling foam concentrate.

Unlike systems with foam concentrate and water, it is generally not practical to resupply ARFF apparatus dry-chemical or clean-agent extinguishing systems during the course of an incident. Once these systems have been expended, they will not be available for the remainder of the incident. These systems have special servicing requirements that only can be achieved safely under unrushed, nonemergency conditions.

One way to approach foam concentrate and water resupply operations is to call the operation "rapid resupply" and plan ways to decrease the vehicle's out-of-service time. The simple use of cam-lock or Storz fittings cuts the time it takes to connect a water or foam concentrate resupply hose (Figure 8.17). Using a pump to transfer foam concentrate from a foam tender or other supply source also makes the operation quicker. Pre-incident planning should prepare for a worst-case scenario fire and establish how much agent is required to extinguish a fire in the largest aircraft that can be expected to land at that airport. Knowing this information helps the department decide what type of resupply system best fits its needs.

Figure 8.17 This hose with cam-lock fittings is easily connected for resupply operations. *Courtesy of Robert Lindstrom.*

Structural Capability

Some ARFF vehicles are designed with structural fire fighting systems, which allow the vehicle to operate in a structural fire fighting mode (Figure 8.18). This has become popular with departments who anticipate making interior attacks with their vehicles and fighting structural fires on the airport. Each manufacturer has a different design for its vehicle's structural fire fighting system. The driver/operator should take the time to learn about the vehicle's structural fire fighting capabilities.

Figure 8.18 Pump operations panel on an ARFF vehicle.

Auxiliary Systems and Equipment

With many vehicles being custom-designed these days, it is important that the driver/operator become familiar with the different auxiliary systems and equipment used. One example of an auxiliary system would be a winterization system. These systems are mini-heating systems used to prevent the vehicle from freezing while standing by in extreme cold weather. These systems are very diverse and require the driver/operator to be familiar with the operation and inspection of the winterization systems.

Another example of equipment used on the vehicle could be an aircraft-skin-penetrating device (Figure 8.19). These devices are used to allow the firefighter to penetrate the aircraft skin and access the fire from a safe environment. These devices are attached to the end of an elevated master stream waterway. They can be designed to use water/foam solutions or other agents through the penetrating nozzle.

Figure 8.19 A firefighter tests an aircraft-skin-penetrating device. *Courtesy of Robert Lindstrom.*

Compressed-Air Foam Systems

One new technology making its way into the ARFF world is the compressed-air foam system (CAFS). The principle behind CAFS is simple. They use either a stored premixed solution of foam concentrate and water or a proportioned foam solution drawn from foam concentrate and water tanks on the apparatus. In either case, compressed air is injected into the foam solution. This causes the foam solution to become highly aerated, with air-to-agent ratios as high as 20:1. This greatly exceeds the normal average expansion rate of 4:1 to 6:1 created by standard ARFF vehicle low-expansion foam systems. The 20:1 ratio means 30 gallons (120 L) of foam solution will produce an end product of 600 gallons (2 400 L) of finished foam.

The ARFF industry is very interested in this process because of the quality of foam blankets and the extended drain (breakdown) times of this highly expanded foam. This higher air ratio also makes the handlines lighter and has a dramatic effect on fire extinguishment times (Figure 8.20). There are manufactured portable systems in use presently that use a premixed container of agent and a compressed air cylinder with a regulation system. These systems work well on small fuel spills and fires on the flight line.

Figure 8.20 Compressed-air foam system.

Electric Generators

Many ARFF vehicles are equipped with onboard electric generators. Some are small, portable generators that produce about 2.0 kW electricity, and others are larger, mounted generators capable of producing 7.5 kW or more. These generators are primarily used for on-scene lighting systems on the vehicles or perhaps for ventilation fans. The driver/operators must be familiar with the daily inspection requirements and all the safety aspects of operating high voltage electrical generators in the presence of water and foam.

Chapter 9

Extinguishing Agents

Job Performance Requirements

This chapter provides information that will assist the reader in meeting the following job performance requirements from NFPA 1003, *Standard for Airport Fire Fighter Professional Qualifications*, 2000 edition. Particular portions of the job performance requirements (JPRs) that are addressed in this chapter are noted in bold text.

3-1.1.1 General Knowledge Requirements. Fundamental aircraft fire-fighting techniques, including the approach, positioning, initial attack, and **selection, application, and management of the extinguishing agents**; limitations of various sized hand lines; use of proximity protective personal equipment (PrPPE); fire behavior; fire-fighting techniques in oxygen-enriched atmospheres; reaction of aircraft materials to heat and flame; critical components and hazards of civil aircraft construction and systems related to ARFF operations; special hazards associated with military aircraft systems; a national defense area and limitations within that area; characteristics of different aircraft fuels; hazardous areas in and around aircraft; aircraft fueling systems (hydrant/vehicle); aircraft egress/ingress (hatches, doors, and evacuation chutes); hazards associated with aircraft cargo, including dangerous goods; hazardous areas, including entry control points, crash scene perimeters, and requirements for operations within the hot, warm, and cold zones; and critical stress management policies and procedures.

3-1.1.2 General Skills Requirements. Don PrPPE; operate hatches, doors, and evacuation chutes; approach, position, and initially attack an aircraft fire; **select, apply, and manage extinguishing agents**; shut down aircraft systems, including engine, electrical, hydraulic, and fuel systems; operate aircraft extinguishing systems, including cargo area extinguishing systems.

3-3.1 Extinguish a 250-ft^2 (23.2-m^2) aircraft fuel spill fire, given PrPPE and a minimum of a 100-lb (45-kg) dry chemical fire extinguisher, so that the agent is applied using the proper technique and the fire is extinguished in 25 seconds.

(a) *Requisite Knowledge*: The fire behavior of aircraft fuels in pools, **physical properties, characteristics of aircraft fuel.**

(b) *Requisite Skills*: **Operate dry chemical extinguishers equipped with a hose line, including removing and operating hose and applying agent.**

3-3.2 Extinguish an aircraft fuel spill fire, given PrPPE, an assignment, an ARFF vehicle hand line flowing a minimum of 95 gpm (359 L/min) of AFFF extinguishing agent, and a fire sized to the AFFF gpm flow rate divided by 0.13 (gpm/0.13 = fire square footage) (L/min/0.492 = 0.304 m^2), so that the agent is applied using the proper techniques and the fire is extinguished in 90 seconds.

(a) *Requisite Knowledge*: The fire behavior of aircraft fuels in pools, **physical properties and characteristics of aircraft fuel, agent application rates and densities.**

(b) *Requisite Skills*: Operate fire streams and **apply agent.**

3-3.3 Extinguish an aircraft fuel spill fire, given PrPPE, an ARFF vehicle turret, and a fire sized to the AFFF flow rate of 0.13 (0492 L/min) divided by the square feet of fire area, so that the agent is applied using the proper technique and the fire is extinguished in 90 seconds.

(a) *Requisite Knowledge*: **Operation of ARFF vehicle agent delivery systems**, the fire behavior of aircraft fuels in pools, **physical properties and characteristics of aircraft fuel, agent application rates and densities.**

(b) *Requisite Skills*: **Apply fire-fighting agents and streams using ARFF vehicle turrets.**

3-3.4 Extinguish a three-dimensional aircraft fuel fire, given PrPPE, an assignment, and ARFF vehicle hand line(s) using primary and secondary agents, so that a dual agent attack is used, the agent is applied using the proper technique, the fire is extinguished, and the fuel source is secured.

(a) *Requisite Knowledge*: **The fire behavior of aircraft fuels in three-dimensional and atomized states, physical properties and characteristics of aircraft fuel, agent application rates and densities**, and methods of controlling fuel sources.

(b) *Requisite Skills*: Operate fire streams and **apply agents**, secure fuel sources.

3-3.7 Attack a wheel assembly fire, given PrPPE, an assignment, an ARFF vehicle hand line and appropriate agent, so that the fire is controlled.

(a) *Requisite Knowledge*: **Agent selection criteria**, special safety considerations, and the characteristics of combustible metals.

(b) *Requisite Skills*: Approach the fire in a safe and effective manner, **select and apply agent**.

3-3.11 Overhaul the accident scene, given PrPPE, an assignment, hand lines, and property conservation equipment, so that all fires are extinguished and all property is protected from further damage.

(a) *Requisite Knowledge*: **Methods of complete extinguishment and prevention of re-ignition**, purpose for conservation, operating procedures for property conservation equipment.

(b) *Requisite Skills*: Use property conservation equipment.

Airport rescue and fire fighting (ARFF) personnel could encounter all four classes of fire in any one incident. To be effective, ARFF personnel must have a full understanding of the chemical and physical nature of fire and the effective use of extinguishing agents. Most airports maintain various types of extinguishing agents, each having a unique use and application. The aircraft fuels, synthetic materials, combustible metals, and other new materials that are constantly being developed and incorporated into modern aircraft all have specific burning characteristics. Fires involving these materials may require the use of specialized extinguishing agents. Firefighters must be familiar with new and existing extinguishing agents and their proper application.

Aircraft Fuels

Because the large volumes of fuel carried on aircraft represent the primary hazard to both occupants and ARFF firefighters, a full understanding of aircraft fuels is necessary.

Types of Fuel

Three basic types of fuels are used in aircraft: aviation gasoline (AVGAS), kerosene, and blends of gasoline and kerosene. The last two are jet fuels, and all cover a broad range in the hydrocarbon series. Jet fuel is used in all jet and turboprop engines; AVGAS is used in reciprocating engines.

Fuels are delivered from bulk storage to aircraft on the ground in two ways: (1) by conventional fuel tank truck used for over-wing gravity refueling (Figure 9.1) and single-point (pressure) refueling; and (2) by underground fuel piping system that uses a fuel service vehicle that does not carry fuel but that pumps the fuel from subsurface connection points.

AVGAS is the same as the gasoline used in automobiles except that AVGAS has a higher octane rating than automotive fuel (87 to 92 in automobile gasoline versus 100 to 145 in AVGAS). The ignition temperature, flash point, flammable limits, and flame spread characteristics are also very similar to that of automotive fuel. The variance in the octane rating does not affect the fire fighting characteristics. Spills of AVGAS and low flash point turbine fuels (Jet-B) over 10 feet (3 m) in any direction and covering an area of over 50 ft^2 (4.6 m^2) or any spill that is of a continuous nature should be blanketed with foam to contain the release of flammable vapors.

Figure 9.1 Gravity refueling is used to refuel this general aviation aircraft.

Jet fuels are divided into two grades: kerosene grades and blends. The kerosene grades, such as Jet-A and Jet A-1 (JP-5, JP-6, JP-8), are the most common. The important characteristics of these types of fuels are listed in Table 9.1. In general, these types of fuels have higher flash points and slower flame spread ratings than does AVGAS.

Foam application should be considered for spills of kerosene grade fuels over 10 feet (3 m) in any direction, spills covering an area of over 50 ft^2 (4.6 m^2), or spills in the presence of an ignition source. ARFF personnel should keep in mind that even though the ambient air temperature may be well below the flash point of the fuel, the ramp/

Table 9.1
Characteristics of Aircraft Fuels

Type of Fuel	Flash Point	Vapor Density	Vapor Pressure	Specific Gravity	Auto Ign. Temp.	Flame Spread	Weight	LEL	UEL	UN No.	DOT Hazard Class	Dot Packing Group
Avgas	-49°F (-45°C)	>1	5.5-7.0 psig	0.702-0.720	n/a	700-800 ft/min (213.4-243.8 m/min)	6.01 lb (2.73 kg)	1.4%	7.6%	UN1203	3	II
Jet A*	100°F-106°F (38°C-41°C)	5.7	0.1 psia at 100°F	0.81	475°F-500°F (246°C-260°C)	100 ft/min or less (30.5 m/min)	7.00 lb (3.18 kg)	0.6%	4.7%	UN1863	3	III
Jet A-1*	100°F-106°F (38°C-41°C)	5.7	0.1 psia at 100°F	0.81	440°F-475°F (227°C-246°C)	100 ft/min or less (30.5 m/min)	7.00 lb (3.18kg)	0.6%	4.7%	UN1863	3	III
Jet A-2	115°F (46°C)		25 mm at 158°F	0.85	442°F (228°C)		7.00 lb (3.18 kg)	0.7%	5.0%	UN1863	3	III
Arctic Diesel	100°F (38°C)	4	2-3 mm Hg/70°	0.77-0.84			7.00 lb (3.18 kg)	0.7%	5.0%	UN1993	3	III
Jet B**	-10°F (-23°C)	2.0-3.0	>21 at 0°	0.75-0.81	470°F-480°F (243°C-249°C)	700-800 ft/min (213.4-243.8 m/min)	6.68 lb (3.03 kg)	1.4%	7.6%	UN1863	3	II
JP 4**	-10°F (-23°C)	2.0-3.0	103-155 mm Hg/70°	0.75-0.81	470°F-480°F (243°C-249°C)	700-800 ft/min (213.4-243.8 m/min)	6.68 lb (3.03 kg)	1.3%	8.0%	UN1863	3	II
JP-6	95°F-145°F (35°C-63°C)	0.1			440°F-475°F (227°C-246°C)	100 ft/min or less (30.5 m/min)	6.50 lb (2.95 kg)	0.6%	4.9%			
JP-8*	100°F-106°F (38°C-41°C)	5.7	0.1 psia at 100°F <5 at 20°C	0.84	410°F-475°F (210°C-246°C)		6.5 lb (2.95 kg)	0.7%	4.7%	UN1863	3	III

* Jet A, Jet A-1, and JP-8 are the same.
** Jet B and JP-4 are the same.

NOTE: The information in the table is incomplete, and the information given is approximate. This table is meant to be used for comparison purposes only. Consult the fuel manufacturer's material safety data sheet for accurate information on a specific brand and type of fuel.

ground surface temperature can be 25°F to 45°F (14°C to 25°C) higher, thus placing firefighters in a dangerous range. (**NOTE:** For more information, see NFPA 407, *Standard for Aircraft Fuel Servicing.*)

Blended fuels, such as Jet-B (JP-4) fuel, are a blend of gasoline and kerosene. They have a lower ignition temperature than Jet-A fuels, which makes blended fuels potentially more dangerous when spilled. ARFF personnel should consider immediately applying foam to any spill.

Conditions of Flammability

When an aircraft crashes, the force of impact often ruptures components and compromises the fuel system. These ruptures are extremely hazardous because of the presence of many ignition sources such as sparks caused by friction, electrical short circuits, hot engine components, and other ignition sources on the ground. If a fire has not occurred, the incident should be approached and considered as if it were on fire. Precautions must be carried out in all phases of the incident, from the initial response to the recovery of the wreckage.

After an accident in which major structural damage has occurred, aircraft fuel may mix rapidly with air and form a mist or vapor. Regardless of the type of fuel involved, this mist is easily ignited, and the resulting fireball acts as an ignition source for other combustibles.

In accidents/incidents involving large amounts of aviation gasoline or jet fuels, reignition (flashback) is a constant threat. ARFF personnel must be aware of the danger of flashback and must completely cover the fuel-saturated areas with foam, reapplying it as necessary to maintain the integrity of the foam blanket.

Extinguishing Agents and Their Application

As mentioned at the beginning of this chapter, it is extremely important for firefighters to understand the extinguishing agents they have at their disposal and the proper application of those agents. In aircraft fire fighting, there are two basic categories of extinguishing agents; primary and auxiliary. Primary agents are those that are designed for mass application and rapid knockdown of a fire. Primary extinguishing agents include the different types of foam, which are discussed later in this chapter. Foaming agents are the primary agents used to combat two-dimensional fires in hydrocarbon fuels such as gasoline, kerosene, heavier oils, and others.

When primary extinguishing agents are applied properly and in sufficient quantities (mass application), they are effective in extinguishing or controlling the flammable liquid fires typical of aircraft accidents/incidents. Occasionally, a three-dimensional fire — one involving a burning fuel flow — may require using auxiliary agents and techniques. This type of fire may prove to be very difficult to extinguish with the use of foam agents alone, but it may be quickly knocked down by using auxiliary extinguishing agents such as dry chemical.

Auxiliary extinguishing agents are agents that are compatible with primary agents and are used in conjunction with the primary agent (usually foam) in fire extinguishment. Common auxiliary extinguishing agents for aircraft fire fighting include dry chemical, halons, and halon replacements. In order for an extinguishing agent to be compatible with the primary agent, the agent's chemical composition must not adversely affect the performance of foam. There are other situations where auxiliary agents may be used as the primary agent on specialized fires including fires in wheel wells, engine nacelles, and fires in interior walls and compartments.

Generally, however, auxiliary agents are not effective as primary agents because they are prone to flashback. Dry chemical, for example, provides rapid knockdown of a flammable liquids fire, but if the fire is not fully extinguished and adjacent ignition sources not sufficiently cooled, the entire fire area will "flash back" unless a vapor-suppressing blanket of foam is applied simultaneously. This is the basic principle of the use of the combined agent vehicle (CAV).

Halons and halon replacements may prove useful for fires that are inaccessible such as engine fires or fires beneath aircraft. Fires involving such combustible metals as magnesium, aluminum powder, and titanium must be extinguished using a dry-powder combustible-metal extinguishing agent such as MET-L-X® or G-1 powder.

This chapter focuses on the extinguishing agents commonly used for aircraft fire fighting. The advantages and disadvantages of each agent, as well as the methods of application, are discussed.

Some of the most common extinguishing agents are:

- Water
- Foaming agents
- Dry chemicals
- Halogenated agents and halon replacements
- Dry powders

Water and Its Application

Water is by far the most commonly used extinguishing agent in the fire service. However, water alone is generally not a suitable extinguishing agent for large aircraft fuel fires, especially ones in deep pools or pits, unless foaming agents or surfactants are added to it. In cases where water may be the only agent available, it can be used to push the fire away from the aircraft. Water is also the preferred extinguishing agent for fires in the interior of aircraft involving Class A materials.

By applying water correctly, firefighters can push burning fuel to an area far enough from the aircraft where it can be contained until it can be extinguished or allowed to burn out. Water also can be used to cool the aircraft fuselage, thereby reducing the likelihood that the exterior fuel fire could spread into the interior of the fuselage. It can be used effectively for controlling spot fires and eliminating reignition sources by cooling hot pieces of wreckage. In addition, water can provide an effective heat shield for aircraft passengers and personnel fighting the fire. Finally, using water to extinguish fires in nearby Class A materials contributes directly to the overall extinguishment effort.

Many procedures for applying water in aircraft fire fighting have been explored. ARFF personnel have been most successful when they have used fog and spray streams. The higher the nozzle pressure, the smaller the water particles become and the more heat the stream absorbs. However, the more finely divided the stream becomes, the more it is subject to the effects of wind and thermal column updrafts; as a result, it may be more difficult to reach the seat of the fire.

When structural apparatus is being used to combat a spilled fuel fire and when water is the only available extinguishing agent, it should be applied from 1½-inch (38 mm) or larger lines in a fog pattern. Firefighters should avoid using straight streams because they tend to churn and splash the fuel, causing the flammable liquids to spread the fire to other exposures. While structural fire fighting apparatus does not normally carry as much water as ARFF vehicles, its water supply may last long enough for ARFF personnel to effect rescue if the water is judiciously applied.

In using water as an extinguishing agent, however, ARFF personnel should also keep in mind that there are certain inherent hazards. Water is an excellent conductor of electricity, and personnel must take precautions to avoid electrical shock. Water extinguishes primarily by absorbing heat in the process of being converted to steam, but the steam can obscure vision and may scald aircraft occupants and ARFF personnel. This is significant in fighting fires in the interior of an aircraft. When water converts to steam, it expands at a rate of as much as 1700:1. This high expansion results in "filling up" the interior of the aircraft with steam (especially if ventilation is not adequate). The result of this expansion will be steam burns, not only to unprotected victims still inside the aircraft but also to ARFF personnel (even with approved personal protective equipment).

Foam and Its Uses

Foam is used to combat fires in hydrocarbon fuels such as gasoline, kerosene, heavier oils, and others. Foam has lower specific gravity than the fuels; therefore, it floats on the fuel surface. Applying a blanket of foam to burning hydrocarbons cools the fuel and prevents flammable vapors from reaching the air. A good-quality foam blanket should be a homogeneous mass of minute bubbles that will be minimally disrupted by wind, thermal updraft, or flame and hydrocarbon attack. It will reseal itself if the established foam blanket is disturbed, and it will flow around objects to gain access to and cover areas that are difficult to reach.

Firefighters must understand the characteristics of foam to maximize its application and effectiveness. As the foam is applied, it breaks down, and its water content vaporizes due to the heat and flames. Because of the loss of water through evaporation, the foam must be applied to a burning surface in sufficient volume, at an adequate rate, and reapplied as necessary to be effective. Applying it in this manner ensures that there is a residual foam layer over the extinguished portion of the burning liquid.

The density and rate of foam application become even more critical when the fire control area is considered. This concept is based on the theory that around the aircraft fuselage there is a specifically defined area within which it is feasible to extinguish or control a fire long enough for ARFF personnel to rescue trapped or immobilized occupants. See NFPA 403, *Standard for Aircraft Rescue and Fire Fighting Services at Airports*, for information on determining this area (Figure 9.2). As indicated in this standard, the most serious problem that ARFF personnel face when using foam is that they must quickly apply large quantities of foam in a manner to form a fire-resistant blanket on the fire. This may be especially difficult on larger flammable liquid spills.

The effectiveness of structural fire fighting apparatus using water for controlling flammable liquid fires can be significantly increased by the addition of foaming agents, such as aqueous film forming foam (AFFF), either by premixing the foaming agent in the vehicle's water tank or by using proportioning systems or devices.

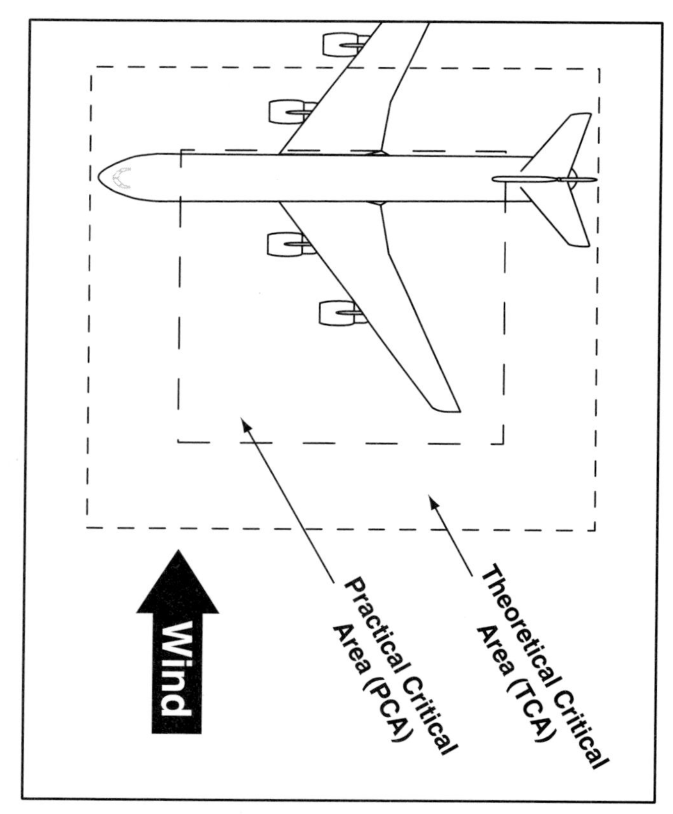

Figure 9.2 The theoretical critical area and practical critical area relative to an aircraft.

Foam Equipment and Systems

The proper understanding of foam and operation of foam fire fighting equipment is a primary responsibility of ARFF personnel. Although the technological advances in foam, its application, and equipment design have made the use of foam somewhat simpler than in the past, foam is not foolproof. ARFF personnel must still understand the basic principles of foam, proportioning, and application if the operation is to be successful.

The following sections examine some of the basic concepts with which airport firefighters should be familiar in respect to foam concentrates, portable foam proportioning equipment, apparatus-mounted foam proportioning systems, and foam application equipment. Because there are many manufacturers of this type of equipment, it is impossible to provide specific operational guidelines for each type of system. However, the information in this chapter provides the principles of each type of system. For more detailed information on foam fire fighting and equipment, see the IFSTA **Principles of Foam Fire Fighting** manual.

Principles of Foam

Foams in use today are of the mechanical type. These must be proportioned (mixed with water) and aerated (mixed with air) before they can be used. To produce quality fire fighting foam, foam concentrate, water, air, and mechanical aeration must be present and blended in the correct ratios (Figure 9.3). Removing any element results either in no foam production or in a poor-quality foam.

To become familiar with types of foams and the foam-making process, it is important to understand the following terms:

- *Foam concentrate* — The raw foam liquid as it rests in its storage container before the introduction of water and air

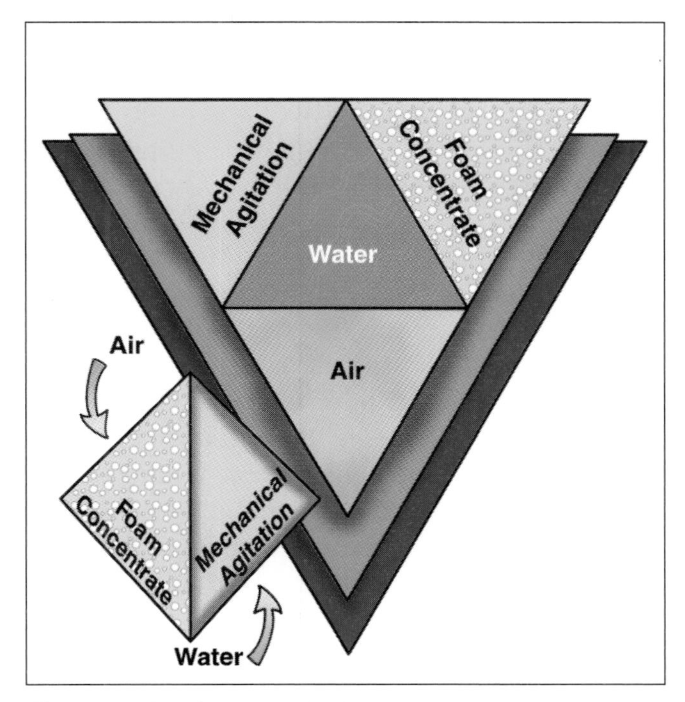

Figure 9.3 The foam tetrahedron.

- ***Foam proportioner*** — The device that introduces foam concentrate into the water stream to make the foam solution
- ***Foam solution*** — The mixture of foam concentrate and water before the introduction of air
- ***Foam*** — The completed product after air is introduced into the foam solution (also known as finished foam)

Aeration should produce an adequate amount of bubbles to form an effective foam blanket. Proper aeration should produce uniform-sized bubbles to provide a long-lasting blanket. A good foam blanket is required to maintain an effective cover over Class B fuels for the period of time desired.

To be effective, foam concentrates must also match the fuel to which they are applied. Class A foams are not designed to extinguish Class B fires. Class B fuels are divided into two categories: hydrocarbons and polar solvents.

Hydrocarbon fuels, such as crude oil, fuel oil, gasoline, benzene, naphtha, jet fuel, and kerosene, are petroleum-based and float on water. Standard fire fighting foam is effective as an extinguishing agent and vapor suppressant because it can float on the surface of hydrocarbon fuels.

Polar solvent fuels, such as alcohol, acetone, lacquer thinner, ketones, and esters, are flammable liquids that are miscible (capable of being mixed) in water. Fire fighting foam can be effective on these fuels but only in special alcohol-resistant (polymeric) formulations. It should be noted that many modern automotive fuel blends, which include gasoline with 10 percent or more solvent additives, should be considered polar solvents and handled as such during emergency operations.

Class B foams designed solely for hydrocarbon fires will not extinguish polar solvent fires regardless of the concentration at which they are used. Many foams that are intended for polar solvents may be used on hydrocarbon fires, but such use should not be attempted unless the manufacturer of the particular concentrate being used specifically says this can be done. Additionally, polar solvent foams cannot be used in crash apparatus because the foam concentrate is too viscous to proportion properly.

CAUTION: Failure to match the proper foam concentrate with the fuel results in an unsuccessful extinguishing attempt and could endanger firefighters.

Specific types of foam concentrates are addressed later in this chapter.

How Foam Works

Foam extinguishes and/or prevents fire by the following methods (Figure 9.4):

- ***Separating*** — Creating a barrier between the fuel and the fire
- ***Cooling*** — Lowering the temperature of the fuel and adjacent surfaces
- ***Suppressing*** (sometimes referred to as smothering) — Preventing the release of flammable vapors and therefore reducing the possibility of ignition or reignition

Figure 9.4 How foam works to extinguish fire.

In general, foam works by forming a blanket on the burning fuel. The foam blanket excludes oxygen and stops the burning process. The water in the foam is slowly released as the foam breaks down. This provides a cooling effect on the fuel and surrounding surfaces in contact with the fuel.

Foam Proportioning

The term *proportioning* is used to describe the mixing of water with foam concentrate to form a foam solution. Most foam concentrates are intended to be mixed with either fresh water or saltwater. For maximum effectiveness, foam concentrates must be proportioned at the specific percentage for which they are designed. This percentage rate for the intended fuel is clearly marked on the outside of every foam container. Failure to proportion the foam at its designated percentage, such as trying to use 6% foam at a 3% concentration, results in poor-quality foam that may not perform as desired.

Most fire fighting foam concentrates are intended to be mixed with 94 to 99.9 percent water. For example, when using 3% foam concentrate, 97 parts water mixed with 3 parts foam concentrate equals 100 parts foam solution. For 6% foam concentrate, 94 parts water mixed with 6 parts foam concentrate equals 100 percent foam solution.

The selection of a proportioner depends on the foam solution flow requirements, available water pressure, cost, intended use (apparatus-mounted or portable), and agent to be used. Proportioners and delivery devices (foam nozzle, foam maker, etc.) are engineered to work together. Using a foam proportioner that is not compatible with the delivery device (even if the two are made by the same manufacturer) can result in unsatisfactory foam or no foam at all. For example, a proportioner that is designed to be used at 95 gpm (380 L/min) must be used with a 95 gpm (380 L/min) nozzle, or the foam will not proportion properly — if at all.

There are four basic methods by which foam may be proportioned:

- Induction
- Injection
- Batch mixing
- Premixing

A variety of equipment is used to proportion foam. Some types are designed for mobile apparatus, and others are designed for fixed fire protection systems. The common types of foam proportioners are covered later in this chapter.

Induction

The induction (eduction) method of proportioning foam uses the pressure energy in the stream of water to induct (draft) foam concentrate into the fire stream. This is achieved by passing the stream of water through a device called an eductor that has a restricted diameter (Figure 9.5). Within the restricted area is a separate orifice that is attached via a hose to the foam concentrate container. The pressure differential created by the water going through the restricted area and over the orifice creates a suction that draws the foam concentrate into the fire stream. In-line eductors and foam-nozzle eductors are examples of foam proportioners that work by this method.

Figure 9.5 Typical in-line foam eductor.

Injection

The injection method of proportioning foam uses an external pump or head pressure to force foam concentrate into the fire stream at the correct ratio in comparison to the flow. These systems are commonly employed in apparatus-mounted or fixed fire protection system applications.

Figure 9.6 A firefighter pours foam concentrate into the apparatus water tank for batch mixing.

Batch Mixing

Batch mixing is the simplest method of mixing foam concentrate and water. With batch mixing, an appropriate amount of foam concentrate is poured directly into a tank of water (Figure 9.6). Batch mixing is commonly used to mix foam within a fire apparatus water tank or a portable water tank. It also allows for accurate proportioning of foam. Batch mixing may not be effective on large incidents because when the tank becomes empty, the foam attack lines must be shut down until the tank is completely filled with water and more foam concentrate is added. Another drawback of batch mixing is that Class B concentrates and tank water must be circulated for a period of time to ensure thorough mixing before being discharged. The time required for mixing depends on the viscosity and solubility of the foam concentrate.

Premixing

Premixing is one of the more commonly used methods of proportioning. With this method, premeasured portions of water and foam concentrate are mixed in a container. Typically, the premix method is used with portable extinguishers, wheeled extinguishers, skid-mounted multiagent units, and vehicle-mounted tank systems (Figure 9.7).

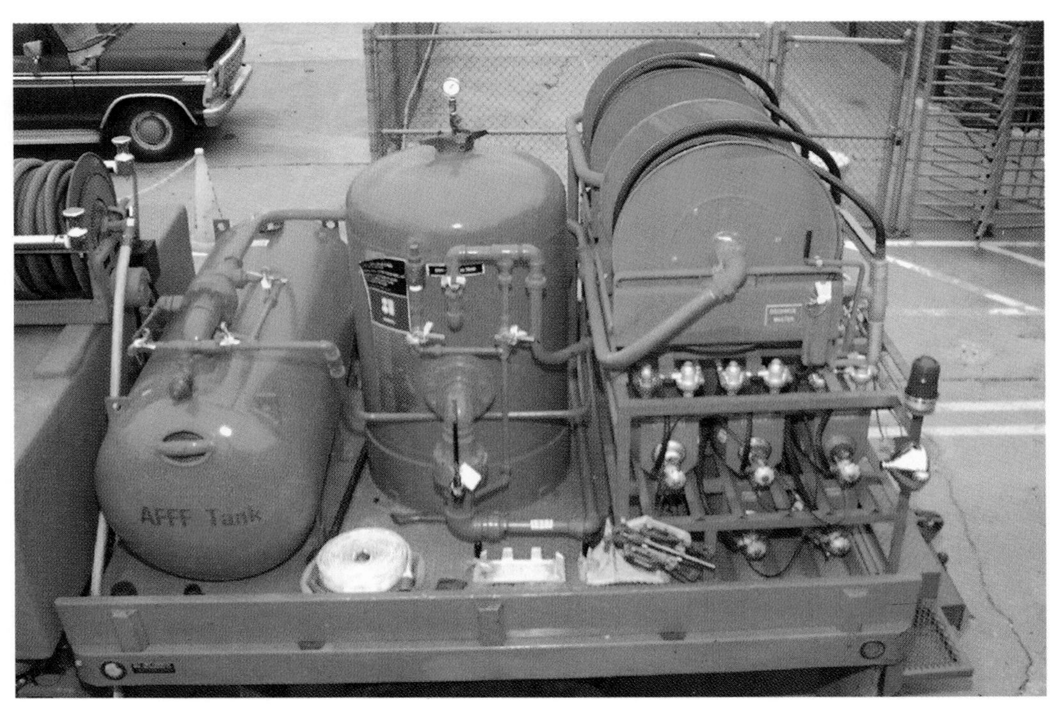

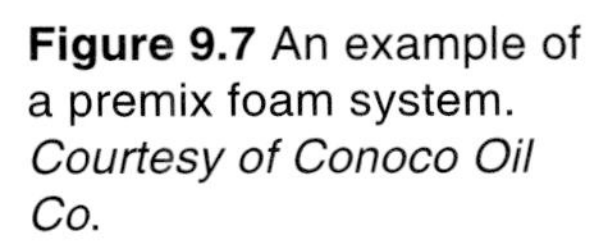

Figure 9.7 An example of a premix foam system. *Courtesy of Conoco Oil Co.*

In most cases, premixed solutions are discharged from a pressure-rated tank using either a compressed inert gas or air. An alternative method of discharge uses a pump and a non-pressure-rated atmospheric storage tank. The pump discharges the foam solution through piping or hose to the discharge devices. Premix systems are limited to a one-time application. When used, they must be completely emptied and then refilled before they can be used again.

How Foam Concentrates are Stored

Foam concentrate is stored in a variety of containers. The type of container used in any particular situation depends on how the foam is generated and delivered. The three common methods of foam concentrate storage are: pails, barrels (drums), and apparatus tanks.

Pails

Five-gallon (20 L) plastic pails are common containers used for shipping and storing foam concentrate (Figure 9.8). These containers are durable and are not affected by the corrosive nature of foam concentrates. Pails may be carried on the apparatus in compartments, on the side of the apparatus, or in topside storage areas. The containers of alcohol-resistant foams must be airtight to prevent a skin from forming on the surface of the concentrate. Foam concentrate may be educted directly from the pail when using an in-line or foam-nozzle eductor.

Barrels

Foam concentrate may also be shipped and stored in 55-gallon (220 L) plastic or plastic-lined barrels (drums) (Figure 9.9). Foam concentrate can then be transferred to pails or to apparatus tanks for actual deployment. Some departments have apparatus that are designed to carry these barrels directly to the emergency scene for deployment. Foam concentrate can be educted directly from barrels in the same manner that it is educted from pails.

Apparatus Tanks

Most ARFF apparatus are equipped with integral, onboard foam-proportioning systems and usually have foam concentrate tanks piped directly to the foam delivery system (Figure 9.10). This eliminates the need to use separate pails or barrels. Foam concentrate tanks can be found also on municipal and industrial pumpers and on foam tenders.

Figure 9.8 Five-gallon foam pail. *Courtesy of Conoco Oil Co.*

Figure 9.9 Foam concentrate in 55-gallon drums. *Courtesy of Conoco Oil Co.*

Figure 9.10 The foam tank vent/fill opening on an ARFF apparatus.

Foam Concentrates

Mechanical foam concentrates can be divided into two general categories: those intended for use on Class A fuels (ordinary combustibles) and those intended for use on Class B fuels (flammable and combustible liquids). The following sections contain information on the concentrates in both of these categories.

Class A Foam

Class A foam has been used since the 1940s; however, only recently has the technology come of age. This agent has proven to be effective for fires in structures, wildland settings, coal mines, tire storage, and other incidents involving similar deep-seated fuels. Because Class A foams are not effective in fighting Class B fires, these concentrates have limited applications in ARFF settings. However, some jurisdictions may choose to use Class A foams to attack interior aircraft cabin fires or other structural-related fires on the airport property.

WARNING

Use Class A foam only on Class A fuels. It is not specifically formulated for fighting Class B fires and will not provide the extinguishing and vapor-suppression capabilities of Class B foams.

For more information on Class A foam concentrates and their use, see the IFSTA **Principles of Foam Fire Fighting** manual.

Class B Foam

Class B foam is used to extinguish fires involving flammable and combustible liquids (Figure 9.11). It is also used to suppress vapors from unignited spills of these liquids. There are several types of Class B foam concentrates; each type has its advantages and disadvantages.

Class B foam may be proportioned into the fire stream via apparatus-mounted or portable foam proportioning equipment. The foam may be applied either with standard fog nozzles (aqueous film forming foam [AFFF] and film forming fluoroprotein foam [FFFP] concentrates only) or with air-aspirating foam nozzles (all types).

In general, different manufacturers' foam concentrates should not be mixed together in apparatus tanks because they may be chemically incompatible. The exception to this would be foam manufactured to U.S. military specifications (mil-spec concentrates). The mil-specs are written so that mixing can be done with no adverse effects. On the emergency scene, concentrates of a similar type (all AFFFs, all fluoroproteins, etc.) but from different manufacturers, may be mixed together immediately before application.

The chemical properties of Class B foams and their environmental impact vary depending on the type of concentrate and the manufacturer. Consult the data sheets provided by the manufacturer for information on a specific concentrate.

Proportioning

Today's Class B foams are mixed in proportions from 1% to 6%. The proper proportion for any par-

Figure 9.11 Firefighters apply Class B foam. *Courtesy of Joel Woods, Maryland Fire and Rescue Institute.*

ticular concentrate is listed on the outside of the foam container. Some multipurpose foams designed for use on both hydrocarbon and polar solvent fuels can be used at different concentrations, depending on which of the two fuels they are used. These concentrates are normally used at a 1% or 3% rate on hydrocarbons and 3% or 6% rate on polar solvents, depending on the manufacturer's recommendations. (**NOTE:** The polar solvent types of foams are not acceptable for use in crash apparatus because they are too thick to proportion properly.)

Foam Expansion

Foam expansion refers to the increase in volume of a foam solution when it is aerated. This is a key characteristic to consider when choosing a foam concentrate for a specific application. The methods of aerating foam solution result in varying degrees of expansion, which depends on the following factors:

- Type of foam concentrate used
- Accurate proportioning of the foam concentrate in the solution
- Quality of the foam concentrate
- Method of aspiration

Depending on its purpose, foam can be described by three types: low-expansion, medium-expansion, and high-expansion. NFPA 11, *Standard for Low-Expansion Foam,* states that low-expansion foam has an air/solution ratio up to 20 parts finished foam for every part of foam solution (20:1 ratio). Medium-expansion foam is most commonly used at the rate of 20:1 to 200:1 through hydraulically operated nozzle-style delivery devices. In the high-expansion foams, the expansion rate is 200:1 to 1000:1.

Rates of Application

The rate of application for fire fighting foam varies depending on any one of several variables:

- Type of foam concentrate used
- Whether or not the fuel is on fire
- Type of fuel (hydrocarbon/polar solvent) involved
- Whether the fuel is spilled or in a tank; if the fuel is in a tank, the type of tank will have a bearing on the application rate

The minimum foam solution application rates for aircraft fuel spill fires are established in NFPA 403, *Aircraft Rescue and Fire Fighting Services at Airports.* Refer to the IFSTA **Principles of Foam Fire Fighting** manual and to NFPA 11, *Standard for Low-Expansion Foam,* for fuel tank fire fighting operations, foam proportioning systems, and foam generating systems.

Unignited spills do not require the same application rates as ignited spills because radiant heat, open flame, and thermal drafts do not attack the finished foam as they would under fire conditions. No specific rate is given by NFPA 11 for unignited spills. In case the spill does ignite, however, firefighters should be prepared to flow at least the minimum application rate for the specified amount of time based on fire conditions.

Specific Foam Concentrates

Numerous types of foams are selected for specific applications according to their properties and performance. Some foams are thick and viscous and form tough, heat-resistant blankets over burning liquid surfaces; other foams are thinner and spread more rapidly. Some foams produce a vapor-sealing film of surface-active water solution on a liquid surface. Others, such as medium- and high-expansion foams, are used in large volumes to flood surfaces and fill cavities. The following sections highlight each of the common types of foam concentrates.

Aqueous Film Forming Foam (AFFF)

AFFF (commonly pronounced "A triple F") is extremely effective in ARFF applications. It is the recommended extinguishing agent for hydrocarbon fuel fires and is the most commonly used foam at airports today. Aqueous film forming foam (AFFF) is a synthetically produced material. AFFF consists of liquid concentrate that is made from fluorochemical and hydrocarbon surfactants combined with high-boiling-point solvents and water with suitable foam stabilizers. Because AFFF has a lower specific gravity than hydrocarbon fuels, AFFF floats on the surface of these fuels and, because of its low viscosity, quickly spreads across the fuel surface to form a vapor-suppressing blanket (Figure 9.12). Furthermore, AFFF has a significant bleeding effect, so there is a cooling film floating on the surface of the fuel continually. Finally, this film is self-sealing when disturbed.

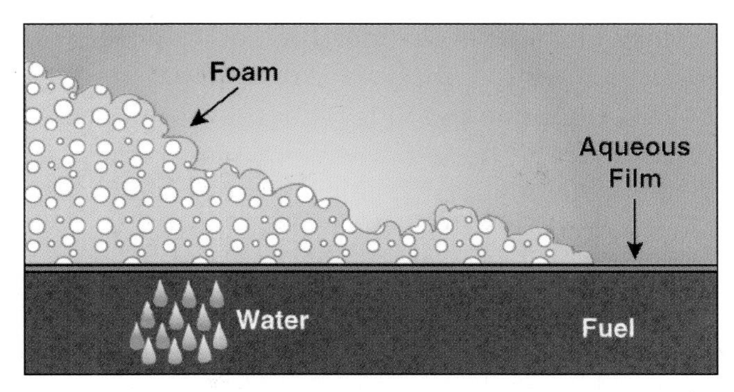

Figure 9.12 The film of AFFF floats ahead of the foam blanket.

When AFFF (as well as FFFP, which is discussed later) is applied to a hydrocarbon fire, three things occur:

- An air/vapor-excluding film is released ahead of the foam blanket.
- The fast-moving foam blanket then moves across the surface and around objects, adding further insulation.
- As the aerated (7:1 to 20:1) foam blanket continues to drain its water, more fire is released. This gives AFFF the ability to "heal" over areas where the foam blanket is disturbed.

It is available in either 1%, 3%, or 6% concentrates to be mixed with water to form the foam solution. AFFF may be used with fresh water, saltwater, or brackish water. It resists breakdown by dry chemicals, making it suitable for use in combination with other agents.

How fast the fire is extinguished depends upon the manner in which AFFF is applied, its application rate, and its density. AFFF may be applied with an aspirating or nonaspirating nozzle.

Alcohol-resistant AFFF is available from most foam manufacturers. On most polar solvents, alcohol-resistant AFFF is used at 3% or 6% concentrations, depending on the particular brand used. Alcohol-resistant AFFF can also be used on hydrocarbon fires at a 1% or 3% proportion, depending on the manufacturer, but is not acceptable for use in ARFF apparatus because it is too viscous.

Regular Protein Foam (PF)

Before the 1970s, protein foam (PF) was used for almost all aircraft fire fighting. However, due to its corrosiveness, the fact that it is not self-sealing, and other limitations, PF is no longer widely used in ARFF applications. It is still used in some industrial settings.

Fluoroprotein Foam (FPF)

Fluoroprotein foam (FPF) is also not widely used in aircraft fire fighting. However, it is widely used in protecting fuel tanks and petroleum processing facilities primarily because its unique fuel-shedding qualities make it highly desirable for subsurface injection applications. Wherever aircraft operate, bulk fuel storage is present, so ARFF personnel need to be aware of this agent and its capabilities.

Film Forming Fluoroprotein Foam

Film forming fluoroprotein foam (FFFP) concentrate is based on fluoroprotein foam technology with aqueous film forming foam (AFFF) capabilities. This film forming fluoroprotein foam incorporates the benefits of AFFF for fast fire knockdown and the benefits of fluoroprotein foam for long-lasting heat resistance. FFFP is available in an alcohol-resistant formulation, but as with alcohol-resistant AFFF concentrates, this is not commonly used in ARFF applications.

Film forming fluoroprotein foam (FFFP) is an effective agent on flammable liquid fires. Similar to AFFF, FFFP forms a self-sealing film on the surface of the fuel, continuously suppressing fuel vapors.

FFFP concentrates are available in 3% and 6% solutions that may be applied with a variety of water spray devices. Both fresh water and saltwater are suitable vehicles for the foam solution. Additionally, dry chemicals may be used in conjunction with FFFP for successful multiagent applications.

As with AFFF, the effectiveness of FFFP depends upon the application rate, density, and blanketing of the fuel. However, FFFP is not as effective as AFFF in maintaining foam stability. After extinguishment, the foam blanket should be monitored and reapplied as necessary to avoid breakdown and possible reignition hazards. Film forming fluoroprotein foam may be applied with aspirating or nonaspirating nozzles.

High-Expansion Foams

High-expansion foams are special-purpose foams and have a detergent base. Because they have a low water content, they minimize water damage. Their low water content is also useful when runoff is undesirable. Fixed high-expansion foam systems are found in some airport hangars.

High-expansion foams have three basic applications:

- In concealed spaces such as basements, in coal mines, and in other subterranean spaces
- In fixed extinguishing systems for specific industrial uses such as aircraft hangars, rolled or bulk paper storage, etc.
- In Class A fire applications

High-expansion foam concentrates have expansion ratios of 200:1 to 1,000:1 for high-expansion uses and expansion ratios of 20:1 to 200:1 for medium-expansion uses. (**NOTE:** Whether the foam is used in either a medium- or high-expansion capacity is determined by the type of application device used.)

Foam Proportioning Systems

The process of foam proportioning sounds simple: Add the proper amount of foam concentrate into the water stream and an effective foam solution is produced. Unfortunately, this process is not as easy as it sounds. The correct proportioning of foam concentrate into the fire stream requires equipment that must operate within strict design specifications. Failure to operate even the best foam proportioning equipment as designed can result in poor-quality foam or no foam at all. In general, foam proportioning devices operate by one of two basic principles:

- The pressure of the water stream flowing through an orifice creates a venturi action that inducts (drafts) foam concentrate into the water stream.
- Pressurized proportioning devices inject foam concentrate into the water stream at a desired ratio and at a pressure higher than that of the water.

This section details the various types of low energy foam proportioning devices commonly found in portable and apparatus-mounted applications. A low-energy foam system imparts pressure on the foam solution solely by the use of a fire pump. This system introduces air into the solution when it either reaches the nozzle or is discharged from the nozzle. High-energy foam systems introduce compressed air into the foam solution before it is discharged into the hoseline. (**NOTE:** High-energy foam systems are described later in this manual.)

Portable Foam Proportioners

Portable foam proportioners are the simplest and most common foam proportioning devices in use today at structural fire departments. The three common types of portable foam proportioners are: in-line foam eductors, foam nozzle eductors, and self-educting master stream nozzles.

In-Line Foam Eductors

The in-line eductor is the most common type of foam proportioner used in the structural fire service; however, it is not commonly used in ARFF applications. This eductor is designed to be either directly attached to the pump panel discharge or connected at some point in the hose lay. When using an in-line eductor, it is very important to follow the manufacturer's instructions about inlet pressure and the maximum hose lay between the eductor and the appropriate nozzle.

In order for the nozzle and eductor to operate properly, both must have the same rating in gpm (L/min). Remember that the eductor — not the nozzle — must control the flow. If the nozzle has a flow rating lower than that of the eductor, the eductor will not flow enough water to pick up concentrate. An example of this situation is a 60 gpm (240 L/min) nozzle with a 95 gpm (380 L/min) eductor.

Foam Nozzle Eductors

The foam nozzle eductor operates on the same basic principle as the in-line eductor. However, this eductor is built into the nozzle rather than into the hoseline (Figure 9.13). As a result, its use requires the foam concentrate to be available where the nozzle is operated. If the foam nozzle is moved, the foam concentrate also needs to be moved. The logistical problems of relocation are magnified by the gallons of concentrate required. Use of a foam nozzle eductor compromises firefighter safety: firefighters cannot move quickly, and they must leave their concentrate behind if they are required to back out for any reason.

Figure 9.13 Foam nozzle eductor.

Self-Educting Master Stream Foam Nozzles

The self-educting master stream foam nozzle is used where flows in excess of 350 gpm (1 400 L/min) are required. These nozzles are available with flow capabilities of up to 14,000 gpm (56 000 L/min) (Figure 9.14). These nozzles are not commonly used for ARFF operations.

Apparatus-Mounted Foam Proportioning Systems

Foam proportioning systems are commonly mounted on structural, industrial, wildland, and fire boats, as well as on aircraft rescue and fire fighting apparatus. The majority of the following foam proportioning systems can be used for both Class A and Class B foam concentrates:

- Installed in-line eductors
- Around-the-pump proportioners
- Bypass-type balanced-pressure proportioners
- Variable-flow variable-rate direct-injection systems
- Variable-flow demand-type balanced-pressure proportioners
- Batch-mixing

Installed In-Line Eductor Systems

Installed in-line eductors use the same principles of operation as do portable in-line eductors. The only difference is that these eductors are permanently attached to the apparatus pumping system. The same precautions regarding hose lengths, matching nozzle and eductor flows, and inlet pressures listed for portable in-line eductors also apply

Figure 9.14 A self-educting master stream foam nozzle.

to installed in-line eductors. Foam concentrate may be supplied to these devices from either pickup tubes (using 5 gallon [20 L] pails) or from foam concentrate tanks installed on the apparatus.

Around-the-Pump Proportioners

The around-the-pump proportioning system consists of a small return (bypass) water line connected from the discharge side of the pump back to the intake side of the pump (Figure 9.15). An in-line eductor is positioned on this bypass line. A valve positioned on the bypass line, just off the pump discharge piping, controls the flow of water through the bypass line. When the valve is open, a small amount of water (10 to 40 gpm [40 L/min to 160 L/min]) discharged from the pump is directed through the bypass piping. As this water passes through the eductor, the resulting venturi effect draws foam concentrate from the foam concentrate tank and into the bypass piping. The resulting foam solution is then supplied back to the intake side of the pump, where it is then pumped to the discharge and into the hoseline.

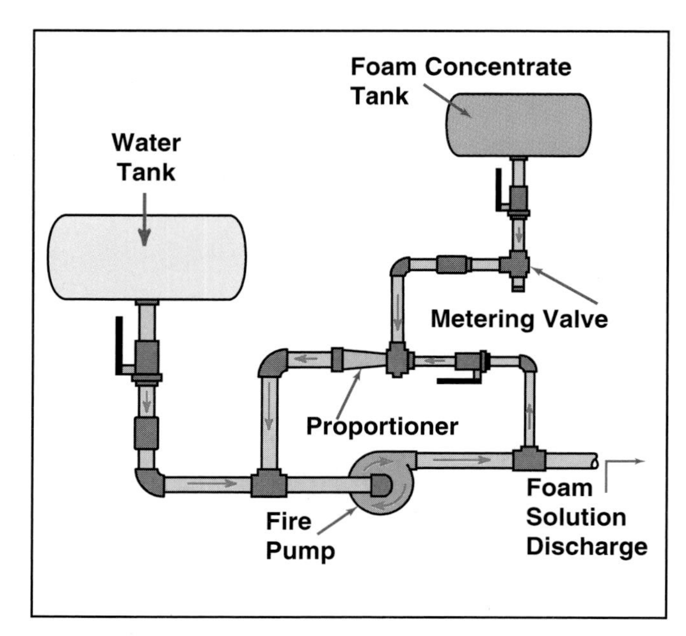

Figure 9.15 An around-the-pump proportioner.

Around-the-pump proportioning systems are rated for a specific flow and should be used at this rate, although they do have some flexibility. For example, a unit designed to flow 500 gpm (2 000 L/min) at a 6% concentration proportions 1,000 gpm (4 000 L/min) at a 3% rate.

A major disadvantage of older around-the-pump proportioners is that the pump cannot take advantage of incoming pressure. If the inlet water supply is any greater than 10 psi (70 kPa), the foam concentrate will not enter into the pump intake. Another disadvantage is that the pump must be dedicated solely to foam operation. An around-the-pump proportioner does not allow plain water and foam to be discharged from the pump at the same time.

Balanced-Pressure and Direct-Injection Proportioners

Most ARFF apparatus use some type of balanced-pressure or direct-injection foam proportioning system. These systems provide the most accurate proportioning of foam concentrate at large flow rates. There are several specific types of systems used in ARFF apparatus, including bypass-type balanced-pressure proportioners, variable-flow demand-type balanced-pressure proportioners, and variable-flow variable-rate direct-injection systems.

Apparatus equipped with a bypass-type balanced-pressure proportioner have a foam concentrate line connected to each fire pump discharge outlet (Figure 9.16). This line is supplied by a foam concentrate pump separate from the main fire pump. The foam concentrate pump draws the concentrate from a fixed tank. This pump is designed to supply foam concentrate to the outlet at the same pressure at which the fire pump is supplying water to that discharge. The pump discharge and the foam concentrate pressure from the foam concentrate pump are jointly monitored by a hydraulic pressure control valve that ensures the concentrate pressure and water pressure are balanced.

The primary advantages of the bypass-type balanced-pressure proportioner are:

- They have the ability to monitor the demand for foam concentrate and to adjust the amount of concentrate supplied.
- They have the ability to simultaneously discharge foam from some outlets and plain water from others.

Limitations of the bypass-type balanced-pressure proportioner include:

- There is a need for a separate foam pump with power take-off (PTO) or other power source.
- Bypass of concentrate in this system can cause heating, turbulence, and foam concentrate aeration (bubble production in the storage tank).

In the variable-flow demand-type balanced-pressure proportioning system, a variable-speed mechanism, which is either hydraulically or electrically controlled, drives a foam concentrate pump. The foam concentrate pump supplies foam concentrate to a venturi-type proportioning device built into the water line (Figure 9.17). When activated, the foam concentrate pump output is automatically monitored so that the flow of foam concentrate is commensurate with the flow of water to produce an effective foam solution.

Advantages of the variable-flow demand-type balanced-pressure proportioning system include:

- The foam concentrate flow and pressure match system demand.
- There is no recirculation back to the foam concentrate tank.
- The system is maintained in a ready-to-pump condition and requires no flushing after use.

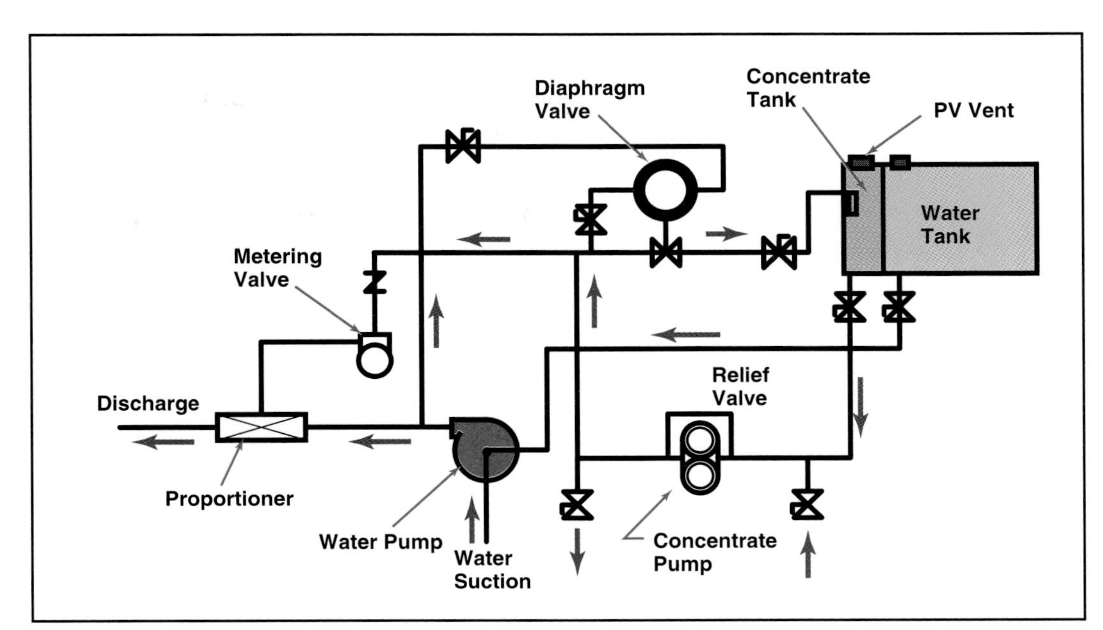

Figure 9.16 A bypass-type balanced-pressure proportioner.

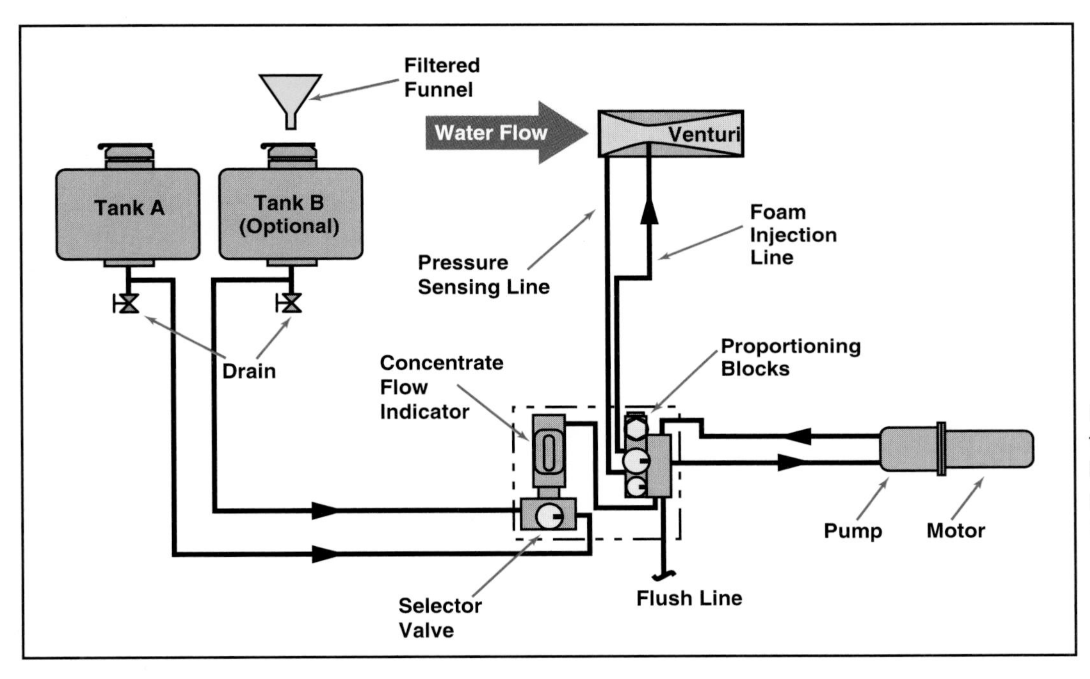

Figure 9.17 A variable-flow demand-type balanced-pressure proportioner. *Courtesy of KK Products.*

- Water and/or foam solution can be discharged simultaneously from any combination of outlets up to rated capacity.

A limitation of these systems is that the fire pump discharges have ratio controllers (which reduce the discharge area); thus, pressure drops across the discharge are higher than those on standard pumpers.

Variable-flow variable-rate direct-injection systems operate off power supplied from the apparatus electrical system. Large-volume systems may use a combination of electric and hydraulic power. The foam concentrate injection is controlled by monitoring the water flow and controlling the speed of a positive displacement foam concentrate pump, thus injecting concentrate at the desired ratio. Because the water flow governs the foam concentrate injection, water pressure is not a factor.

There are several advantages to variable-flow variable-rate direct-injection systems, including:

- They will accurately proportion foam concentrate at any flow rate or pressure within the design limits of the system.

- The system automatically adjusts to changes in water flow when nozzles are either opened or closed.
- Nozzles may be either above or below the pump, without affecting the foam proportioning.
- This system may be used with high-energy foam systems, which are described later in this chapter.

The disadvantage of these systems is that the foam injection point must be within the piping before any manifolds or distribution to multiple fire pump discharges.

Batch Mixing

By far, the simplest means of proportioning foam is to simply pour an appropriate amount of foam concentrate into a tank of water. This method is called batch mixing or the dump-in method. To do this, the ARFF firefighter pours a predetermined amount of foam concentrate into the tank via the top fill opening at the time when foam is needed. The truck is then pumped normally, and foam is discharged through any hoseline that is opened. The amount of foam concentrate needed depends on the size of the water tank and the proportion percentage for which the foam is designed.

In general, batch mixing is used only with regular AFFF (not alcohol-resistant AFFF concentrates) and Class A concentrates. The AFFF concentrate mixes readily with water, and it will stay suspended in the solution for an extended period of time. When batch-mixing AFFF, the water in the tank has to be circulated for a few minutes before discharge to ensure complete mixing.

The disadvantage of this method is that all the water onboard the apparatus is converted to foam solution. This method does not allow for continuous foam discharge on large incidents, as the stream has to be shut down while the apparatus is replenished. It is difficult to maintain the correct concentrate ratio when refilling unless the water tank is completely emptied each time.

High-Energy Foam Generating Systems

High-energy foam systems generally differ from those previously discussed in that they introduce compressed air into the foam solution *prior* to discharge into the hoseline (Figure 9.18). The turbulence of the foam solution and compressed air going through the piping and/or hoseline creates a finished foam. In addition to simply forming the foam, the addition of the compressed air also allows the foam stream to be discharged considerably greater distances than a regular foam or water fire stream.

Figure 9.18 A compressed-air foam system unit on an ARFF vehicle.

This system uses a standard centrifugal fire pump to supply the water. A direct-injection foam-proportioning system is attached to the discharge side of the fire pump. Once the foam concentrate and water are mixed to form a foam solution, compressed air is added to the mixture before it is discharged from the apparatus and into the hoseline. This type of system is commonly called a compressed-air foam system (CAFS).

CAFS systems are most commonly found on structural and wildland fire apparatus. Structural apparatus assigned to an airport facility may have this type of system, but it is not commonly found on other types of ARFF apparatus. For more information on high-energy foam systems, see the IFSTA **Principles of Foam Fire Fighting** manual.

Portable Foam Application Devices

Once the foam concentrate and water have been mixed together to form a foam solution, the foam solution must then be mixed with air (aerated) and delivered to the surface of the fuel. With low-energy foam systems, the aeration and discharge of the foam are accomplished by the foam nozzle, sometimes referred to as a foam maker. Low-expansion foams may be discharged through either handline nozzles or master stream devices. While standard fire fighting nozzles can be used for applying some types of low-expansion foams, it is best to use nozzles that produce the desired result (such as fast-draining or slow-draining foam). This section highlights portable foam application devices. (**NOTE:** Foam nozzle eductors and self-educting master stream foam nozzles are considered portable foam nozzles, but they are omitted from this section because they are covered earlier in the chapter.)

Handline Nozzles

IFSTA defines a handline nozzle as "any nozzle that one to three firefighters can safely handle and that flows less than 350 gpm (1 400 L/min)." Most handline foam nozzles flow considerably less than that figure. The two most common types of handline nozzles used by ARFF firefighters are standard fog nozzles and air-aspirating foam nozzles.

Either fixed-flow or automatic fog nozzles can be used with foam solution to produce a low-expansion, short-lasting foam. It is often referred to as nonaspirated foam. This nozzle breaks the foam solution into tiny droplets and uses the agitation of water droplets moving through air to achieve its foaming action. Its best application is when it is used with regular AFFF. Some nozzle manufacturers have foam aeration attachments that can be added to the end of the nozzle to increase aspiration of the foam solution (Figure 9.19).

Figure 9.19 A nozzle with a foam aeration attachment.

The air-aspirating foam nozzle inducts air into the foam solution by a venturi action. These nozzles provide maximum expansion of the agent. However, the reach of the stream from air-aspirating foam nozzles is considerably less than that of a standard fog nozzle.

Fog and air-aspirating foam nozzles each have advantages and disadvantages in ARFF operations. In general, the following principles should be observed when selecting the nozzle to be used at a particular incident:

- Fog nozzles that produce nonaspirated foam are best-suited for fire attacks on Class B fires. They provide maximum penetration power and rapid extinguishments of these fires.
- Air-aspirating foam nozzles are best-suited for applying a foam blanket to unignited or recently extinguished pools of fuel. The foam produced by these nozzles forms a more effective blanket for long-term vapor suppression than does nonaspirated foam.

Thus, many jurisdictions choose to make a fire attack with standard fog nozzles and then switch to air-aspirating foam nozzles to blanket the fuel once the fire is extinguished.

Turret Nozzles

Turrets may be either aspirating, nonaspirating, or a combination of the two, any of which can be used with great success (Figure 9.20). There are several factors to consider when selecting the type. As with any other type of nozzle, better reach and penetration are achieved with nonaspirating turrets, whereas aspirating types produce better-quality

Figure 9.20 A nonaspirating turret.

foam. Because either type of turret can perform satisfactorily, the type of turret selected is simply a question of preference and local need.

Assembling a Foam Fire Stream

To provide a foam fire stream, ARFF personnel must be able to correctly assemble the components of the system. The following procedure describes the steps for placing a foam line into service using an in-line proportioner. As mentioned earlier, this is one of the most common methods of foam production used in the municipal fire service.

Step 1: Select the proper foam concentrate for the burning fuel involved.

Step 2: Check the eductor and nozzle to make sure that they are hydraulically compatible (rated for same flow) (Figure 9.21).

Step 3: Check to see that the foam concentrate percentage listed on the foam container matches the eductor rating/setting. If the eductor is adjustable, set it to the proper concentration setting.

Figure 9.21 A firefighter checks the flow rating of a nozzle and in-line proportioner.

Step 4: Attach the eductor to a hose capable of efficiently flowing the rated capacity of the eductor and the nozzle.

- Avoid kinks in the hose.
- If the eductor is attached directly to a pump discharge outlet, make sure that the ball-valve gates are completely open. In addition, avoid connections to discharge elbows. This is important because any condition that causes water turbulence adversely affects the operation of the eductor.

Step 5: Attach the attack hoseline and desired nozzle to the discharge end of the eductor. The length of the hose from the eductor to the nozzle should not exceed the manufacturer's recommendations.

Step 6: Open enough buckets of foam concentrate to handle the task. Place them at the eductor so that the operation can be carried out without any interruption in the flow of concentrate.

Step 7: Place the eductor suction hose into the concentrate. Make sure that the bottom of the concentrate is no more than 6 feet (2 m) below the eductor.

Step 8: Increase the water supply pressure to that required for the eductor. Be sure to consult the manufacturer's recommendations for the specific eductor being used. Foam should now be flowing.

There are a number of reasons for failure to generate foam or for generating poor-quality foam. The most common reasons for failure are as follows:

- Failure to match eductor and nozzle flow, resulting in no pickup of foam concentrate
- Air leaks at fittings that cause loss of suction
- Improper cleaning of proportioning equipment that results in clogged foam passages
- Partially closed nozzle control that results in a higher nozzle pressure
- Too long a hose lay on the discharge side of the eductor
- Kinked hose

- Nozzle too far above eductor (resulting in excessive elevation pressure)
- Mixing different types of foam concentrate in the same tank, which can result in a mixture too viscous to pass through the eductor

When using other types of foam proportioning equipment, such as apparatus-mounted systems, ARFF personnel must follow the operating instructions provided by the foam system or fire pump manufacturer. Because the operation of these systems varies widely, it is not possible to give specific operating directions in this manual.

 Foam Application

As part of their primary job duties, airport firefighters may be required to operate a foam handline or master stream on a fire or spill. It is important to use the correct techniques when manually applying foam. Using incorrect techniques, such as plunging the foam into a liquid fuel, reduces the effectiveness of the foam.

Aspirating Versus Nonaspirating Nozzles

Only film forming foams are suitable for nonaspirating application. The application of protein foams requires air-aspirating nozzles. Aqueous film forming foam may be applied with aspirating or nonaspirating turrets and nozzles. However, there are some important factors that ARFF personnel should consider prior to deciding which type of nozzle to use.

During practical application of AFFF, the advantages of using nonaspirating nozzles are evident. The reach of the stream is greater than with aspirating equipment, and larger areas may be covered with conventional variable stream nozzles. In some instances, extinguishing the fire may be quicker with nonaspirating nozzles than with conventional low-expansion devices.

The limitations of using nonaspirating nozzles are not as obvious and are often only realized after laboratory and field testing. Nonaspirating devices do not mechanically draw in the air. The foam produced is largely a function of the foam solution properties, nozzle design, setting selected, droplet size, and impact of the stream on the fuel surface. A low-expansion ratio of 2:1 or 3:1 is usually achieved when using nonaspirating devices. This low-expansion ratio limits the foam's ability to seal a fire surface after the fire is out and reduces its effectiveness against reignition and burnback.

Air-aspirating devices are designed to produce good-quality foam. This design requires that all or nearly all the solution be converted into working foam with good properties such as bubble size, uniformity, stability, water retention, and heat resistance. All these properties are important factors in restricting reignition and burnback. An expansion ratio of 6:1 to 10:1 is commonly associated with low-expansion air-aspirating equipment.

Personnel who must write specifications for new vehicles and equipment or develop fire fighting tactics should be aware of the advantages and limitations of nonaspirating and aspirating nozzles.

Foam Application Techniques

Correct application of any extinguishing agent can be as important as the type of agent selected. In general, large exterior aircraft fuel fires should be approached and application of foam begun at the farthest effective reach of the turret. The principle of "insulate and isolate" explains the general tactics. The initial foam application should be to insulate the fuselage and protect the integrity of the aircraft skin. This will assist in protecting the occupants who may be self-evacuating. The next consideration should be to try to isolate the fuselage from the fire — that is, push the fire away from the fuselage area. While these techniques may have to be modified for any given circumstance or situation, the general principle remains.

Tactically, driver/operators should also consider the correct placement and repositioning of apparatus. It is vitally important to know the effective reach of the turret. Application should begin at the effective reach of the turret. This may be difficult for the driver to judge, having only somewhat of a "one dimensional" view of the accident. If available, an officer or other apparatus, placed at a 45-degree angle can view the effectiveness of the turret's reach. At this point, instructions can be given to the driver regarding positioning for turret effectiveness.

Drivers should also reposition apparatus as necessary to apply the agent to the correct areas. ARFF apparatus are built with the pump-and-roll capability specifically for this purpose. Care should also be taken to use agent sparingly. The turret can be turned off and on as necessary to produce the most efficient application of the available agent.

Roll-On Method

The roll-on method directs the foam stream on the ground near the front edge of a burning liquid pool (Figure 9.22). The foam then rolls across the surface of the fuel. The ARFF continues to apply foam until it spreads across the entire surface of the fuel and the fire is extinguished. It may be necessary to move the stream to different positions along the edge of a liquid spill to cover the entire pool. This method is used only on a pool of liquid fuel (either ignited or unignited) on the open ground.

Figure 9.22 Roll-on method of foam application.

Bank-Down Method

The bank-down method may be employed when an elevated object is near or within the area of a burning pool of liquid or an unignited liquid spill. The object may be a fuselage, wall, tank shell, or similar structure. Firefighters direct the foam stream at the object, allowing the foam to run down onto the surface of the fuel (Figure 9.23). As with the roll-on method, it may be necessary to direct the stream off various points around the fuel area to achieve total coverage and extinguishment of the fuel. This method is used primarily in dike fires and fires involving spills around damaged or overturned transport vehicles.

Figure 9.23 Bank-down method of foam application.

Seat-of-the-Fire Method

For maximum effectiveness, AFFF should be applied with the use of a zero-degree, "seat-of-the-fire" agent delivery angle to maximize agent effectiveness and minimize extinguishment time. Recent tests have shown that the combined results from small-, medium-, and large-scale static pool fire tests indicated that the use of a 40-degree "raindrop" AFFF delivery technique increased the extinguishment times by an average of 70 percent versus application at zero degrees. Equally important, the effect of wind must be considered, used to advantage when approaching an aircraft fire, and continuously considered throughout the fire fighting phase of the operation.

Dry Chemicals and Their Application

The terms *dry chemical* and *dry powder* are often incorrectly used interchangeably. Dry-chemical agents are for use on Class A-B-C fires and/or Class B-C fires. Dry-powder agents are for Class D fires only. Dry-chemical agents are extremely effective for an initial attack and quick knockdown of a fuel fire, brake fires, hydraulic fires, or lubricant fires. They are also effective for extinguishing three-dimensional or running-fuel fires. However, dry chemicals do not have the vapor-sealing properties or the flashback-preventive characteristics of foam, and reignition may occur due to the lack of cooling effect. Once dry chemicals have accomplished a quick knockdown, a blanket of foam should be applied to prevent fuel vapors from reigniting.

When a dry-chemical agent is discharged into a fire or flames, it inhibits the chemical chain reaction, thereby extinguishing the fire. All dry-chemical agents are nonconductive, making them suitable for use on energized electrical equipment.

Dry chemicals used to combat hydrocarbon fuel fires may contain any of a number of different chemical compounds including:

- Sodium bicarbonate
- Potassium bicarbonate
- Urea-potassium bicarbonate
- Potassium chloride
- Monoammonium phosphate

Dry chemicals are compatible with film-forming foams, but dry chemical may degrade a protein foam blanket. Compatibility between primary and auxiliary agents should be confirmed prior to using them together or successively in fire fighting operations.

Many dry-chemical agents are corrosive to metals, so it may be better to use another agent such as halon, halon substitutes or even carbon dioxide on any electronic equipment or aircraft engines.

Dry chemicals are applied by directing the agent at the base of the fire and sweeping the nozzle back and forth over the fire. Guidelines for applying dry chemicals are as follows:

- Apply dry chemicals from a position upwind of the fire when possible.
- Apply dry chemicals so that the agent will blanket the fire.
- Be aggressive in attacking and extinguishing the fire, but do *not* splash or churn the fuel.
- Monitor the fire area for reignition, especially behind the operator, and reapply the agent as necessary.

The dry-chemical agents themselves are nontoxic and generally are considered quite safe to use. However, the cloud of chemicals may reduce visibility and create respiratory problems like any airborne fine particles. Dry chemical can be a minor respiratory irritant; therefore, ARFF personnel should always wear SCBA when applying them.

Dry-Chemical Extinguishers

There are two basic types of dry-chemical extinguishers: regular B:C-rated and multipurpose A:B:C-rated. Regular B:C-rated extinguishers are recommended for fighting aircraft engine fires. Multipurpose A:B:C-rated extinguishers can be used but are not recommended due to the corrosiveness of the agent. Unless specifically noted in this section, the characteristics and operation of both types are exactly the same.

During manufacture, various additives are mixed with the base materials in order to improve their storage, flow, and water-repellent characteristics. This process keeps the agents ready for use even after being undisturbed for long periods, and it makes them free-flowing.

CAUTION: Never mix or contaminate dry chemicals with any other type of dry chemical because they may chemically react and cause a dangerous rise in pressure inside the extinguisher and reduce extinguishing capabilities.

On Class A fires, the discharge should be directed at whatever is burning in order to cover it with chemical. When applied at an adequate rate and in sufficient quantities, dry chemicals may also be used effectively to extinguish flammable liquid fires.

There are two basic designs for handheld dry-chemical extinguishers: stored-pressure and cartridge-operated (Figure 9.24). The stored-pressure type contains a constant pressure of about 200 psi (1 400 kPa) in the agent storage tank. Cartridge-operated extinguishers employ a pressure cartridge connected to the agent tank. The agent tank is not pressurized until a plunger is pushed to release the gas from the cartridge. Both types of extinguishers use either nitrogen or carbon

Figure 9.24 Cartridge-operated and stored-pressure dry-chemical extinguishers.

dioxide as the pressurizing gas. Cartridge-operated extinguishers use a carbon dioxide cartridge unless the extinguisher is going to be subjected to freezing temperatures; in such cases, a dry nitrogen cartridge is used.

Wheeled Units

Dry-chemical wheeled units are similar to the handheld units but are larger (Figure 9.25). They are rated for regular B:C or multipurpose A:B:C based on the dry chemical in the unit.

Operating the wheeled dry-chemical extinguisher is similar to operating the handheld, cartridge-type dry-chemical extinguisher. The extinguishing agent is kept in one tank, and the pressurizing gas is stored in a separate cylinder. When the extinguisher is in position at a fire, the hose first should be stretched out completely. This procedure is recommended because removing the hose can be more difficult after it is charged and because the powder can sometimes pack in any sharp bends in the hose. The pressurizing gas must be introduced into the agent tank and allowed a few seconds to fully pressurize the tank before the nozzle is opened. The agent is applied the same as that described for the handheld, cartridge-type dry-chemical extinguishers.

Figure 9.25 Typical wheeled dry-chemical extinguisher.

CAUTION: The top of the extinguisher should be pointed away from the firefighter or other personnel when pressurizing the unit. Because of the size of the nozzle, the firefighter should be prepared for a significant nozzle reaction when it is opened.

Halogenated Agents and Their Application

Halogenated extinguishing agents are hydrocarbons in which one or more hydrogen atoms in the molecule have been replaced by halogen atoms. The hydrocarbons from which halogenated extinguishing agents are derived are highly flammable gases; however, substituting halogen atoms for the hydrogen atoms results in compounds that are nonflammable and have excellent flame extinguishment properties. The common elements from the halogen series are chlorine, fluorine, bromine, and iodine. While a large number of halogenated compounds exist, only a few are used to a significant extent as fire extinguishing agents. The two most common ones are Halon 1211 (bromochlorodifluoromethane) and Halon 1301 (bromotrifluoromethane).

Halons are either gases or liquids that rapidly vaporize in fire. Although discharged as a mixture of liquid and vapor, halons extinguish fires best as vapor clouds that chemically interrupt the chain reaction of combustion.

Halogenated vapor is nonconductive and is effective in extinguishing surface fires in flammable and combustible liquids. Similar to dry chemicals, they have almost no flashback-preventive capabilities; but because halons easily penetrate inaccessible areas, they are effective for fires involving aircraft engines, electronic gear, and other complex equipment. In addition, halons are clean agents that leave no corrosive or abrasive residue that could contaminate sensitive electronic equipment. However, these agents are not effective on fires in self-oxidizing fuels such as combustible metals, organic peroxides, and metal hydrides and have been shown to react explosively when used on combustible metal fires involving magnesium. Although the halons have long been used for the protection of internal combustion engines, their primary modern-day application is for the protection of sensitive electronic equipment such as computers. They are compatible with dry chemicals and AFFF.

NOTE: Because of their ozone-depletion potential, halogenated extinguishing agents are included in the *Montreal Protocol on Substances that Deplete the Ozone Layer*, which required a complete phase-out of the production of halogens by the year 2000. The only exceptions allowed under the agreement are for essential uses where no suitable alternatives are available. Many fire departments have aggressive programs to replace halon extinguishers and to use halon replacements where possible.

Chapter 10

Aircraft Rescue and Fire Fighting Tactical Operations

Job Performance Requirements

This chapter provides information that will assist the reader in meeting the following job performance requirements from NFPA 1003, *Standard for Airport Fire Fighter Professional Qualifications*, 2000 edition. Particular portions of the job performance requirements (JPRs) that are addressed in this chapter are noted in bold text.

3-1.1.1 General Knowledge Requirements. Fundamental aircraft fire-fighting techniques, including the approach, positioning, initial attack, and selection, application, and management of the extinguishing agents; limitations of various sized hand lines; use of proximity protective personal equipment (PrPPE); fire behavior; fire-fighting techniques in oxygen-enriched atmospheres; **reaction of aircraft materials to heat and flame**; critical components and hazards of civil aircraft construction and systems related to ARFF operations; special hazards associated with military aircraft systems; a national defense area and limitations within that area; characteristics of different aircraft fuels; hazardous areas in and around aircraft; aircraft fueling systems (hydrant/vehicle); aircraft egress/ingress (hatches, doors, and evacuation chutes); hazards associated with aircraft cargo, including dangerous goods; hazardous areas, including entry control points, crash scene perimeters, and requirements for operations within the hot, warm, and cold zones, and critical stress management policies and procedures.

3-1.1.2 General Skills Requirements. Don PrPPE; operate hatches doors, and evacuation chutes; **approach, position, and initially attack an aircraft fire**; select, apply, and manage extinguishing agents; shut down aircraft systems, including engine, electrical, hydraulic, and fuel systems; operate aircraft extinguishing systems, including cargo area extinguishing systems.

3-2.2 Communicate critical incident information regarding an incident or accident on or adjacent to an airport, given an assignment involving an incident or accident and an incident management system (IMS) protocol, so that the information provided is accurate and sufficient for the incident commander to initiate an attack plan.

(a) *Requisite Knowledge*: Incident management system protocol, the airport emergency plan, airport and aircraft familiarization, communications equipment and procedures.

(b) *Requisite Skills*: Operate communications systems effectively, communicate an accurate situation report, **implement IMS protocol and airport emergency plan**, recognize aircraft types.

3-3.4 Extinguish a three-dimensional aircraft fuel fire, given PrPPE, an assignment, and ARFF vehicle hand line(s) using primary and secondary agents, so that a dual agent attack is used, the agent is applied using the proper technique, the fire is extinguished, and the fuel source is secured.

(a) The fire behavior of aircraft fuels in three-dimensional and atomized states, physical properties and characteristics of aircraft fuel, agent application rates and densities, and **methods of controlling fuel sources**.

(b) Operate fire streams and apply agents, secure fuel sources.

3-3.5 Attack a fire on the interior of an aircraft while operating as a member of a team, given PrPPE, an assignment, an ARFF vehicle hand line, and appropriate agent, so that the integrity is maintained, the attack line is deployed for advancement, ladders are correctly placed when used, access is gained into the fire area, **effective water application practices are used, the fire is approached, attack techniques facilitate suppression given the level of the fire, hidden fires are located and controlled**, correct body posture is maintained, hazards are avoided or managed, and the fire is brought under control.

(a) *Requisite Knowledge*: Techniques for accessing the aircraft interior according to the aircraft type, methods for advancing hand lines from an ARFF vehicle, precautions to be followed when advancing hose lines to a fire, observable results that a fire stream has been applied, dangerous structural conditions created by fire, principles of exposure protection, potential long-term consequences of exposure to products of combustion, physical states of matter in which fuels are found, **common types of accidents or injuries and their causes**, the role of the backup team in fire attack situations, **attack and control techniques, techniques for exposing hidden fires**.

(b) *Requisite Skills*: Deploy ARFF hand line on an interior aircraft fire; gain access to aircraft interior; open, close, and adjust nozzle flow and patterns; apply agent using direct, indirect, and combination attacks; advance charged and uncharged hose lines up ladders and up and down interior and exterior stairways; **locate and suppress interior fires**.

3-3.6 Attack an engine or auxiliary power unit/ emergency power unit (APU/EPU) fire on an aircraft while operating as a member of a team, given PrPPE, an assignment, ARFF vehicle hand line or turret, and appropriate agent, so that the fire is extinguished and the engine or APU/EPU is secured.

(a) *Requisite Knowledge*: Techniques for accessing the aircraft engines and APU/EPUs, methods for operating turrets, methods for securing engine and APU/EPU operation.

(b) *Requisite Skills*: Deploy and operate ARFF hand line, operate turrets, gain access to aircraft engine and APU/EPU, secure engine and APU.

3-3.7 Attack a wheel assembly fire, given PrPPE, an assignment, an ARFF vehicle hand line and appropriate agent, so that the fire is controlled.

(a) *Requisite Knowledge*: Agent selection criteria, **special safety considerations, and the characteristics of combustible metals**.

(b) *Requisite Skills*: **Approach the fire in a safe and effective manner**, select and apply agent.

3-3.8 Ventilate an aircraft through available doors and hatches while operating as a member of a team, given PrPPE, an assignment, tools, and mechanical ventilation devices, so that a sufficient opening is created, all ventilation barriers are removed, the heat and other products of combustion are released.

(a) *Requisite Knowledge*: **Aircraft access points; principles, advantages, limitations, and effects of mechanical ventilation**; the methods of heat transfer; the principles of thermal layering within an aircraft on fire; **the techniques and safety precautions for venting aircraft**.

(b) *Requisite Skills*: Operate doors, hatches, and forcible entry tools; operate mechnical ventilation devices.

3-3.10 Preserve the aircraft accident scene, given an assignment, so that evidence is identified, protected, and reported.

(a) *Requisite Knowledge*: Airport emergency plan requirements for preservation of the scene.

(b) *Requisite Skills*: **Preserve the scene for investigators**.

3-3.11 Overhaul the accident scene, given PrPPE, an assignment, hand lines, and property conservation equipment, so that all fires are extinguished and all property is protected from further damage.

(a) *Requisite Knowledge*: **Methods of complete extinguishments and prevention of re-ignition, purpose for conservation**, operating procedures for property conservation equipment.

(b) *Requisite Skills*: Use property conservation equipment.

3-4.1 Gain access into and out of an aircraft through normal entry points and emergency hatches and assist in the evacuation process while operating as a member of a team, given PrPPE and an assignment, so that passenger evacuation and rescue can be accomplished.

(a) *Requisite Knowledge*: Aircraft familiarization, including materials used in construction, aircraft terminology, automatic explosive devices, **hazardous areas in and around aircraft, aircraft egress/ingress (hatches, doors, and evacuation chutes)**, military aircraft systems and associated hazards; capabilities and limitations of manual and power rescue tools and specialized high-reach devices.

(b) *Requisite Skills*: Operate power saws and cutting tools, hydraulic devices, pneumatic devices, and pulling devices; operate specialized ladders and high-reach devices.

3-4.2 Disentangle an entrapped victim from an aircraft, given PrPPE, an assignment, and rescue tools, so that the victim is freed from entrapment without undue further injury and hazards are managed.

(a) *Requisite Knowledge*: Capabilities and limitations of rescue tools.

(b) *Requisite Skills*: Operate rescue tools.

3-4.3 Implement initial triage of the victims of an aircraft accident, given PrPPE, an assignment, and the triage protocol of the airport, so that each victim is evaluated and correctly categorized according to protocol.

(a) *Requisite Knowledge*: **Categories of triage according to the triage protocol of the airport, methods of assessment**.

(b) *Requisite Skills*: **Assess the critical factors of patient condition, label patient for triage category**.

On or off the airfield, aircraft accidents may occur without notice. Any number of problems can develop while the aircraft is in flight or on the ground. This chapter deals with the types of accidents/incidents with which aircraft rescue and fire fighting (ARFF) personnel are faced and the actions they should perform.

Size-Up

As with any other type of incident, size-up at an aircraft incident is one of the most critical parts of the operation. What is done at this stage of the operation sets the tone for the rest of the incident. The initial unit on the scene should transmit a clear report of conditions, summon whatever additional resources may be needed, and describe the plan of action to be implemented. This allows other responding units to envision the scene and prepare for their possible role. The sooner additional companies and/or specialized units are called, the more successful the operation is likely to be. Rapid size-up also provides the responding chief officers with some of the information they need to assume command upon their arrival.

Initial spotting of the first piece of apparatus often dictates the positioning of later-arriving units. With an aircraft crash, immediate emphasis needs to be placed on the rescue of occupants. Fire may burn through an aircraft's skin in as little as 60 seconds, so rapid setup and fire fighting operations need to commence immediately (Figure 10.1). Because fire fighting agent is often in limited supply, emphasis must be placed on conserving agent during suppression operations to ensure that firefighter safety is not compromised.

Figure 10.1 Rapid setup and fire fighting operations need to commence immediately. *Courtesy of Michael T. Defina, Jr., Metro Washington Airports Authority Fire Department.*

Following the size-up, the nature and scope of the problem dictate the actions required to deal successfully with an aircraft accident/incident. All actions should be based on the same operational priorities as any other emergency: rescue, fire control (prevention or extinguishment), and loss control. However, one of the most difficult decisions may be to take *no action* other than beginning to establish a command structure with which to manage the incident. If the nature and/or scope of the incident is clearly beyond the capabilities of the first-arriving unit, it may be more productive for personnel to delay attacking the incident in favor of managing it. This is accomplished by assuming the position of Incident Commander (IC), designating a command post, and directing other incoming units in a planned and coordinated attack.

If the accident/incident is one in which the first-arriving unit may intervene effectively, the ARFF crew should initiate the plan of action they described in their size-up. Assuming the first vehicle on the scene has taken a position that affords the greatest route of safety for the aircraft occupants, the later-arriving units should turn off their emergency lights and audible devices as they approach the scene. History has shown that occupants migrate toward the warning devices. By having only one vehicle, in the safest exit path, with operating lights and sirens, confusion for the occupants is reduced.

Positioning Apparatus

ARFF apparatus and other responding units must be positioned correctly if rescue and fire fighting operations are to be successful. Because ARFF apparatus often respond single file, the first fire apparatus to the accident site often establishes the route for other vehicles and may dictate the approach into their ultimate fire fighting positions. In positioning apparatus, first-arriving crews and the IC should follow certain guidelines:

- Approach the scene with extreme caution so as not to run over any fleeing occupants, wreckage, ground scarring, spilled fuel, or other hazards. Do not drive through smoke that obscures escaping occupants. Driving over aircraft wreckage could also leave a vehicle with multiple flat tires (Figure 10.2).

- Consider the stability of the terrain, slope of the ground, and direction of the wind prior to entering a crash site. Always attempt to position vehicles uphill and upwind to avoid fuels and fuel vapors that may gather in low-lying areas.
- Do not position vehicles so that they block the entry or exit from the accident site for other emergency vehicles.
- Position vehicles so that they may be operated quickly in the event of flash fire.
- Position vehicles so that they can be used to protect the egress route or rescue operations of persons from the aircraft.
- Position vehicles so that they can be repositioned as easily as possible while limiting maneuvers that require backing the vehicle.
- Position vehicles so that turrets and handlines may be used to maintain the route of egress if necessary.

Other factors that must be considered when determining final apparatus placement on the emergency include:

- The number, type, and capabilities of the responding apparatus
- The number and abilities of responding personnel
- The location and condition of the wreckage
- The number and location of survivors
- Fundamentally hazardous areas associated with aircraft emergencies

Figure 10.2 Running over aircraft wreckage can puncture vehicle tires. *Courtesy of Air Line Pilots Association.*

Wind

While rescue and fire fighting operations can be conducted against the wind, doing so is much more difficult and more hazardous for both ARFF personnel and aircraft occupants. When operating against the wind, the smoke obscures vision, the heat is more intense, and it is more difficult to reach the fire with extinguishing agents. ***Attacking a fire from a position downwind should only be attempted when conditions preclude any other approach.*** Operations conducted from upwind are safer and much more efficient because heat and smoke are carried away from the operating area (Figure 10.3). Attacking with the wind enables extinguishing agents to be applied more effectively, thus reducing extinguishing time. Also, upwind paths of egress for the aircraft's occupants are safer due to the exit corridor being free of heat and smoke. ARFF personnel should make every attempt to use the wind to their advantage. This assists in rescue as well as in fire fighting agent conservation.

Terrain

The influences of some ground features are readily apparent. Soft or muddy soils may stop heavy apparatus and equipment (Figure 10.4). Steep slopes may be difficult to traverse or climb. Low or downslope areas may become saturated with fuel. Rough or rocky terrain may be impassable. However, other terrain effects may not be so obvious. For instance, fire apparatus should not be driven into gullies or downslope depressions near the aircraft into which fuel may have drained or in which

Figure 10.3 Firefighters should make every attempt to attack fire from the upwind side. *Courtesy of Michael T. Defina, Jr., Metro Washington Airports Authority Fire Department.*

fuel vapors may have collected. ARFF personnel should also consider terrain when establishing a triage area, tool and manpower staging, as well as rehab areas.

Wreckage

The condition and location of wreckage and any hazards it creates must be evaluated. Different methods of attack may be required according to whether the aircraft is intact, broken open, fragmented, or upside-down (Figure 10.5). More than one apparatus position may be required if the occupied portion of the aircraft is broken and separated into several pieces of wreckage. ARFF personnel should take time to confirm that initial fire fighting efforts are aimed at a portion of the fuselage and not a section of wing or other part of the aircraft that does not contain occupants.

Figure 10.4 Rough terrain may influence response capabilities. *Courtesy of Air Line Pilots Association.*

Figure 10.5 Aircraft may end up broken or fragmented. *Courtesy of George Freeman, Dallas (TX) Fire Department.*

Survivors

The number and location of occupants influences the point at which rescue efforts should begin. If the occupants have not been evacuated and the fuselage is still intact, ARFF personnel should decide upon the rescue entrance (normal loading doors, emergency exits, or emergency cut-in points). If evacuation has begun from the interior of the aircraft, ARFF personnel should protect the exits being used and may be tasked to assist occupants from the escape slides and direct them to safety (Figure 10.6).

Hazardous Areas

The entire area of an aircraft accident should be considered hazardous, and certain specific areas should be avoided whenever possible:

- Stay clear of aircraft propellers as they may be a hazard, even when an engine is not running.

WARNING

Bumping or turning a prop may cause the magneto to fire, resulting in the engine trying to start and the prop rotating.

- Maintain a safe distance from jet and gas turbine engines. The intake and exhaust areas of jet or gas turbine engines are also hazardous because the heat generated is substantial enough to reach the auto-ignition temperature of most aviation fuels. When operating — even at idle — jet en-

Figure 10.6 A firefighter stands ready to assist passengers from an escape slide. *Courtesy of William D. Stewart.*

gine intakes draw large amounts of air into the engine. This suction is substantial enough to draw personnel into the engine (Figure 10.7). At the opposite end of the engine, exhaust temperatures can be found that could severely burn personnel. Depending on the throttle setting, jet engines emit enough blast to easily blow over a large ARFF vehicle.

- Stay clear of the line of fire for guns and rockets and the rear blast areas of missiles and rockets on military aircraft.
- Because a wing structure may collapse without warning, do not set up an operation that requires ARFF personnel or others to walk under wings or other overhanging wreckage.
- Be aware of the hazards of sharp jagged metal, fires involving advanced aerospace materials (composites), and biohazards such as human remains and wreckage contaminated with aircraft lavatory fluids.
- Stay clear of aircraft radar systems, which are often located in the nose of aircraft, because exposure to the waves generated from the system can cause health damage.

Types of Aircraft Accidents/Incidents

Whether announced or unannounced, the general types of aircraft accidents and/or incidents with which ARFF personnel are confronted are as follows:

Figure 10.7 Cautions and warnings marked on this jet engine remind personnel to avoid hazard areas. *Courtesy of William D. Stewart.*

- Ground emergencies
- In-flight emergencies

Each jurisdiction should have standard operating procedures that address the ARFF response to each of these types of emergencies. These standard operating procedures vary from as little as simply being notified of in-flight emergencies to a full response to a major ground emergency. Historically, the vast majority of in-flight emergencies conclude with the plane landing safely and little, if any, assistance being required of ARFF responders. However, ARFF personnel should not become complacent and must always be prepared for an in-flight emergency to become a ground emergency that requires their intervention.

It is also important to understand the difference between an aircraft accident and an aircraft incident. According to NFPA 402, *Guide for Aircraft Rescue and Fire Fighting Operations*, 1996 edition, an aircraft *accident* involves an occurrence during the operation of an aircraft in which any person involved suffers death or serious injury or in which the aircraft receives substantial damage. An aircraft *incident* encompasses an occurrence other than an accident associated with the operation of an aircraft that affects or could affect continued safe operation if not corrected. An incident does not result in serious injury to personnel or substantial damage to the aircraft. By definition, an aircraft accident is obviously more serious than an incident. It is important to carefully observe aircraft operations in an effort to prevent accidents before they occur.

Ground Emergencies

Ground emergencies involve aircraft that are conducting operations on the ground. This type of emergency could involve an aircraft and a ground vehicle, a structure, or another aircraft (Figure 10.8). Operational plans should be developed that address these types of emergencies. These may range from a simple inspection of the aircraft to a multijurisdictional response.

Ground emergencies (from least serious to most serious) that ARFF crews are likely to encounter are the following:

Figure 10.8 An aircraft colliding with airport property may summon a response. *Courtesy of Michael T. Defina, Jr., Metro Washington Airports Authority Fire Department.*

- Overheated wheel assemblies
- Tire/wheel failures
- Combustible metal fires
- Fuel leaks and spills
- Engine fires or APU fires
- Uncontained engine failures
- Aircraft interior (or "cabin") fires

Overheated Wheel Assemblies

On all types of aircraft, the landing gear is of concern during both normal and emergency landings. Brakes and wheels frequently overheat. The problem of overheated gear assemblies can be compounded if combustible metals, such as magnesium or titanium are used in their construction. Burning magnesium metals are a challenge to extinguish and generally require the use of a Class D fire extinguisher. Landing gear on current generation jet transports (the Boeing 767 and 777, for example) are primarily high-strength steel with some titanium components. The landing gear oleo-strut is serviced with nitrogen and therefore is under very high pressure. ARFF personnel should thoroughly understand the hazards created by overheated wheels, brakes, and tires as well as the techniques and equipment needed for coping with landing gear emergencies.

Most aircraft brake pads are constructed of carbonaceous materials. The brake linings frequently smoke when they are new. If a plane comes to rest with an overheated brake or with a wheel that is smoking around the brake housing and tires, the whole assembly should be allowed to cool naturally without any water or other cooling agent being applied. However, wheel assemblies that are smoking when the aircraft stops should continue to be observed because peak wheel temperatures may not be reached until 15 to 20 minutes after the aircraft has come to a complete stop.

Because overheated brakes are common on all types of aircraft, the operating manuals for some larger propeller-driven aircraft recommend that the propellers be kept turning fast enough so that the prop wash will cool the wheels. Wheels on jet aircraft operate under more severe conditions in that there is no airflow to help keep them cool. This is a major factor affecting brake effectiveness and increases the likelihood of fire.

A landing gear fire may be caused by an overheated brake that ignites grease or hydraulic fluid on the wheel assembly. Wheel grease, when ignited, shows up as long flames coming out slowly from the bottom of the wheel. These fires are usually small and can be extinguished quickly. Rubber tires may ignite at temperatures of 500°F to 600°F (260°C to 315.5°C). Once burning, a tire may develop into an extremely hot and destructive fire. Because it may be difficult to determine exactly what is burning, dry-powder extinguishing agents designed specifically for use on combustible metal fires (Class D agents) should be used if this can be done safely. If Class D agents are not available or have already been utilized, mass application of water (using turrets) can also be used for extinguishment. This should only be done once the area is clear of all personnel.

Related hydraulic failures involving aircraft control systems may contribute to the overheating of wheel assemblies. If the landing gear hydraulic system contains petroleum-based fluids, a fire may start around the hydraulic fittings near the wheel. These fires should be controlled immediately as described in the next section. Failure to do so could allow conducted heat to impinge on the fuselage, allowing the fire to spread to the interior. If the hydraulic system contains a synthetic fluid/compound, such as Skydrol®, ignition of the fluid is uncommon. However, if synthetic fluid is released in the form of a mist, it may ignite and sustain fire. If the Skydrol® does ignite, it will thermally decompose at high temperatures, producing toxic vapors. Regardless of what is suspected to be burning during a landing gear fire, ARFF personnel must wear full protective clothing and SCBA.

Tire/Wheel Failures

When excessive heat is conducted from the brake to the wheel, the tire is also heated, increasing the pressure within it. This can cause the tire to deteriorate and, in time, blow out (Figure 10.9). Experiments have shown, however, that the increase in air pressure alone is not enough to cause a good tire to fail. More likely, the combination of increased temperature of the brake/wheel assembly and the increased tire pressure will lead to the disintegration of the wheel assembly, sometimes with explosive force. ARFF personnel, if improperly positioned, may be at risk from the flying fragments of a blown wheel assembly.

Figure 10.9 Excessive heat and increased tire air pressure cause tires to deteriorate and blow out upon landing. *Courtesy of Michael T. Defina, Jr., Metro Washington Airports Authority Fire Department.*

Modern aircraft wheels are commonly equipped with fusible plugs incorporated into the rims. These plugs are designed to melt, automatically deflating the tires when the rim reaches a predetermined temperature, usually from 300°F to 400°F (149°C to 204°C). Releasing the tire pressure reduces the pressure on the wheel, thus diminishing the possibility of wheel collapse and fragmentation. Caution should still be used as incidents have occurred in which the fusible plugs failed to function properly.

Rapid cooling of hot wheels, especially if localized, may cause them to crystallize and shatter explosively. Personnel and equipment should not be in the line of probable fragmentation, which is the area at least 300 feet (100 m) directly to each side of the heated wheel. If it is decided that a wheel assembly must be cooled, water fog may be used but applied only in short, intermittent bursts (5 to 10 seconds every 30 seconds). The heated wheel area should be permitted to cool slowly so that the metal will not crystallize.

A wheel fire also directly threatens the aircraft through the ignition of the magnesium rims employed on some aircraft. Although magnesium does not ignite easily, once ignited, it burns intensely and is difficult to extinguish. Wheel assemblies of current generation jet airliners are primarily constructed of aluminum alloy. Brakes that have been stressed during an aborted takeoff or as a result of consecutive landings without cooling may develop temperatures high enough to initiate a brake or wheel fire. However, wheel fires may not occur until the aircraft is standing still because of the lag time for wheel parts to absorb enough brake heat for ignition.

WARNING

When responding to a hot brake incident or wheel fire, always approach from forward or aft of the wheel assembly while exercising extreme caution. Never approach from the sides in line with the axle. In addition, always wear full protective gear including SCBA. The brakes of some aircraft contain beryllium, which produces toxic fumes and smoke.

Firefighters in full protective clothing, approaching the involved tire/wheel from the front or rear and applying dry-chemical agents usually achieve rapid knockdown if only the tire is burning. Because dry chemicals provide little cooling effect, re-ignition may occur a number of times until the material cools below its ignition temperature. Firefighters should simply apply additional dry chemical each time flame reappears. If the wheel assembly is also involved and combustible metal construction is suspected, dry-powder (Class D) agents should be applied.

A dry-chemical fire extinguisher is recommended for controlling tire fires on all aircraft because it is less likely to create localized cooling of the metal in the wheel parts. When only parts of the wheel are cooled quickly, cracks in the metal may develop as a result of differential contraction. These cracks may spread or enlarge, causing the wheel to separate around the rim when tire pressure increases due to the heat of the fire. Serious accidents have resulted when ARFF personnel have used CO_2 or an improperly applied stream of water on a wheel fire. As long as one or more tires remain inflated, the use of a dry-chemical extinguisher is the preferred method of extinguishment.

If fusible plugs have deflated all the tires or if dry-chemical agents are not available, water may be applied in short, intermittent bursts to extinguish the tire fire. If further cooling is necessary after the fire has been extinguished, the agent should be directed at the brake area only. As always, ARFF personnel should protect themselves by standing either forward or aft of the wheel (Figure 10.10).

Combustible Metal Fires

A variety of metals, some of them combustible, are used throughout modern aircraft. Magnesium and titanium are the two combustible metals found in widespread use in aircraft.

Magnesium is a lightweight, silvery-white metal. Its ignition temperature is generally considered to be close to its melting point of 1,202°F (650°C). It is classified as a combustible metal, even though as a solid it does not ignite easily. Whether it ignites easily depends upon its mass (thickness and shape). Magnesium is used on most large propeller-driven aircraft (Douglas DC-6, for example) and in early transport-type jet aircraft (such as the Boeing 707) landing gear, engine mountings, wheel cover plates, and engine components.

Titanium, a silvery-gray metal, is as strong as ordinary steel but is only about 56 percent as heavy. Some titanium alloys are up to three times as strong as the best available aluminum alloys. Titanium's ignition temperature is generally considered to be near its melting point of 3,140°F (1 727°C). It is used in engine parts and nacelles because of its resistance to heat and fire. It is also used in the landing gear assemblies of modern jet transports (for example, the Boeing 777).

Combustible metals introduce additional problems when they become involved in an aircraft fire. Applying large amounts of water may cool overheated magnesium, but once the metal is burning,

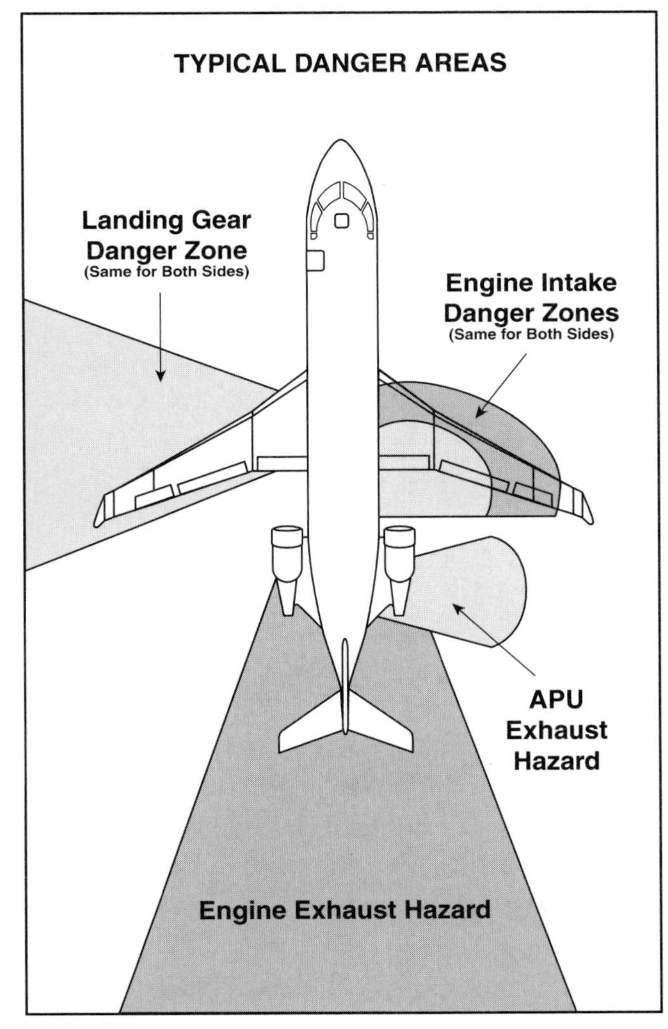

Figure 10.10 Firefighters must stay aware of all danger zones when working around an aircraft.

water increases the intensity of the fire. Water may still be used in copious amounts to cool the uninvolved metal and to protect other nearby combustibles. These metals have such a strong affinity for oxygen that once ignited, they continue to burn — even in atmospheres of carbon dioxide or nitrogen. Because burning magnesium and titanium — both requiring special extinguishing agents — cannot readily be extinguished with water or foam, they present a constant threat where flammable vapors may be present.

Specialized extinguishing agents such as MET-L-X® and G-1 powder are effective in controlling magnesium and titanium fires. However, if these agents are not available, water in heavy, coarse streams provides the next best method of fire control. Initially, such streams will intensify the fire and cause the burning metals to spark and shower significantly. Water application is effective, however, because it causes the burning metal to break loose from the aircraft and prevents the unburned metal from reaching ignition temperature. See Chapter 9, "Extinguishing Agents," for more discussion of extinguishing agents and application techniques.

Fuel Leaks and Spills

In some cases, ARFF personnel respond to incidents where aircraft fuel has spilled or is leaking, but it has yet to ignite. At all such incidents, ARFF personnel should take the following precautions:

- Attempt to shut off the fuel at the source or by utilizing emergency fuel shutoff or transfer valves.
- Avoid actions that could provide an ignition source.
- Evacuate aircraft if the spill poses a threat to occupants.
- Keep all nonessential personnel clear of the area.
- Make sure that fire fighting personnel are wearing full protective clothing, including SCBA.
- If necessary, blanket all exposed fuel surfaces with foam.
- Contain spilled fuel to as small an area as possible.
- Prevent leaking or spilled fuel from entering runoffs, storm drains, sewers, buildings, or basements.
- Keep apparatus and equipment ready to protect rescue operations in case a fire should occur.
- Position the apparatus upwind and uphill from the fuel spill.

Spills can involve significant amounts of fuel. For example, each wing tank of a Boeing 757 holds 14,600 pounds, or just over 2,100 gallons (8 400 L), of fuel.

The procedures for handling fuel spills described in this section are subject to the regulations and procedures established by the authority having jurisdiction. Information from NFPA 407, *Standard for Aircraft Fuel Servicing*, was included in the development of this information.

Each incident may be somewhat unique, but certain general principles apply in all such cases. Every fuel spill involves several variables:

- Size of the spill
- Terrain
- Equipment
- Weather conditions
- Type of flammable liquid
- Aircraft occupancy
- Emergency equipment and personnel available

If a fuel leak develops or a fuel spill occurs during aircraft servicing:

- Personnel must stop the fueling operation immediately.
- Nonessential personnel should leave the area until the hazard is neutralized, repairs are made, and the area is safe.
- Safety personnel should be notified of such incidents so determination can be made to allow operations to remain in progress or terminate them until the problem has been corrected.

During any spill or leak, extreme caution must be exercised to avoid actions that could provide ignition sources for the fuel vapors. If fuel is leaking or spilling from a fuel-servicing hose or equipment, the emergency fuel shutoff valve must be turned off immediately. If the fuel is leaking or spilling from an aircraft at the filler opening, vent line, or tank seam, fuel delivery must stop immediately. All electrical power to the aircraft should be shut down, and the aircraft should be evacuated. Maintenance personnel must thoroughly check the aircraft for damage and for flammable vapors that may have entered concealed wing or fuselage compartments before the aircraft is put back into service. Maintenance records should be kept of each incident or occurrence that describe the cause, corrective action taken by various personnel, and action taken to prevent recurrence. This information should also be contained in the fire department's incident report.

- Small spills involving an area less than 18 inches (450 mm) in any plane dimension normally involve minor danger. Personnel should stand ready until the aircraft departs the area of the spill because engine exhaust could ignite the spill. Local standard operating procedures should be followed. In most cases, small fuel spills are contained by firefighters and can be cleaned up by the responsible airport personnel. These spills contain such a small amount of fuel that they may be absorbed, picked up, and placed easily in an approved container.
- Small or medium static spills — not over 10 feet (3 m) in any dimension nor over 50 square feet (4.6 m^2) in area — should have a fire watch posted. One or more fire extinguishers with at least a 20-B rating should be available. Absorbent materials or emulsion compounds should be used to absorb the spilled fuel, especially if aviation gasoline or low-flash-point fuels are involved. The contaminated absorbent should be picked up and placed in an approved container and disposed. Aircraft fuels will damage some types of ramp surfaces, so personnel should pick up spilled fuel as quickly as possible.
- Large spills — over 10 feet (3 m) in any dimension or over 50 square feet (4.6 m^2) in area — or smaller spills continuing to enlarge should be handled by the fire department. The department should be summoned immediately, and anyone in the spill area should move upwind of the spill at once.

All fuel spills occurring as a result of a collision should be blanketed with foam to prevent ignition and to prevent damage to the aircraft or other exposures.

The safe handling of large fuel spills involves these considerations.

- Evacuate the aircraft.
- Eliminate all sources of ignition.
- Limit available fuel vapors.
- Detect any fuel vapors remaining inside the aircraft.
- Do not allow anyone to walk through the liquid fuel. Any clothing that has been sprayed or soaked with fuel should be removed at once, taking care to avoid additional sources of ignition. Fuel contamination should be washed from skin with soap and water.
- Do not start any aircraft, motor vehicle, or other spark-producing equipment in the area before the spilled fuel is blanketed or removed. Shutting down or moving the equipment may provide a source of ignition. This decision needs to be evaluated carefully. If a vehicle engine is running at the time of the spill, drive the vehicle from the hazard area unless the danger to personnel is judged to be too great. If a vehicle would have to be driven through the spilled fuel to get out of the hazard area, it should probably be left in place without shutting off the engine. Before fuel-servicing vehicles are moved, any fuel hose that was in use or connecting the vehicle and aircraft must be shut down and safely stowed.

If it is decided that a vehicle with an internal-combustion engine must be shut down within a spill area, the engine speed should be reduced to idle before being turned off in order to prevent a backfire.

If any aircraft engine is operating at the time of the spill, move the aircraft from the hazard area unless the size of the spill would be increased or the prop wash or jet blast would increase the extent of the fuel-vapor hazard.

It may be necessary to blanket with foam any spills of aviation fuel that would be considered large spills as previously defined. The severity of the hazard created by a fuel spill depends primarily upon how volatile the fuel is and the proximity to sources of ignition. Aviation gasoline and other fuels with low flash points at normal temperatures and pressures give off vapors that are capable of forming ignitable mixtures with the air near the surface of the liquid. This process is not usually true of kerosene fuels (Jet A or Jet A-1) except where ambient temperatures are at least in the 100°F (38°C) range and the temperature of the fuel has reached the same range.

Local regulations and procedures must be followed. Attention must be given to not allow fuel to enter sewers or storm drains. If fuel has entered sanitary sewers or storm drains, personnel should dam inlets to prevent additional fuel from entering. The responsible utilities supervisor and local environmental health officials should be notified immediately. No further action to dilute or disperse the fuel should be taken until these officials arrive to assess the situation and make recommendations to the IC.

WARNING

Unless ordered to do so by a responsible local official, never flush fuel or other contaminants into sewers or storm drains nor introduce water into these conduits in an attempt to dilute the contaminant. Such actions could increase the possibility of ignition and could expose the airport to significant liability under environmental protection laws.

If sewer or storm drain contamination is extensive, steps should be taken to keep sources of ignition, such as operating vehicles and aircraft, away from manholes or storm drain inlets and outfalls until the atmosphere at these sites can be sampled and determined to be within safe limits.

NOTE: See NFPA 415, *Standard on Airport Terminal Buildings, Fueling Ramp Drainage, Loading Walkways*, for further information on aircraft fueling ramps designed to reduce the danger resulting from spilled fuel by controlling its flow.

Thoroughly inspect aircraft onto which fuel has been spilled to make sure that no fuel or fuel vapors have accumulated in flap well areas or internal wing sections not designed for fuel storage. It is extremely important that cargo, baggage, mailbags, or similar items that have come in contact with fuel be decontaminated before being placed aboard any aircraft.

Engine/APU Fires

In the event of an engine or auxiliary power unit (APU) fire, the cockpit crew may make the first attempt to extinguish the fire by using onboard extinguishing systems. At other incidents fire personnel may deal with an aircraft that is unoccupied; therefore, airport firefighters must be familiar with aircraft shutdown procedures.

When dealing with an engine or APU fire, it is important to realize that directing a stream of water or AFFF into the air inlet will not always extinguish the fire. Although agent will go through the core of the engine or APU, the fire very likely involves the accessory section around the outside core of the engine. The safest method of extinguishment is to operate the engine or APU fire shutdown system from the cockpit or, where provided, from an external fire protection panel. Large-frame aircraft usually have easily identifiable engine and APU fire shutoff handles in the cockpit. Many also have external APU fire protection panels on either the nose landing gear, in the main wheel well, or in the tail. In addition to arming the extinguishing agent bottles, these systems simultaneously shut off the power plant's fuel, hydraulic, electrical, and pneumatic connections. If unable to access the aircraft's fire protection system, responders may be tasked with opening the engine cowlings or APU access panels doors in an effort to fully extinguish the fire. Due to the location and configuration of the access panels, firefighters must exercise extreme caution when performing this task. Hot and burning fluids or engine parts may be trapped inside these areas. These could fall onto the firefighters when the panels are opened. Fire fighting personnel may want to consider using a piercing tool to apply extinguishing agent prior to opening. Some aircraft are equipped with fire extinguishing access ports or knock-in panels, which can be used to apply agent directly to the engine.

Another type of engine fire is a tail cone fire. This fire occurs when too much fuel is ejected into the engine during start-up. This causes fuel and fuel vapor to be emitted through the engine. Upon reaching the tailpipe assembly, it ignites because of the high temperatures. Usually the pilots will shut off the fuel and motor (rev) the engine. This action will blow out the excess burning fuel from the back of the engine, at which time restart procedures can be conducted. Quite often the fire department is not called unless the pilot is unsuccessful in extinguishing the fire. On occasion, burning fuel may drip out onto the ground from the tail cone.

Uncontained Engine Failures

Another type of emergency involving jet engines occurs when fan or compressor blades separate or the turbine section disintegrates (Figure 10.11). When this happens, fragments tear through the engine cowling and can penetrate aircraft structures. A similar problem can occur with propeller-driven aircraft when a propeller blade separates (Figure 10.12). The resulting imbalance can cause disintegration of the engine and loss of control of the aircraft.

The worst-case scenario involves fragments of engine components piercing the fuselage and/or wing structure, causing injuries to occupants, puncturing fuel tanks, severing fuel and hydraulic lines, or damaging the flight-control system. Due to the location and configuration of fuel tanks and lines, this type of incident may result in a three-dimensional flowing fuel fire. This would make it necessary for the flight crew to immediately evacuate the aircraft. Firefighters may be forced to make an aggressive interior fire attack in order to support evacuation or property conservation.

Aircraft Interior Fires

Sometimes an aircraft lands and the flight crew reports a strong odor of something burning. The flight crew and occupants may observe visible smoke. While ARFF personnel are checking the interior of the aircraft, other ARFF personnel should conduct a thorough examination of the exterior, including the wheel wells, for smoke or signs of charring and blistering. In aircraft, any indication of a fire or overheat condition in a wheel well after takeoff is cause for an immediate return and emergency landing. If the air traffic control tower or ARFF personnel confirm that a fire exists, the flight crew will most likely initiate an evacuation once the aircraft is brought to a stop.

Common sources and areas of smoke and the odor of something burning aboard an aircraft are:

- Overheated fluorescent light ballasts
- Food preparation areas
- Lavatories
- Cockpit area
- Avionics and electronic equipment compartments
- Cargo compartments
- Overheated electrical components

Figure 10.11 Aircraft engine damage caused by fan blade separation. *Courtesy of Air Line Pilots Association.*

Figure 10.12 When the propeller blade separated from an engine, it literally cut into the fuselage. *Courtesy of Air Line Pilots Association.*

Overheated ballasts in fluorescent lighting fixtures occur as frequently in aircraft as they do in buildings and are usually not serious. But because the consequences of ignoring overheated ballasts could be serious, flight crew personnel who recognize this characteristic odor must not assume that the problem is minor and dismiss it.

As with commercial and domestic kitchens on the ground, food preparation areas aboard aircraft are frequent sources of smoke. ARFF personnel must thoroughly check this area, including all drawers, storage compartments, and hot plate heating elements. Power switches and circuit breakers for galley equipment are located in the cockpit.

Since 1985, smoke detectors have been installed in all lavatories on commercial aircraft, and they should help pinpoint smoke in this location. However, these detectors sound in the local area only and do not transmit an alarm to the cockpit. The cockpit crew may be unaware that a detector has been activated until notified by flight attendant crew members. This could delay initiation of emergency landing procedures.

In the cockpit area, there may be one or more circuit-breaker panels. If any of the electrical systems throughout the aircraft malfunction, the flight crew should be alerted by a tripped circuit breaker. Because of the sensitivity of aircraft circuit breakers, the flight crew may make several attempts to reset a breaker before taking action to correct the problem. Also because of their familiarity with the aircraft, flight crew members may be able to assist ARFF personnel in locating concealed fires.

Because aircraft interior fires may originate in numerous places in addition to the main passenger cabin, ARFF personnel should understand the structural characteristics of an aircraft fuselage. Fires in concealed spaces may travel between the skin of the aircraft and interior liners and may extend the length or width of the aircraft. It may be difficult, under such conditions, to determine either the source of ignition or the extent of the fire spread. If available, portable infrared heat detectors can be used to locate "hot spots" that indicate concealed fires. Other methods of determining the location of concealed fires are to remove sections of flooring, wall panels, and ceilings. On the exterior of the aircraft, paint blistering or discoloration may assist in locating fire areas. Applying a light water mist and watching for areas where the water turns to steam and evaporates quickly can also pinpoint fire areas.

If there is no sign of evacuation upon landing, ARFF personnel must immediately gain access into the aircraft and begin rescue and fire fighting operations.

WARNING

Exercise extreme caution when gaining entry into the aircraft due to the emergency escape slide systems attached to each door and, depending on the aircraft, to over-wing exits as well. If opened from outside, the escape slide may deploy and can seriously injure or kill unsuspecting emergency personnel.

Caution is also needed because an interior fire may lack only oxygen; opening the exits allows fresh air into the superheated atmosphere, and a flashover or rollover could occur. Because a free-burning fire in an aircraft almost always vents itself by burning through the skin of the aircraft in the early stages of the fire, the conditions conducive to a backdraft are unlikely.

Under ***no*** circumstances should firefighters entering aircraft impede the emergency exit of occupants. However, allowing occupants to exit the aircraft does not prevent firefighters from opening all available exit doors, hatches, and windows in an attempt to ventilate the aircraft.

On most aircraft, over-wing exit size will allow entry for ARFF personnel wearing full protective equipment including SCBA. Once inside, advancing down narrow, restrictive aisles — perhaps congested by escaping occupants and cluttered with loose carry-on baggage — may be difficult. Interior aircraft fires should be fought in the same manner as structural fires. Proper ventilation, followed by an interior attack, should be part of a planned and coordinated operation. Water can be used to fight an interior fire but the extinguishing agent of choice is usually Class A and Class B foam. When water runs out of the aircraft, it has a tendency to dilute the foam blanket that is serving as a vapor-suppressing agent. Other agents, such as clean agents, and dry chemicals, can be used after occupant evacuation or if occupants are not present.

ARFF personnel should attempt to locate and determine the extent of fire involvement before attempting entry. Ventilation should then be established as quickly as possible. Early ventilation is important because applying water to the fire will quickly cause the limited interior space to fill with smoke and steam. This condition compounds the difficulty of search and rescue efforts and puts occupants and ARFF personnel at risk for steam burns. Once ventilation is started, personnel should gain entry, initiate an immediate search of the interior, and begin the fire attack from the unburned side.

Unoccupied Aircraft

Fires in unoccupied aircraft often develop into major incidents because of delayed detection. An unattended aircraft with all doors closed may sustain a smoldering fire for long periods, resulting in a buildup of smoke and potentially explosive gases that may go unnoticed until the aircraft is opened. Opening an aircraft door under such conditions is *extremely* hazardous because of the potential for a flashover, rollover, or in rare instances, even a backdraft. As in structural fire fighting, this situation indicates the need for vertical ventilation. Charged hoselines should be in position to immediately respond to the buildup of the fire that occurs when ventilation has been established. Penetrating nozzles may be used to good effect under these conditions.

Aircraft are often left attached to the jetway during airline arrival and departure operations and during overnight layovers. In many cases, electrical power is supplied to the aircraft via an external power cable from the jetway. A fire in either an occupied of unoccupied aircraft could jeopardize the safety of the terminal as well as airport operations. Special considerations should be given to how fire rescue personnel would handle an incident involving a jetway, structure, or multiple structures on the airport. To prepare for these types of emergencies, ARFF personnel should develop pre-emergency plans for these structures and conduct training to examine the most effective means for dealing with these structures.

Cargo Aircraft

Interior fires in fully loaded cargo aircraft differ significantly from fires in passenger aircraft because of the differences in the number of occupants and in fire load. Hazardous cargo is possible with either type of aircraft; however, cargo aircraft are much more likely to have larger amounts of and more hazardous cargo/dangerous goods than passenger aircraft. (**NOTE:** For more information on dangerous goods, see Chapter 12, "Hazards Associated with Aircraft Cargo.")

In the event of a fire aboard a cargo aircraft on the ground, the flight crew is usually able to exit the aircraft through normal entry doors or through cockpit emergency exits. Once it has been determined that all crew members are out and there is no longer a rescue concern, attention can be focused on fire attack. If the cargo section doors cannot be opened, a conventional interior attack may be difficult. The use of skin-penetrating nozzles may be the best tactic to combat a cargo aircraft interior fire. By using these penetrating nozzles, rescue and fire fighting personnel may locate the hottest point of the fire from the exterior and then penetrate the fuselage at that location. This technique properly applies extinguishing agent onto the fire without exposing ARFF personnel to the hazards of an interior attack.

On most fully loaded cargo aircraft, it is virtually impossible to move through the cargo hold. Clearance of only a few inches often exists between the containers and the fuselage. If a small fire is present, it may be possible to unload the cargo to access the fire. Before making an interior attack, ARFF personnel should attempt to determine the presence, types, and quantities of dangerous goods on the aircraft. Information regarding dangerous goods can be found on the waybill, which can be found in the cockpit or in an area around the main loading door. Other than radioactive materials, dangerous goods have to be accessible by the flight crew and are often stored near the front of the aircraft. Regardless of the amount of dangerous goods presumed to be onboard, a hazardous materials response team should be immediately requested when there is a cargo aircraft emergency. If available, infrared thermal-imaging devices might be used to assist in finding the seat of the fire.

In-Flight Emergencies

In-flight emergencies include fires as well as other problems that may lead to an aircraft accident/incident. These emergencies include the following:

- System failure
- Hydraulic problems
- Engine failure/fire
- Inoperable or malfunctioning flight controls
- Gear failure (gear retracted or otherwise unsafe for landing), either hydraulic or mechanical
- Special military considerations (explosives becoming dislodged, ejection seat activated, canopy becoming unattached, etc.)
- Loss of cabin pressure
- Onboard fire
- Bird strike
- Structural failure
- Low or no fuel
- Lighting strike, turbulence, wind shear, and icing (**NOTE:** Although these items are not emergencies in themselves, their effects can cause emergencies.)

While in flight, aircraft frequently develop minor difficulties that may or may not be cause for alarm. The majority of these in-flight problems go unnoticed by occupants because they are not serious enough to cause the aircraft to operate abnormally. An example of this type of incident is a minor electrical short or malfunction in one of the warning systems. A malfunction of this type can cause a fire warning light on the instrument panel to indicate a problem when none actually exists. When a fire warning light activates, the crew tries to determine whether there is a fire by making instrument checks and visual observations. If the pilot-in-command is satisfied that the aircraft is safe and airworthy after these checks have been made, the flight continues normally. If a problem actually exists and an emergency is declared, air traffic control notifies the airport fire department, and ARFF personnel respond to their pre-designated standby locations and await the aircraft. Upon landing, the in-flight emergency switches to a ground emergency and depending on the severity may require a full-scale emergency response.

Hydraulic/Gear Failures

Incidents such as hydraulic failure or inoperative landing gear may seriously jeopardize the safety of the aircraft and its occupants (Figure 10.13). Depending on the severity, the aircraft may experience a variety of flight control problems both while flying and once on the ground. This type of emergency may affect aircraft steering, braking, and/or stopping (Figure 10.14). ARFF responders may want to

Figure 10.13 Aircraft with a collapsed landing gear. *Courtesy of Air Line Pilots Association.*

Figure 10.14 Upon landing, aircraft with landing gear failure may lose steering control. *Courtesy of Air Line Pilots Association.*

consider alternate standby locations when dealing with an emergency of this nature so that the safety of the ARFF crews is not jeopardized.

In-Flight Fires

An interior fire aboard an occupied aircraft is a true emergency, particularly if the fire occurs in flight. Because of the automatic fire detection systems aboard modern aircraft, interior fires are usually detected in their incipient stage.

If the fire is accessible in flight, the flight crew will usually attempt to extinguish it using onboard fire extinguishers. If the fire cannot be handled with the onboard fire protection equipment or if its location is inaccessible in flight, it may develop into a serious fire and spread rapidly. In this case, an emergency landing will be attempted immediately.

Depending on the amount of time it takes to make an emergency landing, heat, smoke, and toxic gases may accumulate, creating a deadly threat to the occupants of the aircraft. If the toxic gases build to a sufficient level, flashover or rollover can occur when emergency exits are opened. It is vitally important that rescue workers vent the aircraft as quickly as possible.

Emergency Evacuation Assistance

Once the aircraft has landed, the flight crew usually initiates an emergency evacuation. ARFF personnel should not impede the egress of occupants and crew in an attempt to enter the fuselage for rescue and/or fire fighting. Personnel must locate and open any other available exits. Additionally, many occupants may not be able to extricate themselves, so ARFF personnel should be prepared to assist *after* all those who are able have exited. Discharging agent onto an aircraft with an interior fire wastes agent. The importance lies in opening up the aircraft. This can be accomplished by using all available exits and ARFF crew members assisting occupants from the escape slides by positioning themselves to the side of the slide and lifting occupants to their feet as they approach the bottom of the slide. If dealing with an exterior fire, personnel should position ARFF apparatus in an effort to keep the fire away from the exits being used for egress. Upon making an interior attack, they should use hose streams for ventilation as well as for extinguishment.

Low-Impact Crashes

Aircraft crashes that do not severely damage or break up the fuselage are likely to have a large percentage of survivors and are generally referred to as *low-impact crashes*. These types of incidents may involve fuel fires, although nonfire incidents are not uncommon. Regardless, the first priority of ARFF personnel is to ensure the safety of occupants and crew. Although fatalities are possible in low-impact crashes, nonfatal injuries of varying degrees are more likely. While occupants are often able to extricate themselves and walk away from low-impact crashes, rescue operations may have to be performed in conjunction with fire suppression efforts if there are trapped and/or seriously injured occupants.

Even in low-impact crashes, ARFF personnel should initiate extrication operations only after donning full protective clothing and self-contained breathing apparatus. In addition, handline teams should back up rescue personnel for protection from a flash fire. Quite often depending on the size of the debris field, handlines will need to be deployed and used during the initial attack of the fire as the turret nozzle stream will be unable to reach the fire area.

Wheels-Up or Belly Landings

One example of a low-impact crash is the wheels-up or belly landing (Figure 10.15). These may result

Figure 10.15 During a low-impact crash, aircraft such as this one that landed wheels-up, often stay intact. *Courtesy of Air Line Pilots Association.*

from a hydraulic system failure or other cause. Fire is not uncommon, although it is not inevitable in these incidents. When an aircraft scrapes along the ground, fuel tanks often rupture, and tremendous heat and sparks can be generated by friction to provide an ignition source. These hazards are usually greater when the plane lands on airport runways rather than on soft ground. In any event, after a belly landing, suppression efforts to minimize ignition are extremely critical.

In landings of this type, it is almost impossible for the pilot to maintain control of the aircraft. Upon touchdown, the plane may break up or veer off the runway. ARFF personnel should accept the uncertainty of where the aircraft will come to rest and stage apparatus a safe distance from the runway to avoid being struck. The aircraft should be pursued only after it has passed staged vehicles.

Following such a landing, a large aircraft may remain substantially intact, and a majority of the occupants may be able to leave the aircraft on their own. If fire does occur, an aggressive attack to keep the fire clear of the fuselage — especially at the exits — is critical. Egress operations will be hampered due to the final attitude of the aircraft. Escape slides are designed for wheels-down evacuations. When the wheels are up and the aircraft is resting on its fuselage, occupants exiting often crowd together at the bottom of the slide, making egress substantially slower.

Ditching

Another example of a low-impact crash is a wheels-up landing on water, known as *ditching*. In these cases, ARFF personnel can often intervene effectively with rescue boats and personnel trained in water rescue to assist in removing people from the aircraft. Saving occupants from drowning may be a significant challenge for ARFF personnel.

A number of airports have large bodies of water either in their approach/departure patterns or in close proximity. Aircraft accidents/incidents in bodies of water may result when an aircraft skids off a runway, lands short, aborts a takeoff, ditches, or crashes. Such accidents may be dangerous and frustrating for ARFF personnel attempting to extinguish a fire and perform rescue operations. The surface of the water may be covered with fuel, which may or may not be burning. If practical, personnel should apply a blanket of foam to the entire area. If the aircraft is partially in the water and has not ignited, ARFF personnel should remain alert when carrying out rescue operations because fuel rising to the surface may contact heated engine parts and ignite.

ARFF personnel should also be aware that aircraft wreckage might be floating because of pockets of air trapped in the top of the compartments. Making an opening at a point above the water level may permit the air to escape and cause the wreckage to submerge before occupants have been removed.

Rescue personnel need specialized equipment to perform rescue operations in water. Inclement weather, especially in winter, can bring on hypothermia that can quickly disable aircraft occupants and rescue personnel. Depending upon air and water temperatures plus the victim's age, physical condition, and extent of injuries, hypothermia may prove fatal in minutes. In cold water, rescue and fire fighting personnel may use special flotation suits that will support two to three additional people. Neoprene wet suits also may be used by rescue personnel; however, the protection they offer is not as good in cold temperatures as that of the dry suits.

For aircraft accidents in swamps, marshes, and tidal flats inaccessible by conventional rescue boats and land vehicles, airboats may be the best alternative. These flat-bottomed, shallow-draft boats need only a few inches of water to operate efficiently, and they are capable of crossing wide expanses of tidal flats on the wet mud.

Although helicopters may be effective in some water rescue operations, in others, the rotor downdraft may push rescuers and flotation devices away from victims.

Rejected Takeoff with Runway Overrun

This type of low-impact crash may result from a sudden loss of power, slippery runway conditions, or the lack of needed runway to stop the aircraft. Again, this type of emergency often leaves the aircraft intact or in large pieces and is usually survivable. Quick response while protecting the egress route is vital to passenger survival.

Helicopter Crashes

The use of helicopters in general aviation has increased significantly; as a result, accidents involving helicopters have also increased. Because helicopters are of relatively light construction, they do not withstand the violent forces encountered in vertical impact. The undercarriage, rotors, and tail units usually break apart, leaving the wrecked interior of the fuselage as the main debris (Figure 10.16). Rotors, which are usually found close to the passenger area, may continue to spin after a crash. Approaching the aircraft while these are still spinning should be avoided. The main wreckage usually contains the engine and fuel tank and should be approached with caution. The hazards associated with fuel tanks and fuel fires are the same for helicopters as for all other aircraft.

High-Impact Crashes

Aircraft crashes with severe damage to the fuselage (fuselage disintegration) and with a significantly reduced likelihood of occupant survival are generally referred to as *high-impact crashes.* At this type of incident, firefighters should see to scene security, protection of evidence, and protection of exposures. By definition, a high-impact crash is an accident in which the fuselage is substantially damaged; the G forces upon the occupants exceed human tolerance levels; or, the seats and safety belts fail to restrain the passengers during the impact. In this situation, an aircraft will often break apart upon impact with the ground or trees (Figure 10.17). Sometimes, hitting obstructions may cause it to cartwheel. If this happens, the main structural components, such as the wings, tail, and undercarriage, may be torn off and scattered over a wide area in the line of approach. Crew members or occupants may be thrown from the aircraft before it comes to rest, so under these conditions, a thorough and wide-ranging search should be carried out for casualties.

Controlled Flight into Terrain

On occasion an aircraft may crash for no apparent reason. When the aircraft does not encounter any type of mechanical or aerodynamic problem that makes it unable to fly, the pilot often by accident will fly the plane into the terrain. Quite often poor weather conditions are to blame for this type of accident. Improper instrument settings, miscalculated computer settings, or pilot distractions have also been known to cause this type of accident. Because this type of crash may occur anywhere, access to the scene may present a true challenge to the ARFF responder.

Hillside Crashes

Aircraft fires on hillsides are sometimes so difficult to reach that fire fighting may be limited to preventing fire spread and performing a thorough overhaul. The aircraft fuel usually scatters over a wide area and burns out, leaving only pieces of burning wreck-

Figure 10.16 Firefighters must be aware of debris from broken rotors. *Courtesy of Metro Washington Airports Authority.*

Figure 10.17 The remnants of an aircraft after a high-impact crash. *Courtesy of Air Line Pilots Association.*

Figure 10.18 Remote crash site. *Courtesy of Air Line Pilots Association.*

age and vegetation (Figure 10.18). However, because of the slope, mass transport of burning material may spread the fire much faster than would otherwise be expected.

Crashes Involving Structures

A crash into a building obviously creates a more complex problem than an accident involving only an aircraft. The first-arriving fire officer must attempt to accurately assess the situation, transmit a clear description, and employ available resources as appropriate.

The aircraft may break open upon impact, and flying debris may damage surrounding properties. Damage to the roofs and upper stories of buildings may occur, floors and walls may collapse or be on the verge of failure, and people inside and outside the affected buildings may be injured. Rescue personnel should search involved properties and evacuate the entire area. Sightseers should be kept as far away from the area as possible.

Almost certainly, the aircraft's fuel tanks will be severely damaged and the contents dispersed. Just after takeoff, a Boeing 747-400 may be carrying as much as 58,000 gallons (240 000 L) of fuel. As soon as possible, ARFF personnel should take steps to prevent fuel from running into gutters and out of the immediate crash area. If fuel enters sewers or storm drains, the intakes must be dammed, and those responsible for these utility systems must be notified as soon as possible. If fuel has entered a waterway, floating booms may be needed to help contain the contamination. Additional agencies, such as the Coast Guard, Department of Fish and Game, and others, may need to be notified.

Rescue personnel should prohibit smoking and take precautions to eliminate other sources of ignition. Fires may be widely separated and may spread rapidly because of scattered fuel, severed gas lines, and damage to domestic electrical systems.

Response Procedures

Each jurisdiction must have established procedures for responding to all types of aircraft emergencies. All firefighters must understand their role in the overall operation so that all necessary functions are accomplished rapidly and effectively. While response procedures vary from jurisdiction to jurisdiction, this section highlights some of the more common procedures that most agencies should incorporate into their standard operating procedures.

Standard Emergency Response

Runway standby positions for ARFF vehicles in anticipation of an emergency should be predetermined in a standard operating procedure. In the event of an emergency, units should go directly to these positions unless directed elsewhere. Responding units should, if possible, have the following minimum information concerning the accident:

- Make and model of aircraft
- Emergency situation
- Response category
- Amount of fuel on board
- Number and locations of occupants, as well as injured, if known
- Nature and location of any cargo of critical significance
- Location of aircraft (if landing, the runway to be used; if crashed, the site)

While time is essential, ARFF personnel must temper their response with discretion, taking weather, visibility, terrain, and traffic into consideration. Promptness *and* safety are equally important response considerations. The fire department section of the airport emergency plan should include response routes to be used unless unforeseen conditions dictate otherwise. This procedure allows all units to anticipate the actions of other units. The following are considerations for selecting these routes:

- Probable accident sites
- Presently available routes (location of frangible crash gates)
- Possible alternative routes
- Design of apparatus (weight, height, width, etc.)
- Load capacity of bridges, ramps, etc.
- Terrain (rough, even, paved, unpaved, flat, hilly, etc.)
- Effects of weather
- Other obstacles

If, for any reason during an emergency response, a driver's vision becomes obscured, the driver should approach the scene using extreme caution to ensure that he or she does not strike fleeing occupants with the moving vehicle. If two people are on board, one person should get out of the vehicle and sweep the area in front of the vehicle to ensure that it is clear and that occupants will not be run over. If the apparatus driver/operator should lose sight of the firefighter on foot, the driver/operator must stop the apparatus immediately to avoid the possibility of running over the firefighter. Response must not be resumed until visual contact has been reestablished. During night operations, flashlights may be needed to direct apparatus safely onto the scene.

They must also respond in a way that avoids damaging the responding apparatus and equipment. They should avoid running over aircraft debris scattered throughout the accident scene. Preserving the accident scene and safeguarding evidence is a responsibility of all ARFF personnel.

Response time to aircraft accidents is critical to initiating an effective rescue effort. The authority having jurisdiction for their respective airport may require the primary airport ARFF apparatus be able to respond from the station to the midpoint of the most distant runway and begin application of extinguishing agent within three minutes of notification. Additional apparatus must be able to respond and begin extinguishment within four minutes. In any case, ARFF personnel should be aware of the standards of response based on the authority having jurisdiction for each respective airport.

Unannounced Emergency Response

An *unannounced emergency* is one that occurs without prior warning. With in-flight (announced) emergencies, ARFF personnel are usually given certain pre-approach information before the aircraft attempts to land. However, in either case, the available information may be sketchy, such as "Aircraft on fire on the approach end of runway one-seven."

Size-Up

The first firefighter or company officer to arrive at the scene should perform a quick size-up. This initial Incident Commander should develop an action plan that allows for the best possible fire attack on the aircraft while remembering egress routes, wind direction, terrain and aircraft attitude. More information on size-up is described earlier in this chapter.

Initial Attack/Fire Control

Existing fire and crash conditions govern the placement of fire fighting apparatus for the initial attack. The main objective during this attack is the rescue of occupants trapped within the aircraft. Fires threatening these areas should be extinguished as soon as possible. Other nonthreatening fires may be left for later-arriving units. At times, it is difficult to distinguish between rescue and extinguishment activities because they are interrelated and are often performed simultaneously.

Two important factors in the initial fire attack and rescue size-up are whether survivors are being evacuated before the fire fighting apparatus arrives and whether the aircraft fuselage is intact. If the flight crew has begun evacuating the occupants, the first-arriving unit should establish a safe exit to permit evacuation to continue and to make sure that the escape chutes remain intact and free of fire. If the fuselage is not intact, more than one rescue area may have to be established. Utilization of extending booms may assist in extinguishing fires in the confined areas of a crash scene. Application methods consisting of low sweeping patterns and the conservation of agent are critical to ensuring both occupant and firefighter safety.

Quickly controlling an area of fire to establish a safe egress area involves initial mass application of an extinguishing agent. In the case of specially designed aircraft fire fighting apparatus, turrets and ground sweeps should be used to control the fire around the exterior of the fuselage. Handlines should be used for backup, interior attack, and overhaul. The initial attack begins during the approach of the fire fighting vehicles. Roof turrets, bumper turrets, and ground sweeps should be used as soon as the vehicles are within range of the aircraft's occupied sections. However, because limited quantities of extinguishing agents are carried on apparatus, turrets should be used only when the agent can be applied without being wasted. The initial discharge of foam should be made along the fuselage in order to prevent fire from impinging on it and to begin to create an exit.

Although structural apparatus may lack specialized delivery systems, they can still be effective on aircraft fires by using aqueous film forming foam (AFFF). Given an adequate supply of AFFF and additional water available from hydrants, relays, or drafting sources, structural apparatus can sustain an effective attack as long as necessary. Wide coverage and considerable heat absorption can be achieved by using larger handlines and master stream appliances with appropriate fog nozzles.

During the control phase of an exterior fire, all efforts should be directed at insulating and isolating the occupied portions of the aircraft. This process is done by concentrating the extinguishing agents on the occupied portions of the aircraft and the surrounding areas. When conditions permit, ARFF personnel should position the apparatus at the nose or tail of the aircraft so they can apply agent on either side of the aircraft (Figure 10.19). Thus, they may keep the aircraft interior acceptable for occupants to survive while rescue personnel remove them through a controlled exit to safety. If fire is confined to the engine nacelles or wings, personnel should attempt to stop the fire at the wing root or engines. If fuel is leaking from fuel tanks and spreading on the ground, personnel should attempt to keep the fire from the fuselage and egress areas at least until the occupants have been evacuated or rescued.

In accidents involving fire or a high likelihood of fire, the initial attack is usually made with one or more units operating both roof and bumper turrets while additional units perform handline operations and interior attack in an attempt to establish a safe area in and around the aircraft exits. This attack is followed up with rescue personnel, protected by handlines, making entry into the aircraft. Water supplies must be adequate to support interior fire fighting operations. ARFF personnel should be familiar with the cabin length of the aircraft prior to entry to ensure that ample hose line is taken into the aircraft. In nonfire accidents, the same basic procedures should be followed. Instead of fighting fire, however, firefighters must blanket fuel spills with foam and charged handlines. At the same time, vehicle turrets must be kept ready in case fire erupts.

Ventilation

An aircraft with an interior fire may be ventilated the same as any other enclosed structure. Proper ventilation clears smoke, provides a safer atmosphere for effective rescue, helps firefighters locate deep-seated fires, and facilitates overhaul operations.

Conventional methods of ventilation for fire fighting also apply to aircraft ventilation. Cutting ventilation openings on a fully loaded cargo aircraft with an interior fire is a good tactic to reduce horizontal fire spread and can be done safely from an elevated platform. However, cutting ventilation openings in an aircraft fuselage is not recommended because it is time-consuming, may be dangerous, and should be considered as a last resort (Figure 10.20). Using existing openings, such as doorways and emergency exits, for ventilation is much faster and more efficient if they provide enough open area. If the doors and hatches are inaccessible, personnel must consider using other emergency access or cut-in areas.

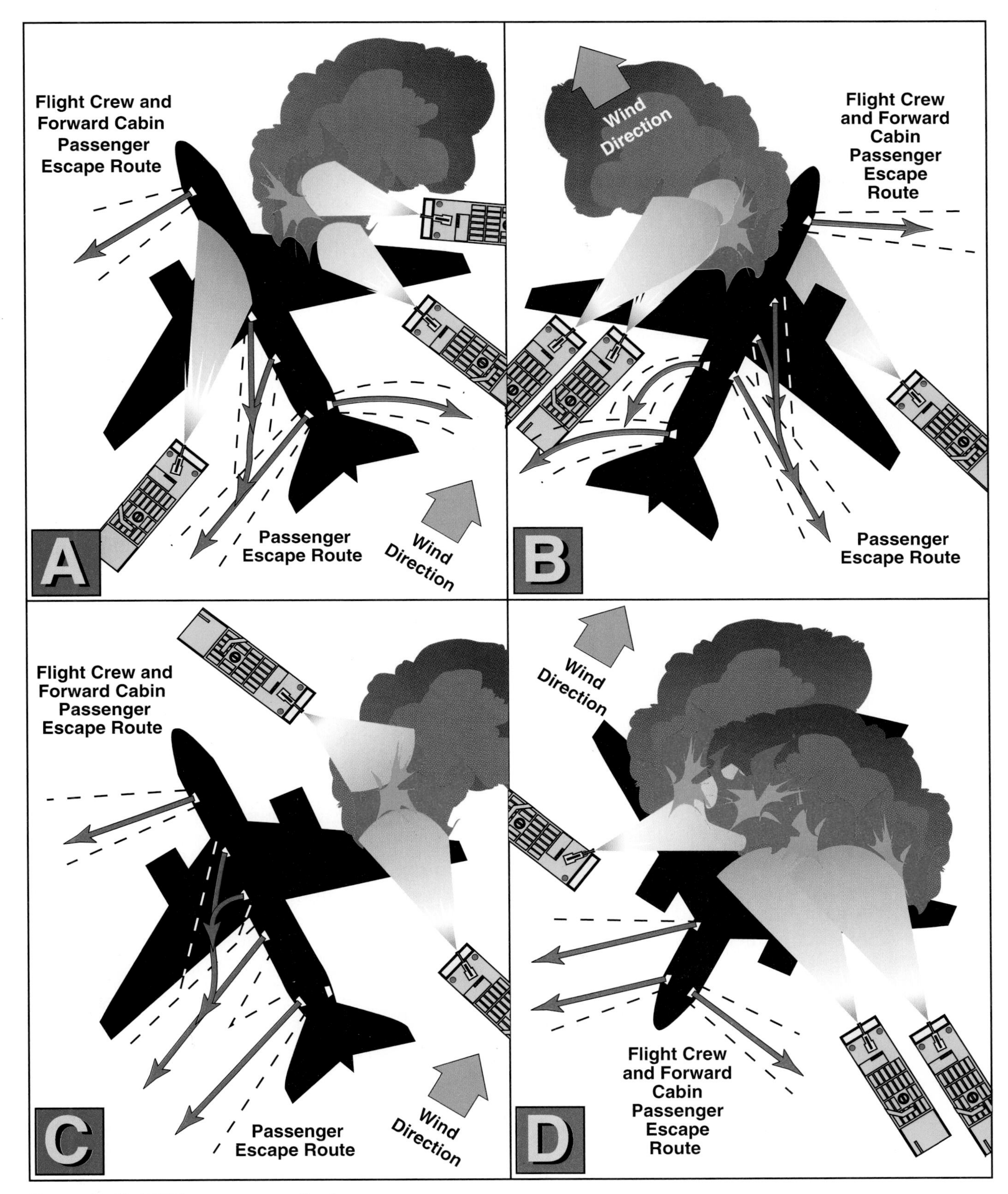

Figure 10.19 ARFF apparatus positioning.

Figure 10.20 Using gas-powered saws can be time-consuming and dangerous. *Courtesy of Jim Nilo, Richmond International Airport.*

WARNING

Beware of the potential hazards of high-pressure hydraulic lines; compressed-gas cylinders; pneumatic lines; and on military aircraft, unexploded ordnance, if it is necessary to penetrate the skin of any aircraft in areas not marked as cut-in areas.

Positive-pressure ventilation (PPV) is an effective means of clearing heat and smoke from inside the aircraft. For more information on positive-pressure ventilation techniques, see the IFSTA **Fire Service Ventilation** manual.

If the situation makes PPV impractical, negative-pressure ventilation may have to be used. Fans placed at various doors and windows are used to draw smoke from the aircraft interior with this method.

Personnel working in the aircraft interior should have a charged hose line with combination nozzle on hand to deal with any re-ignition or spot fires that could occur as a result of ventilation operations or self-ventilation.

Rescue

Initially, all resources at an aircraft incident should be directed toward rescuing occupants of the aircraft. After rescuing passengers and crew, ARFF personnel can concern themselves with saving property and equipment. Rescue apparatus should be positioned so that all rescue and forcible entry equipment is as near to the probable point of entry as possible without endangering the vehicle. This positioning will facilitate taking additional equipment from the apparatus to the accident as needed.

The easiest and quickest way for rescue personnel to gain access to an aircraft is through normal doors and hatches. These openings usually have external releases. The same rule of forcible entry in structural fire fighting also applies to aircraft: *try before you pry* — try the normal means of opening the door before attempting to force it open. If occupants are attempting to exit the aircraft, these portals should be left to them. ARFF personnel should enter by other means, usually through the emergency exits.

CAUTION: ARFF personnel inside an aircraft must have effective communications with those working on the exterior. Conditions on the outside of the aircraft may quickly change, such as a sudden exterior flashback, and personnel operating on the inside could be jeopardized.

As discussed in Chapter 4, "ARFF Firefighter Safety," the two-in/two-out rule also applies to aircraft rescue and fire fighting. A minimum of two rescue personnel should be used for all rescue operations at any single point of entry. Rescue personnel should approach and enter the aircraft by way of the exit area established by the turrets and/or handlines. They should be extremely alert when attempting to remove survivors from an aircraft that is surrounded by widespread fire because flashbacks may cut off the route of escape.

WARNING

During rescue, do not move crash survivors from a clean survivable atmosphere into one in which they cannot survive without protective clothing.

All larger commercial passenger aircraft and some military aircraft are equipped with escape chutes or slides (Figure 10.21). When deployed, these slides either inflate automatically or can be

Figure 10.21 Cargo aircraft also are equipped with escape slides. *Courtesy of Air Line Pilots Association.*

manually activated to inflate by a crew member or passenger. If these chutes or slides are provided and are in use when ARFF units arrive, they should not be disturbed unless they have been damaged by use or are threatened by fire exposure. The slides normally provide a much faster means of evacuation than do steps or ladders.

Occupants of civilian airliners are held in their seats by one lap belt. Releasing this belt frees the occupant of any restraint. Crew members, on the other hand, have a combination lap belt and shoulder harness. The shoulder harness straps fit into the release mechanism of the lap belts. To release the lap belt and shoulder harness straps, simply operate the lap belt release mechanism. If the mechanism fails to release, the straps may be cut with a belt cutter. Disconnect or cut all straps or retention devices holding the occupant; disconnect all personal service connections, and remove the victim.

If the pilot is wearing a parachute, the standard harness may be removed by releasing the strap across the wearer's chest and the strap across each leg at the thigh. When all straps have been released or cut, the harness will slip off the wearer's shoulder. If the wearer has a suspected back injury, the parachute may be left on to support the wearer's back. Whenever possible, using an extrication board or backboard is preferable. On some ejection seats, the parachute is part of the seat. In this case, the firefighter should leave the harness on the pilot as it provides handholds that can be used during extrication.

Once rescue operations are under way, other operations may begin as resources allow. Fire prevention and/or extinguishment efforts should take place, and the aircraft's batteries should be disconnected to reduce sources of ignition. Do not rely upon cockpit switches to achieve a "power off" condition. Batteries should be disconnected and the terminals taped so there is no danger of accidental contact if the aircraft is moved. Most aircraft batteries can be connected or disconnected by simply turning a small handwheel referred to as a *quarter-turn quick disconnect.* These wheels are clearly marked, and the directions are given on adjacent panels. Accident investigators need to know the position of all switches at the time of the accident, so if ARFF personnel do flip any switches, they should note it in their incident report.

Extinguishment

The extinguishment phase is merely an extension of the fire control phase because the control phase includes maintaining an escape exit from fire and, whenever possible, completely isolating the occupied portion of the aircraft. As additional resources become available, either because personnel are no longer needed for rescue operations or because additional units have arrived, the area already secured should be expanded outward to the perimeter of the fire area.

This phase is the final effort prior to overhaul; therefore, extinguishment of surface fires must be complete in order to avoid further fire damage and to secure the area for overhaul. Complete fire extinguishment should not be attempted if evacuation and rescue operations would have to be reduced. However, conducting the extinguishment phase concurrently with rescue may be justified by the situation and the amount of apparatus and manpower available.

Extinguishment involves the elimination of all surface fire, whether on the ground or inside the aircraft. Even foamed areas should be examined and additional foam applied wherever the foam blanket has been compromised. *Crash debris should only be moved if absolutely necessary for rescue. If possible, it should be photographed or at least documented for future reference before it is moved.*

During this phase of the operation, reserve apparatus and equipment may be pressed into service. Additional mobile water-supply vehicles or structural pumpers in relays may be used to replenish depleted water supplies. Special lighting and air-supply units may be needed. After the responsible investigator provides authorization, wreckers and heavy equipment may be used to move parts of the wreckage to ensure thorough extinguishment. Special-purpose vehicles designed to carry mass quantities of medical supplies may also be needed at this time.

Overhaul

After every aircraft incident/accident, a thorough overhaul inspection must be conducted, regardless of whether fire was apparent or not. Because of the possibility of a toxic atmosphere existing within the aircraft and the possibility of a sudden flash fire occurring, personnel should wear protective clothing, including SCBA, during overhaul operations.

The largest aircraft in operation, such as the 747, 767, A300, A340, 777, L-1011, DC-10, MD-11, and C5, present unique problems with their very large areas and extreme heights. The extraordinary size of these aircraft increases the opportunity for concealed-space fires in their interiors. Physically locating and reaching the seat of the fire may become a problem.

As always, the on-scene investigating authority should be consulted before overhaul operations begin. During overhaul, ARFF personnel must make sure that all fire is completely extinguished. This phase of aircraft interior fire fighting is one of the most difficult and is also one of the most hazardous. Because toxic gases and fumes are concentrated and other hazards may be present, firefighter safety is a major concern. To protect themselves, ARFF personnel should wear SCBA until the atmosphere in which they are working has been checked with the appropriate gas and particulate detectors and declared safe. In addition, a charged handline must be kept close at hand.

Because of the configuration of aircraft interiors, carpeting, wall panels, partitions, and ceiling coverings may need to be removed to get to deep-seated, concealed fires. During the overhaul phase of the operation, interior crews need to exercise extreme caution to ensure that any fire that has extended in the void space over the ceiling panels is not allowed to extend and come down behind them.

Care should be taken to preserve as much of the interior in its original configuration as is reasonably possible. This process will assist in determining the origin and cause of the fire and will facilitate the investigation. If ARFF personnel must remove wall panels or disturb other items, they should make descriptive notes or take photographs to indicate the original position of the items.

While visible smoke and/or steam usually indicates the location of hot spots, it may be necessary to use the back of the hand on the surface of the aircraft to find hot spots if infrared heat detectors are not available. Firefighters may have to open some parts of the wreckage for complete extinguishment. Whenever the skin of the aircraft is penetrated, however, ARFF personnel should consider the potential hazards of cutting into high-pressure hydraulic lines, compressed gas cylinders, pneumatic lines, and unexploded ordnance on military aircraft. All hot spots should be cooled until extinguishment is complete and re-ignition no longer occurs.

During overhaul, personnel should avoid disturbing any evidence that may aid investigators in determining the cause of the accident or the extent of damage while ensuring personal protection against bloodborne pathogens. Overhaul personnel should move only those parts of the aircraft that are absolutely essential to complete fire extinguishment. If the aircraft or its parts and controls must be moved because they present a direct hazard to human life, every effort must be made to preserve physical evidence and record the original condition and location of whatever was moved. ARFF personnel should be familiar with their fire department's SOPs that cover this area of operations. The FAA furnishes general guidance for preservation of evidence in Advisory Circular 150/5200-12B, *Fire Department Responsibility in Protecting Evidence at the Scene of an Aircraft Accident*. Also, the National Transportation Safety Board (NTSB) provides general guidelines for handling civil aircraft accidents with which ARFF personnel should be familiar.

Only authorized personnel should remove bodies that remain in the wreckage after the fire has

Figure 10.22 Identifying victim remains is a critical part of the investigation. *Courtesy of Air Line Pilots Association.*

been extinguished. Prematurely removing bodies may interfere with identifying them and may destroy evidence required by the medical examiner, coroner, or other investigating authority. If it is absolutely necessary to remove a body prior to the arrival of the medical authority, ARFF personnel should tag each body with a number or secure a stake to note where the body was found (Figure 10.22). They should note on the tag the location from which the body was removed and also record that information on a drawing of the aircraft accident site in their incident report. This information will be critically important in the accident investigation.

Response to Accidents Involving Military Aircraft

Upon notification that an accident involving military aircraft has occurred, the nearest military installation should be contacted. After being notified, the military will dispatch assistance teams that usually include the following personnel:

- Base fire department personnel
- Explosive ordnance disposal (EOD) personnel
- Military police
- Medical personnel
- Bioenvironmental personnel
- Mortuary personnel
- Information officer
- Accident Investigation Board
- Legal officer
- Heavy-equipment personnel

When military officials arrive at the scene of an aircraft accident, they will need to obtain the following information from witnesses to assist in the investigation of the accident:

- Time of the accident
- Direction in which the aircraft was headed
- Weather conditions at the time of the accident
- Whether anyone was seen parachuting from the aircraft
- Whether there was an explosion in the air prior to the crash

A military crash site contains hazards identical to those of a civilian crash site. Military aircraft should be considered somewhat more hazardous due to the additional systems and devices found onboard.

One added hazard involves the fuels used to power and operate these aircraft. Many military aircraft use a varied mixture of jet fuel, which has a flash point significantly lower than civil aviation fuel. Another hazard associated with military aircraft and fuel systems encompasses a fuel described as hypergolic fuel. Military aircraft may be equipped with an emergency power unit (EPU) instead of an APU. The EPU may utilize hydrazine, a hypergolic fuel, as the fuel supply for the emergency power unit instead of normal jet fuel. Firefighters may not know whether or not the system was used prior to or during the response. Inhalation, ingestion, and absorption hazards may be present when working with these alternative fuels. *Hypergolics* are substances that ignite spontaneously on contact with each other (such as hydrazine with an oxidizer). For example, the F-16 uses H-70, which is 70 percent hydrazine and 30 percent water. Those aircraft that use hydrazine carry a minimum of 7 gallons (28 L). The need for a highly reliable and quickly responsive way of obtaining emergency electrical and hydraulic power aboard aircraft is likely to increase the use of hydrazine. Hydrazine has an odor similar to ammonia, is toxic in both liquid and vapor form, and may explode. It is a strong reducing agent and is hypergolic with some oxidizers such as nitrogen tetroxide and the metal oxides of iron, copper, and lead. Auto-ignition may occur if hydrazine is absorbed in rags, cotton wastes, or similar materials.

WARNING

Always wear full protective clothing when dealing with hydrazine emergencies as it may be absorbed through the skin. Even short exposures may have serious effects on the nervous and respiratory systems.

NOTE: Refer to Technical Manual TO 00-105E-9, which is accessible at the website, http://www.robins.af.mil/ti/tilta/documents/to00-105e-9.htm (case-sensitive address). Chapter 11, "Airport Emergency Plans," also contains additional information regarding procedures to follow if faced with a military aircraft accident.

Aircraft Accident Victim Management

When considering a high-impact crash, emergency medical services (EMS) most likely will not need to be addressed. Injuries will be localized to rescuers accessing the scene while performing body recovery and accident investigation. Depending on the time of year, either heat-related or cold-related medical problems may need to be prevented.

A low-impact crash presents rescuers with the greatest challenge in treating and transporting what could be a very large number of victims. To perform this function efficiently, a system should be utilized that allows rescuers the ability to triage, treat, and transport victims in a short period of time. Factors that weigh into the system include time of day, time of year, location of accident, and availability of resources. The time of day dictates the number and type of resources that rescuers can plan on being available. The time of year relates to responding agencies preparing for the type of resources they will need to protect occupants and firefighters from unpredictable weather conditions.

When treating victims, personnel should first ensure personal protection against bloodborne pathogens and then initiate a triage system that can be performed quickly. A colored ribbon or tag may be attached to the victim. The level of urgency is indicated by the color of the ribbon or tag. Green represents low priority or walking wounded, yellow is medium priority, and red is high priority. Victims who are deceased should be marked with a black ribbon or tag so that they are not rechecked at a later time.

Once triaged and tagged with the level of urgency, victims should be moved to a treatment area, with the high-priority victims being moved first (Figure 10.23). The treatment area should be located upwind and uphill from the crash scene to avoid drifting smoke and running-fuel hazards. At the treatment area, the patient should be re-evaluated and placed into the appropriate area for treatment to begin. Means of transporting victims to the hospital should be specified in the airport emergency plan and may consist of helicopters, buses, and ambulances (Figure 10.24).

Figure 10.23 Patient triage is an important part of victim management. *Courtesy of William D. Stewart.*

Figure 10.24 Patient transportation should be arranged by the triage/transportation officer. *Courtesy of William D. Stewart.*

An area should be identified where the green-tagged victims or walking wounded can be transported and treated. Benefits to removing them from the scene include preventing them from returning to the crash site and isolating them from the crash scene for mental health reasons. Victim isolation is also important to ensure the victims are not bothered by members of the press or by attorneys seeking to represent them in legal matters.

An important resource during the entire operation may be the flight crew of the aircraft. If physically able, the flight crew can provide needed information regarding the aircraft, cabin, and occupants. Flight crews are often trained in emergency first aid along with other life-saving training.

Airports are required to develop an AEP or Airport Emergency Plan, which outlines the following resources needed for either type of crash. These include: food services for the rehab area, portable lighting, drinking water, rest room facilities, coroner and mortuary services, blankets and medical supplies, communication facilities, and a procurement system that has been pre-established. On-scene operations can often extend for days. Airport management must take appropriate measures to rotate personnel. An aviation disaster will test any emergency response system, but with proper logistical planning most of the challenges can be planned for and resolved prior to responding. This pre-incident planning will prove crucial in providing the resources needed to save lives as well as protect the emergency response worker. Chapter 11 discusses the airport emergency plan in more detail.

Another critical part of emergency preparedness involves critical incident stress and the effects that personnel may experience during and after an aircraft disaster. Chapter 4 covers with detail the precautions and procedures for dealing with these types of issues.

Chapter 11

Airport Emergency Plans

Job Performance Requirements

This chapter provides information that will assist the reader in meeting the following job performance requirements from NFPA 1003, *Standard for Airport Fire Fighter Professional Qualifications*, 2000 edition. Particular portions of the job performance requirements (JPRs) that are addressed in this chapter are noted in bold text.

3-2.2 Communicate critical incident information regarding an incident or accident on or adjacent to an airport, given an assignment involving an incident or accident and an incident management system (IMS) protocol, so that the information provided is accurate and sufficient for the incident commander to initiate an attack plan.

(a) *Requisite Knowledge*: Incident management system protocol, the **airport emergency plan**, airport and aircraft familiarization, communications equipment and procedures.

(b) *Requisite Skills*: Operate communications systems effectively, communicate an accurate situation report, implement IMS protocol and airport emergency plan, recognize aircraft types.

3-3.10 Preserve the aircraft accident scene, given an assignment, so that evidence is identified, protected, and reported.

(a) *Requisite Knowledge*: **Airport emergency plan requirements for preservation of the scene.**

(b) *Requisite Skills*: Preserve the scene for investigators.

In preparation for the occurrence of an aircraft accident/incident, ARFF personnel should create a plan of operation to develop appropriate procedures and identify needed resources. Such a plan addresses the need for a coordinated response to emergency situations within the airport property along with the surrounding local jurisdictions. The plan should be as complete and detailed as possible to ensure that all involved agencies are aware of their roles and responsibilities under various conditions (Figure 11.1).

Figure 11.1 Emergency response plans should address incidents both on and off the airport. *Courtesy of Michael T. Defina, Jr., Metro Washington Airports Authority Fire Department.*

Developing an airport emergency plan (AEP) is not an end in itself, nor is it a guarantee for an effective emergency response. However, airport emergency planning helps reduce the confusion that often exists during emergency operations. The efficiency with which an emergency operation is handled may depend upon how well the involved agencies have planned. Beyond that planning, the success of the operation depends upon how well those involved understand the plan and how well they execute it. The plan will only be successful if training is designed and conducted to address each part of the plan. This is best accomplished by conducting small exercises that focus on each specific task of response coordination. Upon completing the small exercises, a large-scale drill should be conducted to exercise the complete plan, and discrepancies should be noted.

Structural fire departments with little or no direct airport responsibility should also plan for the possibility of an aircraft accident in their response area. Their plans should reflect all conceivable contingencies within their boundaries, as well as their role in mutual aid agreements with airport fire departments. Because of the many variables, all agencies involved must recognize the need for validating their plan at least annually through joint training exercises. Structural firefighters providing protection to areas around airports must be knowledgeable in aircraft rescue and fire fighting. They should train with aircraft rescue firefighters on a regular basis and be included in airport emergency planning and exercises.

The airport emergency plan also addresses rescue and fire fighting operations, guidelines for communicating with the news media, the legal obligations of personnel involved, and the joint training necessary to implement and maintain the plan. Once a plan is written, it can be tailored to address such emergencies as severe weather, threats of terrorist activities, hijacking and bomb threats, as well as major structural emergencies. This chapter discusses airport emergency planning and the many considerations involved in AEP development. A joint action plan involving numerous agencies may be based on the information provided in this chapter plus information available in the following publications:

- NFPA 424M, *Airport/Community Emergency Planning*
- NFPA 402, *Aircraft Rescue and Fire Fighting Operations*
- NFPA 1561, *Fire Department Incident Management System*
- ICAO *Airport Services Manuals*, Part 1, Part 5, and Part 7
- *FEMA Disaster Planning Guidelines for Fire Chiefs*, Federal Emergency Management Agency (FEMA)
- FAR Part 139.325, *Airport Emergency Plan*
- FAA Advisory Circulars 150/5200-12B, *Fire Department Responsibility in Protecting Evidence at the Scene of an Aircraft Accident*, 150/5200-31A, *Airport Emergency Plan*, 150/5210-2A, *Airport Emergency Medical Facilities and Services*, 150/5210-7C, *Aircraft Rescue and Firefighting Communications* and 150/5210-13A, *Water Rescue Plans, Facilities, and Equipment.*
- Air Line Pilots Association (ALPA) *Guide to Accident Survival Factors*

In many cases, a single, detailed plan can serve the needs of all agencies involved. In others, a single, all-inclusive plan may be too cumbersome. In these cases, a parent document supplemented by a number of individual elements or annexes may be required, each defining the functions of all agencies under a given set of conditions or the responsibilities of a given agency under all conditions. All plans, whether simple or complex, should identify and reflect potential needs, local resources, and available mutual aid.

While a number of different formats serve equally well, some general considerations are common to all plans. Sections that should be common to all plans include:

- Introduction
- Aircraft incidents and accidents
- Bomb incidents
- Structural fires
- Mass-casualty and/or fatality incidents
- Natural disasters
- Sabotage/terrorism/hijacking
- Hazardous materials emergencies
- Disabled aircraft removal
- Civil disturbances
- Water rescue situations
- Unauthorized aircraft movements

The plan should identify the steps needed to mitigate each of these emergencies. In preparing a plan or plans, planners should answer the questions of who, what, when, where, why, and how. Some of the variables that must be considered when planning for possible accidents/incidents include the following:

- Types of aircraft common to the areas under consideration
- Possible types of accidents (risk analysis)
- Possible accident/incident sites (on and off the airport)
- Accident site accessibility
- Climatic considerations
- Emergency response notification (recall of off-duty firefighters)
- Mutual aid support
- Available apparatus, equipment, and water supply
- Rehabilitation
- Critical incident stress debriefing
- Communications
- News media
- Government agencies
- Rescue and fire fighting operations
- Accident reporting
- Training for mutual aid support personnel
- Support agencies and organizations
- Joint training exercises
- Accidents involving military aircraft

Each jurisdiction may organize its plan according to local needs.

Types of Aircraft

The type and number of aircraft involved in an accident, to a large extent, determine the kinds and quantities of ARFF resources required. ARFF personnel handle accidents/incidents involving general aviation aircraft in the same way they do those involving commercial aircraft — with the exception that there may be a significant difference in the number of passengers and the amount of fuel aboard. However, incidents/accidents with agricultural aircraft, which are in the general aviation category, may dictate an automatic haz-mat team response because of the potential of a chemical hazard. Military aircraft can pose some unique hazards regarding specialized types that often carry munitions. ARFF personnel should be familiar with the military aircraft that fly in their area, along with base locations and contact numbers in the event of an aircraft accident.

Types of Accidents/Incidents

As specifically defined in Chapter 10, "Aircraft Rescue and Fire Fighting Tactical Operations," ARFF responses vary depending on the type of emergency. Understanding the difference between an incident and an accident will help personnel define what equipment is needed and the operations that will need to be conducted. ARFF personnel respond to aircraft accidents and incidents that are either declared or undeclared prior to the aircraft landing. In most cases, aircraft incidents do not create an immediate risk to the occupants of the aircraft or to ARFF personnel. Some accidents may involve survivors; others will not. In a low-impact crash, if egress is not blocked by fire, fatality rates tend to be low (Figure 11.2). In high-impact crashes, major aircraft structural damage results, and fatality rates are significantly higher (Figure 11.3). In both types of accidents, ARFF personnel should concentrate on the number one goal — rescue. If ARFF personnel are unable to rescue occupants, their next goal

Figure 11.2 A low-impact crash such as this one may result in few or limited injuries. *Courtesy of Air Line Pilots Association.*

Figure 11.3 Survivors of a high-impact crash are unlikely, as this debris from a high-impact crash illustrates. *Courtesy of Air Line Pilots Association.*

is to provide fire control in an effort to provide survivors with a means of escape. (**NOTE:** See Chapter 10, "Aircraft Rescue and Fire Fighting Tactical Operations," to better understand the tactics used to accomplish this task.) Providing effective fire control aids in protecting possible survivors while preserving crash-scene evidence.

A large percentage of aircraft accidents result in fire because of the large quantities of fuel and numerous ignition sources that are present. Because aircraft fires spread rapidly, ARFF personnel must respond quickly and take appropriate action to increase the chances of successful rescue and fire fighting operations. As discussed in Chapter 12, "Hazards Associated with Aircraft Cargo," aircraft accidents/incidents sometimes involve hazardous materials such as fuel, chemicals, radiological and/or etiological agents, and explosives. ARFF personnel must be prepared to deal with these situations as well as those involving threats of bombs, sabotage, and hijacking.

Possible Accident Sites

Although accidents may occur at any time or place, statistics show that the greatest potential for accidents is during the landing/takeoff phase of the flight. A high percentage of all aircraft accidents/incidents occur on or near airport property and generally in the threshold or departure area of the runway. In these areas, aircraft may be taxiing, taking off, approaching, and landing while at the same time, other aircraft may be moving about, being fueled, and/or serviced. Accidents/incidents on airport property must be anticipated and appropriate plans developed. There is also a need to plan for accidents/incidents occurring away from the airport and in urban, suburban, or rural areas (Figures 11.4 a and b).

Figures 11.4 a and b Reaching a crash in a rural location is less difficult if such areas are considered during emergency planning. *Courtesy of Air Line Pilots Association.*

Grid maps of the airport and surrounding areas should be prepared as part of airport emergency planning. All concerned with aircraft emergency service rescue and fire protection must have up-to-date, standardized grid maps of the airport and surrounding areas within a 5- to 15-mile (8 km to 24 km) radius. The maps should show access routes and key locations such as water supplies, medical facilities, staging areas, and heliports. They should also identify roads, bridges, perimeter gates, and any other pertinent features of the terrain that could prevent or delay response.

Accident Site Accessibility

Personnel may determine the areas in which accidents/incidents are likely to occur by studying the arrival and departure traffic patterns. In a planning survey of potential accident sites, ARFF personnel should check access to these areas. Obstacles to apparatus response and access to primary and auxiliary water supply sources can be identified through ground and air surveys. Alternative response routes may then be developed to facilitate rescue and fire fighting operations.

The size and condition of roads and bridges may restrict access. Underpasses and bridges may impose height, width, and weight restrictions. There also may be various types of fences and gates that restrict access. For example, apparatus would have difficulty reaching a location that is next to the airport but behind a railroad embankment. Without advance planning, vehicles might have to detour several miles to reach the location and lose valuable time in the process.

A variety of terrain features may exist that may impede response capabilities (Figure 11.5). Routes should also be planned to bypass ditches, fenced-in areas, wooded areas, streams, marshes, and swamps that may block or impede the passage of heavy apparatus. The nature of the ground surface plays an important part in rescue and fire fighting operations. The local terrain may also influence how personnel perform rescue operations, as well as which extinguishing methods of agent application they choose.

Advanced planning should include possible routes over areas in which no roads exist. In some cases, personnel can make minor changes during the pre-planning stage to permit vehicles to access a site. For instance, grading out an approach and stabilizing the streambed may make it possible for vehicles to cross shallow streams. In other situations, it may be necessary to construct roads into inaccessible areas that are likely to be accident sites.

Figure 11.5 Rough terrain can hamper the delivery of adequate water supply and resources. *Courtesy of Metro Washington Airports Authority.*

Climatic Considerations

The effects of various weather phenomena peculiar to the area, especially on substandard roads and off-the-road areas, should be considered. Climatic conditions such as wind, rain, sleet, and snow may delay or prevent response to the accident site (Figure 11.6). Terrain that normally accommodates heavy vehicles may turn into a quagmire, causing rescue and fire fighting operations to be bogged down or otherwise hampered. In some cases the use of heavy vehicles and equipment may become impossible.

Figure 11.6 Weather conditions, such as the snow at this rural location, can delay response vehicles. *Courtesy of William D. Stewart.*

During plan development, considerations should be made for protecting aircraft occupants along with all emergency responders from harsh weather conditions. Portable shelters may need to be erected at the crash site if extended operations are anticipated. Depending on the conditions, rescue personnel may need to be transported between the crash scene and a climate-controlled rehabilitation area. If weather conditions dictate, blankets should be available to occupants, especially if delays in transportation from the site are encountered.

Emergency Response Notification

Planners should identify the methods for alerting emergency response and support personnel: horn, siren, or other audible alarm system; telephone or phone pagers; or radio, including portable and/or mobile. Those responsible for completing airport emergency plans should compile an emergency contact list with emergency telephone numbers for key personnel assigned to both the primary or secondary response. Automated dialing systems can be used to initiate an automatic recall of all off-duty fire fighting personnel. Telephone numbers (including pagers) should include those that can be used to reach these personnel at all hours of the day and night, including weekends and holidays. It is critical to keep the personnel recall roster updated to ensure that personnel are able to be located in the event of an emergency.

Primary Response

The primary response often incorporates those agencies that are notified on the initial call from the air traffic control tower. When the crash phone or crash network is activated, each of these agencies initiate response operations. The nature and magnitude of the emergency often dictate their level of response. At the very least, the primary response should include the following personnel and their telephone numbers:

- Aircraft rescue and fire fighting services
- Police departments
- Emergency medical services
- Air carrier/owner (that is, the tenants concerned or a representative of the aircraft operator)

Police

Crowds of unauthorized persons frequently collect near the site of an aircraft accident/incident to watch emergency operations, to take pictures, to collect parts of the wreckage as souvenirs, and also to inquire about family members or friends who may have been involved. Because the presence of these individuals may hinder rescue and fire fighting operations, traffic and crowd control are major concerns. A primary concern is to cordon off the immediate crash site to ensure that only authorized personnel are permitted to enter. Quite often, commercial aircraft carry very large sums of money, which emphasizes the need for law enforcement assistance (Figure 11.7). In addition, law enforcement personnel may have primary responsibility for conducting large-scale evacuation. Therefore, it is important that local, county, and state or provincial law enforcement agencies participate in airport emergency planning to help define and clarify roles and to standardize operating procedures. Law enforcement agencies may also provide specialized equipment and personnel, such as bomb-disposal units.

Figure 11.7 State and local law enforcement may be needed at a crash site. *Courtesy National Transportation Safety Board.*

Emergency Medical Services

Well-trained emergency medical personnel may be needed for triage, emergency care, and transportation of the injured; therefore, planning for emergencies, such as aircraft accidents, requires participation by hospitals and medical response personnel, too. The airport emergency plan should identify the abilities and limitations of all participating medical facilities to accept patients in various injury categories. Arrangements might be made to use an airport facility, such as a hangar or terminal building, to accommodate passengers and crew members with minor injuries. If such a facility is available, it should be secured, with access limited to authorized persons and family members.

Patient transportation. At accidents/incidents involving large passenger aircraft, there may not be enough ambulances immediately available to transport injured persons to hospitals. Identifying the location and availability of mutual aid ambulances—including air ambulances—during airport emergency planning can be critical to the survival of crash victims (Figure 11.8). Local military bases are often excellent resources for providing transport helicopters (medevacs) and medical transport vehicles. Plans should also identify available buses and other non-emergency vehicles that could be used to transport ambulatory or uninjured persons.

Equipment. Airport emergency plans should identify the needed quantities of the following items that should be available for responding emergency medical teams:

- ***Medical and first-aid supplies and equipment.*** Incidents that involve large numbers of victims will require more medical supplies than are typically carried on ambulances and fire vehicles. Most airports have a cache of medical supplies in a designated trailer or vehicle that can be transported to the scene when needed.
- ***Rope or barrier tape for cordoning off an area.*** In addition to keeping bystanders and the media out, this equipment can be used to designate various incident-specific areas, such as triage, treatment, CISD, rehab, and other areas. Some jurisdictions carry different colors of boundary tape for these purposes.

Figure 11.8 Air ambulances play a critical role in transporting patients to the hospital. *Courtesy of Michael T. Defina, Jr., Metro Washington Airports Authority Fire Department.*

- ***Wooden/metal stakes or folding barricades.*** These are used to mark body parts, deceased victims, or other important evidence.
- ***Folding cots and blankets.*** These may be used to treat victims who are awaiting transportation to a medical facility.
- ***Body bags.*** These are needed to remove deceased victims at the appropriate time.
- ***Triage tags and marking pens.*** These are necessary early in the incident for classifying victims according to their treatment priority.

Field hospitals. The emphasis in patient care following most aircraft accidents is on immediate medical stabilization of the injured and timely transportation to the nearest medical facility. However, planners should consider the possibility that an accident may involve such large numbers of injured that immediate transport is not possible. In these cases, it might be necessary to establish a temporary field hospital at the scene to stabilize and maintain patients until transportation is available.

Such a facility would be organized and equipped to provide care beyond that of basic first aid and would be staffed with qualified medical personnel who could perform triage and prioritize transportation of the injured to hospitals. Because the possibility of a multiple aircraft accident must be considered, a field hospital unit should have at least enough supplies to handle an accident involving two of the largest aircraft that use nearby airports. Once again, if available, the military is often an excellent resource for providing transportable field hospital facilities and personnel to support their deployment. Emergency planners will need to involve local emergency management representatives to establish setup procedures and define resource requirements. Local hospitals should also be involved in the planning process to help make decisions about the best use of their resources in such emergencies.

Air Carrier/Owner

Airline personnel can provide concise information regarding the number of occupants, along with known quantities of hazardous materials. In addition, many air carrier organizations have developed family-assistance resource teams that can be deployed and used as part of an emergency response. Establishing a resource directory of all airline station managers will expedite the contact process.

Regardless of the size and magnitude of a disaster, in order for an emergency plan to be effective, it is critically important that the emergency notification list is kept up-to-date. Individuals on the list move or are reassigned, phone numbers change, and different people assume responsibility for various aspects of the response system. Unless some means is developed for monitoring and disseminating these changes, the list will inevitably become outdated.

Secondary Response

As the primary response units arrive on the scene, the secondary response network is activated. Much like the primary response, the secondary response network should expand and contract as dictated by the magnitude of the emergency. At the minimum, the secondary response should include:

- Mutual aid fire resources if needed
- Airport officials and maintenance personnel, where applicable
- Appropriate governmental agencies having responsibilities (such as the FAA, NTSB, Canadian Transportation Accident Investigation and Safety Board (CTAISB), Federal Bureau of Investigation, Post Office Department, Department of Defense)
- Emergency management agencies (Red Cross, for example)
- Coroner/medical examiner
- Critical incident stress debriefing (CISD) team
- Clergy

Mutual Aid Support

Airports should have written mutual aid agreements with surrounding fire and police departments and other emergency service organizations. In an aircraft accident — especially one involving fire — time is of the essence in rescue and fire fighting operations. The amount of resources available quickly may make a significant difference in the final outcome of such an incident (Figure 11.9). In order to coordinate the efforts of all those who may be involved, planners should meet with representatives of all entities as a group to define roles, identify resources, and develop procedures.

Through pre-arranged signed agreements, mutual aid fire fighting services can become part of the primary response notification list. As part of the primary response, dispatch and response are automatic. Otherwise, ARFF personnel will need to contact a mutual aid dispatch center and formally request additional manpower and equipment.

In order to coordinate the efforts of all those who may be involved, planners should meet with representatives of all entities as a group to define roles, identify resources, and develop procedures. One of the most important products of such meetings

Figure 11.9 Planned mutual aid support facilitates the coordination of resources at the incident. *Courtesy of Metro Washington Airports Authority.*

should be a comprehensive list of the resources available from the various entities, along with the key emergency phone numbers. The resource list should include highly specialized items such as bulldozers, cranes, and other heavy equipment and the sources from which they are available (Figure 11.10). All such mutual aid agreements and lists must be periodically reviewed and revised to meet changing conditions and to keep them current.

Available Apparatus, Equipment, and Water Supply

In airport emergency planning, the written plan should designate which types of vehicles should respond to each type of accident/incident. Planners should consider the terrain to be encountered when designating heavy, specialized apparatus. Additional considerations are possible needs and sources for other heavy equipment such as bulldozers, cranes, and forklifts plus special-purpose equipment such as lighting, cutting/welding tools, towing/lifting tools, and boats. To ensure availability of such tools and equipment, contractual information should be part of the airport emergency plan so that such agreements do not have to be made after an accident happens. Special emphasis should be placed on identifying possible water supply sources on or near the airfield. Relay pumping or water shuttle operations may need to be considered if hydrants or a sufficient static supply source is not available near the incident.

Rehabilitation

During intense aircraft rescue and fire fighting operations, firefighters exert a great deal of energy. Sufficient resources are needed to ensure that rescue personnel are provided with an area to where they can retreat from on-scene operations. ARFF personnel must exercise proper decontamination procedures so as not to contaminate the rehab area. The area should be well isolated from the emergency site while staying clear of triage, staging, and other operational areas. In addition, this area should provide shelter, if required, along with a place to sit or lie down, and warm or cold drinks and fresh fruit should be available.

Regardless of the role or task they have performed, all response personnel should be required to spend time in the rehab area to allow nourishment and hydration. Arrangements should be made with competent vendors, canteen providers, or other similar organizations so that a rehab area can be established and supplied throughout an incident. (**NOTE:** For more information on this topic, see the FPP/Brady *Emergency Incident Rehabilitation* manual.)

Figure 11.10 Multiple lifting devices often are needed to move aircraft. *Courtesy of Air Line Pilots Association.*

Critical Incident Stress Debriefing

While working at an accident site, ARFF responders are subjected to a tremendous amount of psychological pressure. This pressure is created when responders are called upon to rapidly handle numerous patients, presenting a variety of injuries, while working in a hazardous environment. Depending on the type of incident, wreckage entrapping many of the victims may require extensive extrication operations. Many injuries are often quite traumatic with responders working in an environment where fatalities may also be present. Rescue workers placed into this type of environment can feel overtaxed and overwhelmed due to the large number of tasks that need to be accomplished.

Both fire department and airport operations should have as part of their resource directories a list of qualified critical incident stress debriefing teams. These teams should provide the ability to respond twenty-four hours a day, seven days a week. They should also be cleared to obtain access to the rehab area in order to provide on-site assistance to all rescue and operations personnel. Confidential emotional support groups should be available and accessible for all persons involved with conducting rescue operations both during and after the incident. Depending on the severity of the accident, these debriefing teams may need to be contracted to provide counseling services for an extended period of time after an incident. It is important to ensure that all personnel are provided the opportunity and encouraged to take part in the group CISD sessions.

Other Logistical Support

Government Agencies

Only an investigation by the National Transportation Safety Board (NTSB) or the Canadian Transportation Accident Investigation and Safety Board (CTAISB) can determine whether a crime was involved; therefore, it is important that the entire area within the established perimeter be treated as a crime scene until declared otherwise (Figure 11.11). Because even the smallest piece of crash debris may be important evidence, media personnel should be allowed into the scene only with an escort designated by Command, and they must not be allowed to disturb anything.

To reduce anxiety among relatives and friends of the occupants of an involved aircraft, the investigative authorities may authorize release of the names of those not seriously injured as soon as possible. However, the names of individuals killed or seriously injured are always withheld until next of kin have been notified, and members of the news media are ethically bound to cooperate with this policy.

Figure 11.11 A representative of the National Transportation Safety Board surveys the restricted area of an incident. *Courtesy of William D. Stewart.*

Military Assistance

Military services, particularly for aircraft and other aerospace vehicle protection, are not necessarily limited to the boundaries of their installations. Public law in the United States allows military funds to be spent for nonfederal, civil, and private interests if the expenditures are in the direct interest of federal agencies or if public disorders or disasters are involved. Military agencies may enter into reciprocal mutual-assistance agreements with surrounding community fire protection organizations to provide many of the services already mentioned. Medevac helicopters, field hospitals, medical support personnel, and additional firefighter personnel are just some of the services that can be prearranged (Figure 11.12). Experience in civil air accidents near military facilities has shown the need for and value of cooperative planning prior to such disasters.

American Red Cross

American Red Cross chapters at the local, state, national and international levels will provide their assistance and services (Figure 11.13). Working in conjunction with the air carrier disaster response team, the Red Cross can provide mental health and counseling services for hospitalized and nonhospitalized survivors. They can help ensure that families are not outnumbered and overwhelmed by well-intentioned organizations and individuals. They can provide mental health resources within the local area while providing child-care services for families who bring young children. The Red Cross coordinates with appropriate representa-

Figure 11.12 Military assistance in providing helicopters is critical when moving large numbers of patients. *Courtesy of William D. Stewart.*

Figure 11.13 Serving refreshments is one of the many ways in which the American Red Cross can assist. *Courtesy of William D. Stewart.*

tives of the coroner/medical examiner's office to assist with death notifications and consults with families to arrange suitable nondenominational memorial services just days following the crash, as well as future memorial services for the burial of unidentified remains. The Red Cross can also support a disaster response by providing canteen services, lodging referrals, along with a list of other valuable resources.

Mortuary Assistance

Plans for coping with a major disaster should include provisions for a temporary morgue. In a major aircraft accident, the number of fatalities may overwhelm local morgue facilities; planning for some alternate means of maintaining dignified custody of remains is necessary. Representatives of the local coroner, morgue, and mortuaries should meet with fire department planners and representatives of the NTSB/CTAISB to decide on mutually acceptable procedures. Arrangements should be made with a resource that can supply refrigerated trucks as a means of a temporary morgue. On the airport there may be facilities available with equipment that may serve the same purpose.

Other Operational Considerations

Communications

In order for multi-agency operations to be successful, clear communication is essential. Although each agency or group of agencies is assigned a specific radio channel for transmitting and receiving during day-to-day, routine activities and emergency activities, all agencies concerned must have one or more common channels for mutual aid operations. In addition, they should have multichannel scanning capability in order to monitor local radio channels for critical traffic. The use of radio codes or esoteric terminology must be suspended during joint operations. A code "10-10" may mean one thing to a fire department, something entirely different to the police, and absolutely nothing to another type of agency. The use of Clear Text or Plain Text, as specified in the Incident Management System (IMS), helps to eliminate confusion. Strict radio discipline must also be exercised to facilitate the proper and efficient use of shared radio channels.

News Media

By establishing good working relationships with news media personnel before aircraft accidents/incidents occur, ARFF personnel can avoid logistical difficulties during these airport emergencies. Recognizing the importance of media coverage, fire department personnel should meet periodically with representatives of the various news media to discuss mutual concerns about necessary scene security versus the public's right to know. At the same time, they can develop clear procedures that allow both disciplines to perform their respective functions without undue interference from the other.

Newspaper, radio, or television representatives may arrive at an accident scene before investigative authorities arrive. They should be directed to a predetermined site designated for the media. Emergency personnel should see that representatives from the media stay clear of the danger area and should refer them to the Public Information Officer (PIO).

The PIO should inform media personnel of any areas where there is a legitimate safety concern, such as those involving hazardous materials or unexploded ordnance, and any other access restrictions by reason of an investigation that is pending or in progress. Beyond that, ARFF personnel should cooperate fully with media personnel as long as it does not interfere with rescue or fire fighting efforts. Should evacuation of the surrounding area be deemed advisable, the broadcast media can be very helpful in notifying the public.

Experience has shown that the public interest is best served when only accurate, factual information is released to the news media. Therefore, ARFF personnel should refrain from making statements and should refer all news media representatives to the public information officer. The PIO should not editorialize or speculate about the incident but should give the media only information that has been confirmed. Should a commercial or military aircraft be involved, the release of information to the media becomes the responsibility of representatives of the carrier and/or the investigative agency once they are on scene.

The law permits photographs to be taken of anything at the scene of a civil aircraft accident as long as no physical evidence is disturbed in the process. Photographs may also be taken at the scene of a military aircraft accident/incident unless classified material is exposed. In that case, military personnel will attempt to cover or remove the material. If they cannot, the PIO should advise photographers that photographs are not permitted. If, after being informed, photographers persist, they will be asked to surrender the film to law enforcement or military personnel. If they refuse, they will be advised that they are subject to penalty under federal law. Willfully retaining a negative that compromises national security is punishable by a fine, imprisonment, or both.

Response to Accidents Involving Military Aircraft

Upon notification that an accident involving military aircraft has occurred, the nearest military installation should be called or the National Response Center should be notified. (**NOTE:** The telephone number for the National Response Center is 1-800-424-8802.) In addition, the regional Federal Emergency Management Agency (FEMA) office should be contacted. After being notified, the military will dispatch assistance teams that usually include the following personnel:

- ***Base fire department personnel.*** Depending on the terms of any existing mutual aid agreement, the ranking officer responding from the base fire department will meet with the civilian IC and either become part of a unified command, act as a technical advisor to the IC, or assume command if the incident is declared a National Defense Area (NDA). While most civilian agencies having mutual aid agreements with federal fire departments usually agree to relinquish control of incidents involving military aircraft, a *unified command* is often appropriate, especially when collateral damage has occurred (Figure 11.14).
- ***Explosive ordnance disposal (EOD) personnel.*** EOD personnel disarm, remove, and recover weapons, parts, and residue.
- ***Military police.*** Military police assist the local law enforcement agency as needed; however, they have no peace-officer authority outside a military installation unless the incident is declared a National Defense Area. In that case, the military police then have full authority over the scene/area control.
- ***Medical personnel.*** Military base medical personnel assist with both military and civilian casualties.
- ***Bioenvironmental personnel.*** Bioenvironmental personnel help with the management of radioactive materials and with decontamination of personnel and equipment.
- ***Mortuary affairs personnel.*** Base mortuary personnel assist with the recovery and identification of human remains.

Figure 11.14 Mobile command post vehicles are often needed to direct response personnel. *Courtesy of Jeff Riechmann.*

- ***Information officer***. The information officer either serves as or works closely with the incident PIO to release information to news media representatives.
- ***Accident Investigation Board personnel***. Unless the accident also involves civilian aircraft, only the military Accident Investigation Board investigates the cause of the accident, determines the nature and extent of hazards created by the crash, and attempts to mitigate them.
- ***Legal officer***. The legal officer advises and assists citizens in their claims against the federal government.
- ***Heavy-equipment personnel***. Military heavy-equipment personnel remove the wreckage of military aircraft; however, the owners of any civilian aircraft involved are responsible for removing that wreckage.

When military officials arrive at the scene of an aircraft accident, they will need to obtain the following information from witnesses to assist in the investigation of the accident:

- Time of the accident
- Direction in which the aircraft was headed
- Weather conditions at the time of the accident
- Whether anyone was seen parachuting from the aircraft
- Whether there was an explosion in the air prior to the crash

Training for Mutual Aid and Support Personnel

Few, if any, fire departments can afford to maintain sufficient numbers of on-duty personnel to handle every contingency within their boundaries without assistance. Most departments are staffed and equipped to handle the usual, day-to-day accidents/incidents; but in a major event or with multiple simultaneous events, mutual aid is often needed. All parties to mutual aid agreements should participate in airport emergency planning, training, and drills.

Because fire department response times may be more critical in an aircraft accident than in other types of more commonly encountered emergencies, mutual aid fire departments and airport support personnel should be sufficiently trained to perform their fire fighting duties quickly and efficiently. This performance can be ensured only through frequent training and evaluation. Nearby structural fire departments should participate with airport fire departments in order to become familiar with the airport and its aircraft. Joint training exercises, drills, and tests should be conducted at the airport (Figure 11.15). Participation in combined training exercises may help to evaluate airport emergency plans.

Personnel assigned to stations near an airport should become familiar with the airport and with the aircraft that commonly use it. Mutual aid companies should be familiar with the runways, taxiways, apron areas, hydrants and other water sources, and access routes to various airport areas. They should also know airport terminology, control-tower light signals, and other information peculiar to an airport operation.

Mutual aid training should emphasize the use of structural apparatus and equipment to combat aircraft fires, both in support of airport fire forces and alone if necessary. Firefighters should practice aircraft fire and rescue operations under conditions that are as realistic as possible. Structural firefighters should practice interior aircraft fire fighting operations and should learn to adapt structural techniques to aircraft fire fighting. Conversely, airport fire fighting forces should be equally familiar with the areas surrounding the airport

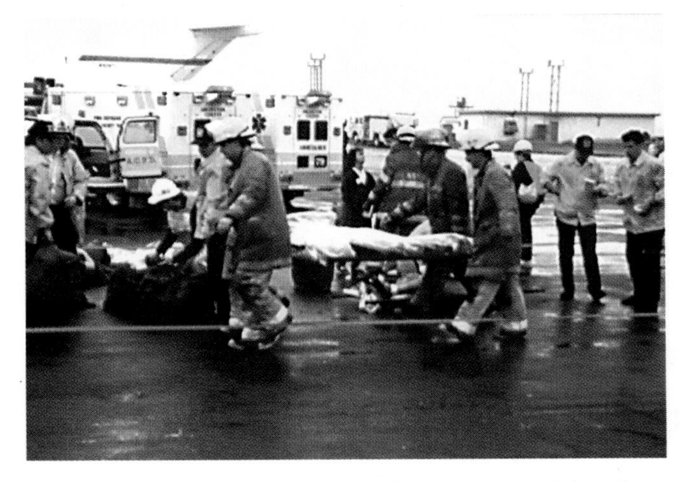

Figure 11.15 Disaster drills provide hands-on training that is critical in preparing response personnel for multicasualty incidents. *Courtesy of William D. Stewart.*

and with how their apparatus and equipment can be used to the best advantage in support of structural fire fighting operations.

If airport support personnel are properly trained, they may be an effective adjunct to regular fire prevention forces in many areas of the airport. Classes for all airport employees should acquaint them with the use of fire extinguishers, fire reporting procedures, and evacuation procedures. With basic fire extinguisher training, support personnel may extinguish incipient fires in aircraft, terminals or other large buildings, fueling areas, hangars, and similar locations.

Joint Training Exercises

Regardless of the amount of thought and effort invested in developing a plan, one or more joint training exercises are needed to test it. Personnel should participate in several full-scale training exercises before they can feel comfortable that the plan will function successfully. As discussed in Chapter 10, "Aircraft Rescue and Fire Fighting Tactical Operations," the operational priorities in aircraft rescue and fire fighting are the same as in other types of emergencies: rescue, fire control, and loss control. The emphasis here should be on tactical considerations among the participating agencies. Successful execution of joint operations depends on airport emergency planning and cooperation. Identifying and reducing differences in apparatus, equipment, terminology, procedures, and operational styles or philosophies can be of critical importance in the disposition of an actual emergency.

After every exercise, all those involved should participate in a comprehensive, nonthreatening critique. All facets of the plan should be reviewed objectively, and any deficiencies that emerge should be corrected. The entire plan should be exercised once a year by conducting at least a tabletop exercise (Figure 11.16). A full-scale disaster drill should be conducted as required by the authority having jurisdiction to ensure that participants are familiar with their roles. Prior to an incident or accident, a thorough plan is the most effective tool. Taking time to develop a complete comprehensive plan has proven to save lives. It is important that the plan encompasses all aspects of on-scene and off-scene operations that will be needed. Contacts must be made to ensure the needed materials are available or can be obtained in a short period of time when needed. Constant review and training will help all personnel become more familiar with the plan along with fine-tuning the plan to meet changing needs.

Figure 11.16 Tabletop exercises define multijurisdictional roles and responsibilities. *Courtesy of William D. Stewart.*

Chapter 12

Hazards Associated with Aircraft Cargo

Job Performance Requirements

This chapter provides information that will assist the reader in meeting the following job performance requirements from NFPA 1003, *Standard for Airport Fire Fighter Professional Qualifications*, 2000 edition. Particular portions of the job performance requirements (JPRs) that are addressed in this chapter are noted in bold text.

3-1.1.1 General Knowledge Requirements. Fundamental aircraft fire-fighting techniques, including the approach, positioning, initial attack, and selection, application, and management of the extinguishing agents; limitations of various sized hand lines; use of proximity protective personal equipment (PrPPE); fire behavior; fire-fighting techniques in oxygen-enriched atmospheres; reaction of aircraft materials to heat and flame; critical components and hazards of civil aircraft construction and systems related to ARFF operations; special hazards associated with military aircraft systems; a national defense area and limitations within that area; characteristics of different aircraft fuels; hazardous areas in and around aircraft; aircraft fueling systems (hydrant/vehicle); aircraft egress/ingress (hatches, doors, and evacuation chutes); **hazards associated with aircraft cargo, including dangerous goods; hazardous areas, including entry control points, crash scene perimeters,** and requirements for operations within the hot, warm, and cold zones, and critical stress management policies and procedures.

3-2.4 Perform an airport standby operation, given an assignment, a hazardous condition, and the airport standby policies and procedures, so that unsafe conditions are detected and mitigated in accordance with the airport policies and procedures.

(a) *Requisite Knowledge:* **Airport and aircraft policies and procedures for hazardous conditions.**

(b) *Requisite Skills:* **Recognize hazardous conditions and initiate corrective action.**

The development of the overnight air cargo business has created large airline companies that are dedicated solely to the transportation of cargo. The cargoes of these aircraft may vary widely. In addition, all commercial aircraft carry various quantities of cargo with each flight.

The term *hazardous materials* is a United States fire service term. In the aviation industry (as well as the Canadian fire service), however, the term *dangerous goods* (DG) is used to describe hazardous substances. Large quantities of dangerous goods are transported by aircraft daily throughout the world, and any air cargo may contain dangerous goods. However, emergency response to incidents involving these aircraft usually remains unchanged from normal aircraft fire fighting procedures.

Because of the materials carried aboard an aircraft for its operation (fuel, hydraulic fluid, etc.) and the combustible metals used in its construction, any aircraft crash could be considered to involve dangerous goods. Technology has advanced aircraft design and construction, incorporating advanced aerospace materials (composites). Exposure to these composites, even in small amounts, may be hazardous to responders, to bystanders, and to electrical/electronic equipment at the scene. Any aircraft subjected to the dynamics of a crash and subsequent fire may release highly harmful substances. Consideration of the possible types and amount of cargo aboard an aircraft suggests a potential for an even greater hazard. ARFF personnel must use proper procedures in response, size-up, and operations to ensure that they are protected from the effects of dangerous goods.

This chapter focuses on the laws and regulations that govern the transportation of dangerous goods by air, product identification procedures, and the personal protective equipment needed by ARFF personnel in situations involving dangerous goods. Also discussed are the safe mitigation and disposition of dangerous goods incidents and the aircraft used in agriculture for applying chemicals.

Laws and Regulations

Dangerous goods shipments by civilian aircraft are regulated by the Code of Federal Regulations (CFR) Title 49, Part 175, *Carriage by Aircraft*, in the United States and by the International Air Transport Association (IATA) regulations for international shipments. However, dangerous goods may be shipped within the United States following IATA regulations instead of Title 49.

While thousands of chemicals are considered to be hazardous if released from their containers, those chemicals considered hazardous in transport are listed in Table 172.101 in Title 49. This table also provides information concerning reportable quantities (RQ) of dangerous goods. Even though the transportation of dangerous goods by air is highly regulated, almost all dangerous goods — except for Class A explosives and poisonous gases — can be and are shipped by air.

Air carriers are required to inspect packages and documents prepared by the shipper to ensure compliance with all appropriate regulations. These procedures, however, do not guarantee that only proper shipments are on board an aircraft. There are many examples of illegal "undeclared" shipments of hazardous materials involving a wide array of dangerous substances that should not be transported by air that are being discovered on all types of aircraft. Responders must be aware that hazardous materials may be involved whether the accident aircraft is a small private plane or a passenger-carrying jetliner.

NOTE: The required paperwork describing dangerous goods being shipped by aircraft may have a red-and-white candy-striped border. This border should alert ARFF personnel to the presence of these materials aboard (Figure 12.1). Required paperwork is more fully discussed in the Product Identification section.

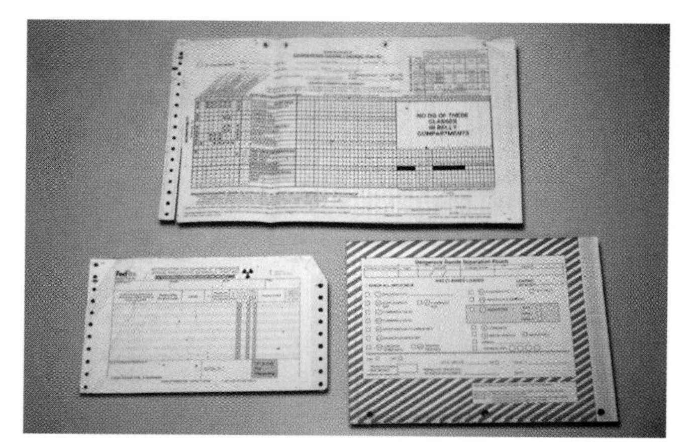

Figure 12.1 Shipping papers, including papers for dangerous goods.

Classification of Dangerous Goods

Dangerous goods may be elements or compounds and may be found as gases, liquids, or solids or in a combination of these physical states. The U.S. Department of Transportation (DOT) defines a *hazardous material* as "a substance that poses an unreasonable risk to the health and safety of operating or emergency personnel, the public, and/or the environment if it is not properly controlled during handling, storage, manufacture, processing, packaging, use, disposal, or transportation." According to NFPA 402, *Guide for Aircraft Rescue and Fire Fighting Operations*, dangerous goods are classified by the UN, ICAO, IATA, and the U.S. DOT as follows:

Class 1 Explosives

Class 2 Gases: compressed, liquefied, dissolved under pressure, or deeply refrigerated

Class 3 Flammable liquids

Class 4 Flammable solids: substances susceptible to spontaneous combustion; substances that, on contact with water, emit flammable gases

Class 5 Oxidizing substances

Class 6 Poisonous (toxic) and infectious substances

Class 7 Radioactive materials

Class 8 Corrosives

Class 9 Miscellaneous dangerous goods

Warning labels that have been approved by the U.S. Department of Transportation using the United Nations labeling system are used on shipments within the United States and internationally (Figure 12.2).

Hazard Class	Name	Label
Class 1	Explosives	EXPLOSIVES 1
Class 2	Flammable Gases	FLAMMABLE GAS 2
	Non-Flammable Gases	NON-FLAMMABLE GAS 2
	Poison Gases	POISON GAS 2
Class 3	Flammable Liquids	FLAMMABLE LIQUID 3
Class 4	Flammable Solids	FLAMMABLE SOLID 4
	Spontaneously Combustible Materials	SPONTANEOUSLY COMBUSTIBLE 4
	Dangerous When Wet Materials	DANGEROUS WHEN WET 4
Class 5	Oxidizers	OXIDIZER 5.1
	Organic Peroxides	ORGANIC PEROXIDE 5.2
Class 6	Poisons	POISON 6
	Keep Away From Foodstuffs (less toxic than above)	HARMFUL STOW AWAY FROM FOODSTUFFS 6
	Infectious Substances	INFECTIOUS SUBSTANCE 6
Class 7	Radioactive I	RADIOACTIVE I 7
	Radioactive II	RADIOACTIVE II 7
	Radioactive III	RADIOACTIVE III 7
Class 8	Corrosive	CORROSIVE 8
Class 9	Miscellaneous	9

Figure 12.2 Warning labels and the hazards they represent.

Shipment of Dangerous Goods

On cargo-carrying aircraft, hazardous freight is usually placed in containers called *unit load devices.* These include aircraft containers, aircraft pallets with a net, or aircraft pallets with a net over an igloo (Figures 12.3 a and b). These containers are then loaded aboard the aircraft. Sometimes loading them requires special equipment, depending upon the size and weight of the particular container. Some air carriers use specially modified unit load devices for transporting certain dangerous goods on the main deck of freighter aircraft. These containers may have special markings and include an integral fire suppression capability (Figure 12.4). Special discharge nozzles located inside the container are coupled to a Halon 1211 portable extinguisher by a connection on the exterior of the unit (Figures 12.5 a-c). Personnel can manually discharge extinguishing agent into the container without having to open it.

Figures 12.3 a and b Two different aircraft containers.

Figure 12.4 This container for shipping dangerous goods has integral fire-suppression capabilities.

Figure 12.5a At the top of the container is a connection for the extinguishing agent hose.

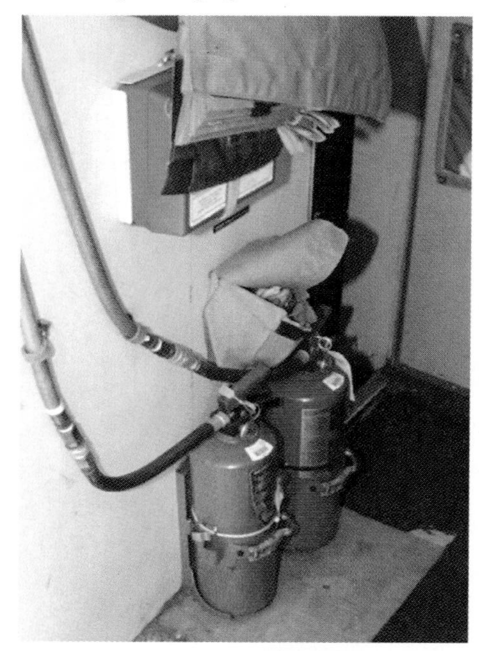

Figure 12.5b The extinguishers and hoses inside the aircraft that connect to the container.

Figure 12.5c The hose stretches from the extinguisher to the cargo area, and the hose nozzle fits into the connection on the container.

Certain dangerous goods must be accessible to the crew in flight in case of a leak or fire. As a general rule, most dangerous goods on the main deck of cargo aircraft are loaded in the most forward location. This is not required, however, if a path or aisle is maintained allowing crew members access throughout the main deck. Also, shipments of radioactive material are usually loaded as far away from the flight crew as possible.

Cargo aircraft have restraining nets or bulkheads to prevent hazardous cargo from shifting. Certain hazardous cargoes are not always transported in specialized containers. These materials, which may be stowed in any aircraft cargo compartment, include dry ice and magnetized materials. Passenger aircraft may have dangerous goods shipments loaded in any of the cargo holds.

CAUTION: Use caution when attempting a rescue through the forward area of cargo-carrying aircraft because of the area's close proximity to any hazardous cargo.

One possible hazard firefighters might encounter at the scene of an aircraft incident or accident is the presence of undeclared dangerous goods cargo. Undeclared dangerous goods cargo, for whatever reason, is cargo that is not packaged properly, does not have shipping documentation, or has not been handled with the safety precautions required of hazardous shipments. This type of dangerous goods cargo may appear in several forms:

- Dangerous goods improperly shipped through the mail
- Dangerous goods transported in passenger luggage
- Dangerous goods illegally shipped as normal cargo to avoid hazardous cargo shipping charges

Product Identification

One of the most important elements of managing a dangerous goods incident is the proper identification of the product involved. This identification may be challenging in air transport situations because of the wide variety of circumstances in which dangerous goods may be encountered. For example, if an aircraft were involved in a high-impact crash, the probability of the presence of dangerous goods might be high, but rapid identification could be nearly impossible.

NOTE: Certain substances may be shipped only on cargo aircraft and not on passenger aircraft. A package labeled "CARGO AIRCRAFT ONLY" or "DO NOT LOAD ON PASSENGER AIRCRAFT" found aboard an aircraft should alert ARFF personnel that the material is extremely hazardous (Figure 12.6).

Identification

All dangerous goods to be shipped by air are supposed to be packaged in accordance with specific guidelines. If they are, and if the packaging is still intact, the recognition and identification of a hazardous shipment are made much easier. There are several means of identifying dangerous goods in air transportation. Some of the means of identification and verification are as follows:

- Package markings
- Labels
- UN/NA number (United Nations/North American Systems)
- Container type
- Material safety data sheet
- Shipping papers
- Name of shipper
- Name of receiver

Figure 12.6 Packages with this label should be shipped in cargo aircraft only.

- Name of air carrier
- Sample analysis of material

In addition, flight crews are required to be notified when dangerous goods have been loaded onto the aircraft. This notification documentation (shipping papers) includes important information such as the proper shipping name, the UN number, hazard class, quantity, and its location on the aircraft. The document(s) is maintained on the flight deck or, on cargo aircraft, may be in a pouch near an exit door — usually the main entry door just aft of the flight deck.

Verification

The initial identification should be verified through multiple sources to ensure accuracy. An error in product identification could be critical and could produce devastating results if the material is consequently handled inappropriately. Therefore, it is recommended that at least three separate sources be used in the identification and verification process. For example, is the product described on the *shipping papers* consistent with the *type of container* and with the *labels* on the container?

Information Gathering (Consulting)

After initial identification and verification have been accomplished, the product must be researched to determine the hazards associated with it. This information gathering helps in the development of a mitigation plan. As in the identification and verification steps, no less than three separate sources of information should be consulted. This procedure helps to ensure the gathering of all information necessary to determine the hazards, select the personal protective equipment to be used, and devise a mitigation plan. Some of the common sources of information include the following:

- *Chemical Hazard Response Information System (CHRIS)* — Superintendent of Documents, United States Government Printing Office
- *Hawley's Condensed Chemical Dictionary*—John Wiley and Sons
- *Dangerous Goods Initial Emergency Response Guide* — Transport Canada
- *Sax's Dangerous Properties of Industrial Materials* — Van Nostrand Reinhold Company, Inc.
- *2000 Emergency Response Guidebook* — U.S. Department of Transportation (DOT)
- *Emergency Handling of Hazardous Materials in Surface Transportation* — Bureau of Explosives (BOE), division of the Association of American Railroads (AAR)
- *The Firefighter's Handbook of Hazardous Materials* — Baker, Maltese Enterprises, Inc.
- *Fire Protection Guide on Hazardous Materials*— National Fire Protection Association
- *Manual for Spills of Hazardous Materials* — Canadian Government Publishing Centre
- *Pocket Guide to Chemical Hazards* — National Institute for Occupational Safety and Health (NIOSH)

A library of information publications should be conveniently and quickly available to response units. Additional data on known chemicals can be obtained on-scene by calling the Chemical Transportation Emergency Center (CHEMTREC®), (800) 424-9300, in the U.S. and calling Canadian Transport Emergency Centre (CANUTEC), (613) 996-6666, in Canada.

Once the materials involved are identified and verified and all their properties and characteristics are known, a mitigation plan may be devised for the disposition of the problem. Based on the known materials regularly transported through the facility, the preparation of pre-emergency plans for specific products or specific situations is recommended. See the IFSTA **Hazardous Materials for First Responders** manual for detailed assistance in developing pre-emergency plans. For more detailed information on handling an actual incident, refer to **Hazardous Materials: Managing the Incident**.

Personal Protective Equipment

Personnel responding to aircraft dangerous goods emergencies need to be protected from the effects of these substances. The authority having jurisdiction determines the type of personal protective equipment (PPE) to be used during the responses. The responders must ensure they are familiar with the type of PPE being used prior to using it in an emergency situation. The authority having jurisdiction must ensure compliance with NFPA 1500, *Standard on Fire Department Occupational Safety and Health Program*, and ensure that all steps of identification, verification, and consultation are completed before determining the appropriate level of protection. In situations involving unknown materials, the role of ARFF personnel may be limited to isolating the contaminated area and denying entry until a hazardous materials response team (HMRT) can obtain a sample of the material for analysis.

There are three basic levels of protection for the emergency responder (see Chapter 4, "Firefighter Safety"):

- Structural fire fighting clothing (see NFPA 1971, *Standard on Protective Ensemble for Structural Fire Fighting*)
- Proximity clothing (see NFPA 1976, *Standard on Protective Clothing for Proximity Fire Fighting*)
- Chemical protective clothing
  - — Vapor protective clothing (see NFPA 1991, *Standard on Vapor-Protective Suits for Hazardous Chemical Emergencies*)
  - — Liquid splash protective clothing (see NFPA 1992, *Standard on Liquid Splash-Protective Suits for Hazardous Chemical Operations*)

The U.S. Environmental Protection Agency (EPA) also has a system for classifying protective clothing for use at dangerous goods incidents. The proper level can be selected only after enough information about the hazardous substance has been gathered to make that determination. The classifications according to EPA are the following:

- Level A — Vapor/gas protection; totally encapsulating suits
- Level B — Liquid protection; nonencapsulating suits
- Level C — Liquid/airborne particulate protection; nonencapsulating suits
- Level D — No specific protection; nonencapsulating suits

Structural firefighter protective clothing (SFPC), as defined by the U.S. Department of Transportation (DOT), consists of full structural fire fighting turnouts (bunkers) and positive-pressure SCBA. Even though it provides a thermal barrier and affords a higher level of respiratory protection than the EPA Level C, which allows filter-type masks, SFPC does not provide any substantial splash protection; thus it only meets the requirements for EPA Level D, the same as an ordinary work uniform. However, according to the *2000 Emergency Response Guidebook*, after weighing all factors, the incident commander (IC) may decide that the risk/benefit of subjecting firefighters wearing SFPC to a "quick in-and-out" mission of critical importance may be appropriate.

Protective equipment for any dangerous goods response should be selected following established SOPs and based upon the nature of the incident and the resources available to the department. Under no circumstances should ARFF personnel be assigned tasks or duties for which they do not have adequate protective clothing or training.

Dangerous Goods Operations

The first responsibility of units responding to dangerous goods incidents is to isolate the scene and deny entry (Figure 12.7). This procedure will stabilize the scene and allow for a detailed risk assessment, including determining what rescue

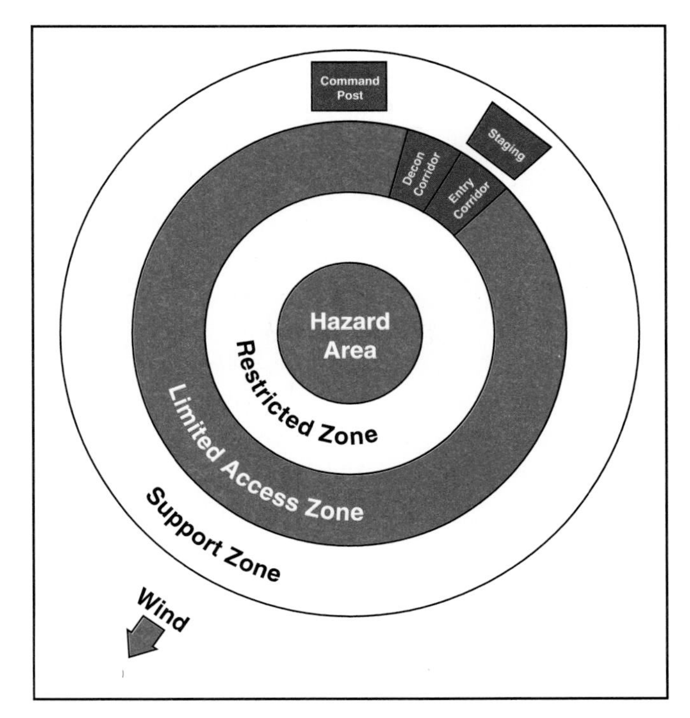

Figure 12.7 Hazardous area control zones.

efforts may be needed. It is necessary to secure the area, establish control zones, and exclude nonessential personnel as rapidly as possible. Through the use of a systematic size-up procedure and after consulting appropriate references, the incident commander should be able to determine whether evacuation beyond the immediate area of the incident is necessary, and if so, to what extent. ARFF personnel are likely to be involved in the mitigation of the hazardous material release and/or engaged in other essential activities in and around the aircraft. Therefore, a large-scale evacuation, if necessary, will probably become the responsibility of law enforcement or other personnel.

If rescue efforts are necessary, the amount of risk to which ARFF personnel would be exposed must be considered. While ARFF personnel may be placed at some degree of risk to effect rescues, their being put in jeopardy to recover bodies is inappropriate. The use of a risk/benefit model should help in determining to what extent responders should be exposed. If the benefit is low and the risk is high, it would not be a good decision to take action at that point. However, if the risk is low and the benefit is high, the operation would be considered a reasonable risk. To quote one adage: We will assume reasonable risk to protect savable lives; we will assume inherent risk to protect savable property; we will assume no risk to protect lives or property that are already lost.

Anytime dangerous goods are involved in aircraft emergencies, a situation that may already be complex and dangerous can become significantly more hazardous for ARFF personnel. The dangerous goods that might be found on or around aircraft are the same as those that could be found on the highway or in a fixed facility. The quantities involved in most aircraft accidents tend to be smaller than those in other environments. However, exposure to even minute amounts of some materials may be extremely hazardous to personnel, so all factors involved and the risk/benefit model should always be considered. For more information on dangerous goods incidents, refer to the IFSTA **Hazardous Materials for First Responders** manual or to **Hazardous Materials: Managing the Incident**.

Response to accidents/incidents involving military aircraft is discussed in Chapter 10, "ARFF Tactical Operations."

Agricultural Application

Another significant use of dangerous goods in aircraft is in the application of agricultural chemicals. These chemicals range from relatively innocuous fertilizers to highly toxic pesticides. Some agricultural chemicals are applied as liquid sprays and others as powders. They are usually shipped to and stored at the aircraft loading point as liquids in drums or as powder in heavy, plastic-lined bags. Different pesticides, herbicides, or fertilizers are sprayed at different times of the year, depending on the life cycle of the pest or the growth stage of the crop.

Agricultural chemicals can be applied with fixed-wing or rotary-wing (helicopter) aircraft. Fixed-wing aircraft are usually loaded at an airport, road, or landing strip as close to the job site as possible. Helicopters are often trucked to and loaded at the job site. Agricultural spraying or crop-dusting aircraft usually have one or more support vehicles in attendance. These vehicles contain fuel, chemical concentrates, water, mixing hopper, and loading equipment. Helicopters often use saddle tanks to hold the chemical solutions. Fixed-wing aircraft usually have a tank between the engine and the pilot's compartment. The quantities of chemicals carried can range up to several hundred gallons (liters).

These chemicals must be applied at very low altitudes to limit overspray and losses due to wind. Pilots often must fly these aircraft extremely close to buildings, trees, power lines, towers, or other types of vertical obstructions. If an accident occurs, the crash site may be very difficult to reach with fire apparatus, and the incident will be complicated by the likelihood of dangerous goods being involved. There is usually no indication on the aircraft of what is being carried. Responders should look for application equipment in the wreckage as an indication that the aircraft was in fact used for spraying/crop-dusting. To determine the chemical being carried, ARFF personnel can contact the owner of the aerial spraying business (the owner may not be the pilot), the owner of the land where the spraying is being conducted, or the local agricultural chemical supplier.

Appendix A

Ground Vehicle Guide to Airport Signs and Markings

Advanced Composites/Advanced Aerospace Materials (AC/AAM): Mishap Risk Control and Mishap Response

Written By: John M. Olson
January 2000

Overview

Although advanced composites and advanced aerospace materials provide several benefits over other materials options, they do present some important environmental, safety, and health concerns. In their final design state these materials are generally considered safe, inert, and biologically benign. However, when damaged by fire, explosion, or high-energy impact, these materials can be characterized by unique hazards and concerns that require timely and appropriate responses.

The following guidelines are provided as recommended precautions and procedures for dealing with a composite mishap response. However, the hazards are dependent upon the type, quantity, damage extent, and mishap scenario. In most cases, the concentration of the materials present drives the level of risk for the potential injection, inhalation, ingestion, and absorption hazards.

These guidelines address all phases of an aircraft mishap response, including fire fighting, investigation, recovery, clean up, and disposal; however, they can be universally applied to any application or situation involving these materials. Due to the infinite variability of mishap scenarios, this information is general in nature, and should be more specifically tailored to individual mishap scenarios as required. Nevertheless, the purpose of these guidelines is to serve as the basis for the development of consistent and effective procedures and policies throughout the world in order to maximize risk control and minimize the environmental, safety, and health hazards caused by composite mishaps. Ultimately, the user is urged to supplement this information with new and updated research and operational guidance as it becomes available, although the conservative measures outlined within this document are the best course of action in the absence of specific or concrete data.

Composite mishap hazards can, in most cases, be efficiently and effectively mitigated with proper training, precautions, and preparation.

Definitions

Composite Material: A physical combination of two or more materials, generally consisting of a reinforcement and a "binder" or matrix material. Generally, the reinforcements, or load-bearing elements, are fibers, while a resin forms a matrix to hold the fibers and fill the voids. The reinforced matrix structure thereby allows fiber to fiber stress transfer. Composite materials generally consist of laminates of several layers in varying directions. In many cases, a honeycomb core material is sandwiched between two of the laminates. The name of the composite describes its physical makeup: type of fiber/type of resin.

Examples: Fiberglass (Glass/Epoxy, Glass/Polyester)

Advanced Composite Material: A composite material comprised of high-strength, high-stiffness reinforcement (i.e., fibers) in a matrix (i.e., resin) with properties that can include low weight, corrosion resistance, unique thermal properties, and special electrical properties. Advanced Composites are distinguished from traditional composites by their increased relative performance, cost, complexity, and mishap hazard potential.

Examples: Graphite/Epoxy, Boron/Epoxy, Aramid (Kevlar)/Epoxy, FQuartz/Cyanate Ester

Advanced Aerospace Material: A highly specialized material fulfilling unique aerospace construction, environment, or performance requirements.

Examples: Beryllium, Depleted Uranium (DU), Radar Absorbent Material (RAM)

It is essential that a clear distinction be made between Advanced Composites and Advanced Aerospace Materials because of several very specific and unique hazards.

Hazard: A condition or changing set of circumstances that presents a potential for injury, illness, or property damage. Likewise, it can be described as the potential or inherent characteristics of an activity, condition, or circumstance, which can produce adverse or harmful consequences.

Given this definition, the hazards associated with mishap damaged advanced composites and advanced aerospace materials will be addressed with a risk control emphasis.

Risk Control: The process of minimizing accidental and other extraordinary losses by anticipating and preventing unplanned events. It emphasizes the complexities of exposures and encompasses broad areas of risk, which are indicative of a mishap scenario. Effective risk management is comprised of both risk control and risk financing in order to control exposures through knowledge, training, preparation, and an understanding of the factors involved. Loss avoidance must be both a pre- and post-mishap effort.

Background

Damage to Advanced Composites and Advanced Aerospace Materials (AC/AAMs) caused by fire, explosion, and/or high-energy impact in a mishap presents unique environmental, safety, and health hazards. In typical aircraft fires, temperatures reach between 1000°-2000°C. Organic matrix materials (i.e., resins and polymers) burn off around 400°C, creating toxic combustion products and liberating the reinforcement (i.e., fibers). Depending upon the type of composite or aerospace materials, the associated material dynamics and response can vary greatly. For example, glass or aramid fiber reinforcements tend to melt under the extreme heat, whereas the heat can oxidize carbon or graphite fibers, thereby altering their size, shape, porosity, and other characteristics.

The intense thermal and mechanical forces in a mishap generally cause degradation, debonding, and/or "explosive" fracture of the advanced composite structures. While absorbing this fracture energy, the reinforcement, usually stiff and strong, may be broken into particulate fibers, turned to dust, or reduced to a cloth-like consistency. AAMs can produce highly toxic oxides or heavy metal concentrations. Liberated carbon fibers can readily penetrate human skin due to their stiffness, whereas boron fibers can penetrate bone. Furthermore, the adsorbed and absorbed pyrolysis and combustion products (generally toxic) on activated, oxidized fibers can be a very potent injection and inhalation hazard because the toxins can be readily placed and retained in the body. This phenomenon is particularly critical in mishaps involving bloodborne pathogens (i.e., HIV, Hepatitus B) that are present in the debris. In almost all cases, the type, amount, and extent of damage controls the concentration of AC/AAMs at a mishap site, which in turn determines the extent of the hazards. The prevailing weather conditions can also greatly affect the extent of the dispersion of the damaged materials within the vicinity of the mishap.

Fire, coupled with heat, shock, and fragmentation, produces several different types of damage in Advanced Composites (ACs). Effects can range from a simple reduction in strength, to a loss of Low Observable (LO) performance, delamination, debonding, charring, melting, burning, and vaporization. The impact upon AAMs can be just as broad but is highly dependent upon the type of material. For example, Depleted Uranium and Beryllium both produce highly toxic oxides when subjected to intense heat (>700°-800°C).

Although AC/AAMs represent only one of the many hazards associated with an aircraft mishap (i.e., fuels, lubricants, exotic metals, and weapons), they do merit increased awareness and informed precautions because of their increased hazard potential, increasingly widespread use, and persistence or durability. Exposures to potentially harmful vapors, gases, particulates, and airborne fibers generated in a composite mishap need to be controlled because of the combined effects of the dispersion forces and the complex chemical mixtures.

Hazard exposure routes for damaged AC/AAMs include absorption (contact), inhalation (breathing), injection (puncture and tearing wounds), and ingestion (eating, drinking, and smoking). The toxicology of respirable particulates and their disease producing potential is a function of three main variables: 1) the dose or amount of particulates in the lung; 2) the physical dimensions of deposited particulates; and 3) the durability (time) in the lung. Fire-exposed carbon fibers tend to break into shorter lengths and split into smaller diameters with sharp points, thereby increasing their probability for respiration and ease of transport. Dry and windy conditions at a mishap site increase the chances for re-dispersion of particulates. Likewise, whether inhaled or injected, ACs are not easily removed or expelled because of their shape, sharpness, and stiffness. Other potential health and environmental effects from AC/AAMs include dermal and respiratory problems, toxic and allergic reactions, contamination, and radiation exposure for AAMs. These impacts may be acute or chronic, as well as local or systemic, depending upon the circumstances. Mechanical injection or cuts are the most common skin hazard, although sensitization (local and systemic) can occur. Irritation to the respiratory tract is also common, much like a nuisance dust irritation hazard. Off-gassing, toxic products in the smoke plume, smoldering debris, and oxidized (fire-damaged) particulates are the primary respiratory hazards.

Mishaps involving AC/AAMs that are electrically conductive (i.e., graphite or carbon fiber) may present electrical shorting or arcing hazards if very high concentrations exist (usually at the immediate mishap site only). Although rare, this may result in electrical equipment degradation or failure, including communication interference. Research has shown that widespread electrical failure due to environmental release and plume dissipation is highly unlikely, except at the mishap site. Disseminated carbon or graphite fibers are also influenced by the presence of high voltage areas and reduce the local dielectric properties of free air, which could in turn cause equipment malfunctions.

Justification

Given the existing and projected increases in usage and applications of AC/AAMs, it is critical to develop realistic policies and procedures that focus on risk control and minimizing the environmental, safety, and health hazards associated with an advanced composite or aerospace material mishap. As the associated knowledge base grows, procedures and guidelines can be situationally optimized in terms of cost, safety, and performance.

Based upon both existing and uncharacterized mishap hazards associated with AC/AAMs, risk reduction measures are necessary. Administrative controls, including adequate personal protective equipment (PPE), training, and safe practices need to be implemented immediately, as the dynamic field environment is not conducive to engineering controls. Conservative, although situationally optimized, risk control measures are essential. Care and common sense approaches are the best course of action. Because aircraft mishaps occur under extremely diverse weather, terrain, and location conditions, with widely varying degrees of damage, a universally applicable set of risk control precautions is not practical. The many variables require conservative protective measures with a complete material lifetime or “cradle-to-grave” mentality of responsibility. This is true for all phases of a mishap response, ranging from first response and fire fighting, to investigation, recovery, clean up, and disposal.

Major Issues

The major issues currently affecting mishap response which involve Advanced Composites/Advanced Aerospace Materials (AC/AAMs) are:

1. Fiber dispersion and re-dispersion

   —Includes the mishap dynamics, effective response procedures, and hold-down material (fixant) suitability.

2. Synergistic material and combustion effects

—The combined effects of multiple materials and varying damage extents.

3. Concentrations and compatibility

   —Exposure limits aren't specifically defined. Equipment, procedure, and fire suppression agent compatibility issues also exist.

4. Adsorbed and absorbed pyrolysis and combustion products

   —Impact and extent of the toxin hazard.

5. Site and equipment contamination and decontamination

   —Procedures for effectively and realistically addressing the hazards.

6. Clean-up and disposal complications (including potential classifications as Hazardous Materials [Haz-Mat] depending upon the type and damage of the materials)

   —Determine proper disposal methods and classifications of waste debris.

7. Bloodborne pathogens

   —Examine the potential for Hepatitis B and HIV transmission from injection by contaminated debris.

8. Radiation exposure effects

   —Evaluate extent and results of potential exposures to Depleted Uranium and other radioactive AAMs.

9. Acute toxic exposures to beryllium oxide and other highly dangerous AAMs

   —Evaluate protective measures and protective equipment.

Personal Protective Equipment (PPE) Guidelines

As the normal first responders, firefighters are considered the primary response group to a mishap with AC/AAMs and are therefore subjected to the greatest hazard exposures from the materials; however, they are the best protected in all but the most extreme cases. As such, all personnel in the immediate vicinity of an AC/AAM mishap, as well as all personnel subject to the concentrated smoke plume, must wear bunker or proximity suits and Self-Contained Breathing Apparatus (SCBA) until the composite material fires have been completely

Personal Protective Equipment (PPE) Requirements

AC/AAM Mishap Condition	PPE Recommended
Burning or Smoldering Materials	1. Full Protective Fire Clothing 2. Self-Contained Breathing Apparatus (SCBA) 3. Do NOT use rubber gloves
Broken, Dispersed, or Splintered Materials (Post-Fire, Explosion, or High-Energy Impact)	1. Protective overalls (Tyvek suit) – coated with hood and foot coverings 2. Full-face respirator with High Efficiency Particulate Air (HEPA) filter (Note: A gas mask with a similar filter may be substituted if equipment or respirator-trained personnel are not present) 3. Hard-soled work boots (steel toe and shank are best) 4. Leather work gloves over nitrile rubber gloves [no surgical gloves]
Minimal or Peripheral Area Material Exposure	1. Long sleeves and long pants for durable work clothing 2. Nuisance dust filter or mask 3. Adequate eye protection (goggles or safety glasses) 4. Hard-soled work boots (steel toe and shank are best) 5. Leather work gloves over nitrile rubber gloves [no surgical gloves]

extinguished, and cooled to a temperature at or below 300°F (149°C) with no intense smoldering. It is important to note that the potential exposure to hazards associated with AC/AAM mishaps may be more severe for secondary exposure groups, including all of the subsequent response operations, than for the initial fire-fighting activities because of the duration of exposure and generally reduced levels of protection. However, the hazard exposures are minimal if Personal Protective Equipment (PPE) is properly used and good mishap response procedures are diligently followed. All affected personnel need to know both the hazards and the proper response for effective mishap risk control. This makes coordination and communication critical for everyone involved. Preparatory knowledge and training, accompanied by common sense, good judgment, and quick decision making are crucial for success.

Mishap Risk Control Guidelines for AC/AAMs

Immediately after a mishap, the situation should be assessed by answering the following questions:

1) Does the mishap scenario involve advanced composite/advanced aerospace materials?
2) If yes, where are they located, and are they damaged?
3) Who can provide information about the type and content of these materials, and can they be reached for specific questions?
4) How does the environment affect the situation? For example, does the weather impact the response, or does the local geography change the response strategy?

Once these basic questions have been addressed, the following steps should be accomplished. The specific response should be tailored to match the extent of the hazards.

1. First responder(s) [usually firefighters] shall conduct an initial mishap site survey for:
   a. Signs of fire, explosion, or high-energy impact damaged Advanced Composite/Advanced Aerospace Materials (AC/AAMs)
   b. Presence of loose/airborne fibers and particulates
   c. Prevailing meteorological conditions/wind direction, including smoke plume assessment (if any)
   d. Degree of site exposure to fire/impact/explosions
   e. Local/proximal equipment/asset damage and hazards, including the debris pattern
   f. Exposed personnel and environmental contamination routes

   Essentially, the first responder will determine the extent of any additional AC/AAM hazards associated with the mishap.
2. Establish control at site with a clear and direct chain of command. If properly protected personnel are not present, avoid the mishap site until appropriately trained and equipped personnel arrive at the scene.
3. Evacuate personnel from areas in the immediate vicinity of the mishap site affected by direct and dense fallout from the smoke plume, along with easily mobile and critical equipment. Continually move fire fighting equipment in order to avoid the smoke plume, especially in larger scale mishaps involving greater amounts of AC/AAMs. Restrict ALL unprotected personnel from assembling downwind of the mishap site. Use of over-pressurized cab equipped fire vehicles is essential if unable to avoid the smoke plume; however, this will require greater decontamination requirements. Some modern AC/AAM form combustion products that are permeable to the protective membranes for ventilation in contemporary fire suit ensembles. Accordingly, contamination, decontamination, and protection become very important concerns for a small percentage of mishaps, usually involving stealth aircraft. If exposed to either the smoke plume, open fire, or smoldering off-gassing of burned AC/AAMs, firefighters should monitor their bodies as part of the whole system. This would include checking for any potential chemical burns/irritation at heavy perspiration areas such as the armpits and groin. However, this phenomenon is very dependent upon rare material concentrations in confined-space-type environments. Nevertheless, it warrants consideration.
4. Alter or move aircraft and flight operations within the immediately exposed mishap and fallout

areas. No ground or flight operations (specifically helicopters) are to be permitted within 500 ft above ground level (AGL) of the site and within 1,000 ft horizontally. (This footprint may be increased depending on concentrations and local conditions).

5. Normal fire fighting mishap response procedures should initially be followed. Once control of any fires is established, the special precautions associated with AC/AAMs should be implemented. Depending upon the type of materials involved, some equipment-related problems might arise. These include: dulling of penetrator tools due to the hardness of some advanced composites, inability to penetrate some areas unless the hard-points and emergency penetration points are known, and internally insulated (imbedded) fires that are difficult to suppress. Complex engine inlets and imbedded exhaust areas are particularly challenging for fire suppression.
6. Extinguish fire and cool AC/AA Materials to below 300°F (149°C). This can be accomplished by spraying a light mist of water or foam on the affected materials once the major fires are extinguished. In some cases, fire suppression agent compatibility will be an issue. For example, dry chemical fire suppressant can destroy some advanced composite components, so care should be exercised with small and isolated fires in order to minimize peripheral material damage. In more extreme cases, known fire suppression agents are somewhat ineffective at extinguishing fires on exotic AC/AAMs. Extreme caution should be exercised in these very rare scenarios.
7. ONLY firefighters equipped with Self-Contained Breathing Apparatus (SCBA) are authorized in the immediate vicinity of a burning/smoldering mishap site until the fire chief declares the area both fire safe and smolder/off-gas hazard safe. If possible, care should be taken to avoid high-pressure water or foam applications due to the high potential for breakup and dispersal of the AC/AA Materials.
8. Avoid dragging fire hoses through mishap debris or contaminated areas, as there is a high potential for abrasion and/or equipment contamination. Gear, including the bunker suits and water/air lines may be snagged by sharp and jagged AC debris. Boots are particularly prone to cuts and penetration where jagged and stiff AC debris is damaged. Likewise, they are big potential sources for contamination from particulates, as well as transfer of these contaminants to other areas beside the immediate mishap vicinity.
9. Cordon or rope off the mishap site and establish a single entry/exit point. Only adequately protected personnel are authorized at the immediate mishap site and peripheral area (contamination reduction zone). The fire chief and bioenvironmental engineer, or the on-scene commander designates the peripheral area in a coordinated effort. As a guide, the peripheral area should be defined as more than 25 feet away from any damaged composite parts, although it will vary based upon local meteorological and geographical conditions.
10. If personnel other than those at the mishap site have been directly and significantly exposed to material and smoke hazards, consult medical personnel for evaluation and tracking. If possible, inform the medical personnel of the type and extent of exposures. Advise and inform the otherwise unthreatened populace of the applicable precautions to take in affected mishap site surroundings or in plume fallout areas. Track patient treatment and outcomes for those involved in the mishap.
11. Coordinate with the on-scene or Incident Commander (IC) to provide necessary access to the mishap site for more thorough survey and investigation. For larger scale mishap response scenarios, especially involving modern, highly unique AC/AAMs, use of a Hazardous Materials Response unit is recommended, because of the added levels of experience and increased capability to control the situation.
12. If possible, toxicology and area studies for dust, inhalable and respirable particulates, and fibers should be conducted by a qualified industrial hygienist or bioenvironmental expert as soon as practical. However, all research personnel must be sufficiently protected. The survey protocol should include a visual observation, personal air, ground, and water sampling, and evaluation of

the engineering controls and PPE in use at the scene.

13. Identify specific aircraft and material hazards as soon as possible by inspection of the debris and consultation with applicable, knowledgeable personnel/sources (i.e., crew chief, system managers, reference documents, web sites, contractors, or aircraft specialists). Indicate or point out AC/AAM locations and concentrations to all response personnel, as appropriate.

14. Minimize airborne dispersion of particulates/fibers by avoiding excessive disturbance from walking, working, or moving materials at the mishap site. This includes fire suppression equipment whenever possible.

15. Locate, secure, and remove any radioactive AAMs by using a Geiger counter to find any applicable debris or particulates. Contact relevant authorities and dispose of in accordance with strict policies.

16. Monitor entry/exit from the single Entry Control Point (ECP). The following guidelines apply:

    a. When exiting the mishap site, personnel should follow clearly defined decontamination procedures. Use of a High Efficiency Particulate Air (HEPA) filtered vacuum system is highly desired. If possible, remove AC/AAM contaminants from outer clothing, work gloves, boots, headgear, and equipment. If this type of vacuum is unavailable, efforts must be made to rinse, wipe, or brush off as much particulate contamination as possible.

    b. Clean sites (i.e., tent or trailer) for donning and removal of Personal Protective Equipment (PPE) should be set up as soon as practical.

    c. No eating, drinking, or smoking is permitted within the exclusion and contamination reduction zones, or as otherwise determined by the Incident Commander. Personnel must be advised to wash hands, forearms, and face prior to eating, drinking, or smoking.

    d. Contaminated protective clothing should be properly wrapped, sealed, and disposed.

    e. Personnel should shower in cool water prior to going off-duty to prevent any problems associated with transfer of loose fibers or particulates. Portable showers may need to be provided.

    f. When practical, contaminated outergarments from victims/response personnel should be removed at the mishap site decontamination area in order to protect the subsequent medical staff. Any ill effects believed to be related to exposure to AC/AAMs should be reported immediately. Likewise, the local medical staff should be advised of the incident, along with the potential hazards. Symptoms of effects could include:

        —Respiratory tract irritation and reduced respiratory capacity

        —Eye irritation

        —Skin irritation, sensitization, rashes, infections, or allergic reactions

    g. All contaminated footwear should be cleaned to limit the spread of debris into clean areas and support vehicles.

    h. Materials Safety Data Sheet (MSDS) should be made available to qualified personnel.

    i. Security restrictions may require additional control measures during emergencies.

17. Secure burned/mobile AC/AAM fragments and loose ash/particulate residue with plastic, a gentle mist of water or fire-fighting agent, fixant material, or a tent-like structure in order to prevent redispersion.

18. Consult the specific aircraft authority and/or the investigators before applying a fixant or hold-down material. However, safety concerns at the immediate mishap site may override any delayed application. Fire-fighting equipment should be available during fixant/stripper application, aircraft breakup, and recovery. Also, any fires must be completely out and the materials cooled to below 300°F (149°C). Two types of fixants are generally used: one for burned AC/AA Materials and debris, and the other for land surfaces. Fixant is usually not needed for open terrain and improved surfaces (concrete or asphalt) unless very high concentrations exist.

19. Obtain and mix (if necessary) the fixant or hold-down solution such as Polyacrylic acid (PAA) or

acrylic floor wax and water. Light oil is not recommended because it may become an aerosol and collect on equipment, hamper material investigations, and present a health hazard of its own. Generic acrylic floor wax, which is widely available, should be mixed in an approximate 8:1 or 10:1 ratio, although this may vary.

20. Apply (preferably spray) a moderate coating of the fixant solution on all burned/damaged AC/AA Materials and to any areas containing scattered/settled particulate debris. Completely coat the material until wet to ensure immobilization of the material, then allow the coating to dry.
21. **NOTE:** Strip-ability of the fixant coating is required where coatings are applied to debris that must later undergo microscopic chemical and material analysis by incident investigators. Care must be exercised in the use of stripping solutions since they can react with some materials and the process of stripping may damage the parts. PAA may be removed by a dilute solution of household ammonia (about 1% by volume of ammonium hydroxide in water) or trisodium phosphate (approximately one 8 ounce cup of trisodium phosphate per 2 gallons of water).
22. If deemed necessary, agricultural soil tackifiers may be used to hold materials on sand or soil. Most solutions can be sprayed onto the ground at a rate of 0.5 gal/sq yd.
23. Improved hard surfaces (i.e., concrete and asphalt) should be vacuumed (with an electrically protected vacuum) if possible. Sweeping operations should be avoided as they re-disseminate the particulates. The effluent from any runoff should be collected via plastic or burlap coated trenches or drainage ditches. **NOTE:** The entire impact or mishap site must be diked to prevent runoff of fire fighting agent (to avoid additional cleanup or environmental contamination).
24. All fixant application equipment should be immediately flushed/cleaned with a dilute solvent to prevent clogging for future use. Likewise, all fire fighting vehicles and equipment must be decontaminated, to the maximum extent possible, at the mishap site. Water and HEPA vacuums may be used.
25. Pad all sharp projections on damaged debris that must be retained so that injuries during handling and analysis can be avoided.
26. Carefully wrap the coated parts and or material with plastic sheeting/film or place them in a plastic bag of approximately 0.006 inches (6 mils) thick. Generic garbage bags are generally inadequate unless they are used as several plies.
27. Conduct all material disposal according to local, state, federal, and international guidelines. Consult with appropriate agencies for relevant procedures and policies for materials that do NOT require mishap investigation analysis or repair. Ensure all parts are released before disposal is authorized. All AC/AAM waste should be labeled appropriately with the type of material followed by the words: "Do Not Incinerate or Sell for Scrap."
28. Complete all necessary soil and surface restoration as required at the mishap site.
29. Place all hazardous waste material in appropriate containers and dispose of properly according to all applicable regulations.
30. If aircraft were subjected to the concentrated smoke plume or debris areas, the following should be accomplished:
    a. Vacuum the air/ventilation/cooling intakes with an electrically protected, HEPA vacuum cleaner.
    b. For internally affected smoke areas, visually and electronically inspect all compartments for debris and vacuum thoroughly.
    c. Prior to flying, perform electrical and systems checks, as well as an engine run-up.
31. For significantly affected structures and equipment, thoroughly clean all antenna insulators, exposed transfer bushings, circuit breakers, and any other applicable electrical components. Inspect air intakes and outlets for signs of smoke or debris and decontaminate if necessary.
32. Continue to monitor affected personnel, equipment, and mishap site.

Rapid Response Checklist

- ❑ Conduct the Initial Mishap Site Survey
- ❑ Establish Control at the Mishap Site with a Clear Chain of Command
- ❑ Evacuate Personnel From the Immediate Mishap Site Vicinity. *Restrict ALL unprotected personnel from assembling downwind of the mishap site.*
- ❑ Restrict Ground and Flight Operations As Appropriate for Conditions
- ❑ Extinguish Fire and Cool AC/AA Materials to below 300°F (149°C)
- ❑ Cordon Off the Mishap Site and Establish a Single Entry/Exit Point
- ❑ Consult Medical Personnel for Evaluation and Tracking of Exposed Personnel
- ❑ Coordinate a Thorough Survey of the Mishap Site with an Incident Commander (IC)
- ❑ Conduct Expert Toxicology and Area Studies With Survey Protocols
- ❑ Identify Specific Aircraft and Material Hazards
- ❑ Avoid Excessive Disturbance of the Mishap Site
- ❑ Locate, Secure, and Remove Radioactive AAMs; Contact Relevant Authorities and Dispose of In Accordance with Strict Disposal Policies
- ❑ Monitor Entry/Exit from the Single Entry Control Point (ECP)
- ❑ Secure Burned/Mobile AC/AAM Fragments and Loose Ash/Particulate Residue
- ❑ Consult Aircraft Authorities Before Applying Fixants or Hold-Down Materials
- ❑ Obtain and Mix a Fixant or Hold-down Solution
- ❑ Apply/Spray the Fixant Solution on Burned/Damaged AC/AA Materials
- ❑ Use Strippable Fixant Coating Where Coatings are Applied
- ❑ Use Agricultural Soil Tackifiers If Necessary
- ❑ Vacuum Improved Hard Surfaces
- ❑ Flush/Clean the Fixant Application Equipment With Dilute Solvent
- ❑ Pad All Sharp Projections On Damaged Debris
- ❑ Wrap Coated Parts and/or Material with Plastic Sheeting/Film
- ❑ Conduct Material Disposal According to Local, State, Federal, and International Guidelines
- ❑ Complete Soil and Surface Restoration
- ❑ Dispose of Hazardous Waste Material Appropriately
- ❑ Continue to Monitor Affected Personnel and Sites

References

"A Composite Picture." *Safety and Health*. Nov 1991. P 38-41.

A Composite System Approach to Aircraft Cabin Fire Safety. NASA Technical Memorandum. Apr 1987.

Advanced Composite Repair Guide. NOR 82-60. Prepared by Northrup Corporation, Aircraft Division, for USAF Wright Aeronautical Laboratories, Wright-Patterson AFB, OH. Mar 1982.

"Aircraft Fire Fighting Procedures for Composite Materials." US Navy/Marine Corps Training Film #112769. Naval Education and Training Support Center, Atlantic. Norfolk, VA. 1993.

American Conference of Governmental Industrial Hygienists. Threshold Limit Values for Chemical Substances and Physical Agents, ACGIH, Cincinnati, OH. 1998.

Baron, P.A. and K. Willeke. "Measurement of Asbestos and Other Fibers." *Aerosol*

Measurement Principles, Techniques, and Applications. Van Nostrand-Rheinhold, New York, NY. 1993.

Bickers, Charles. "Danger: Toxic Aircraft." *Janes Defence Weekly*. 19 Oct 1991.

Brauer, Roger L. *Safety and Health for Engineers*. Van Nostrand-Rheinhold, New York, NY. 1990.

Code of Federal Regulations, 29 *CFR* 1910.1000, *Air Contaminants*.

Composite Aircraft Mishap Safety and Health Guidelines. Project Engineer: Capt Keller. USAF Advanced Composites Program Office, McClellan AFB, CA. 18 Jun 1992.

Composite Aircraft Mishap Safety and Health Guidelines. ASCC ADV PUB 25/XX. Air Standardization Coordinating Committee, Washington, DC. 16 Sep 1992.

Composite Material Protective Equipment and Waste Disposal. Memo from 650 MED GP/SGB to 411 TS/CC, Edwards AFB, CA. 14 Oct 1992.

Conference on Advanced Composites, 5-7 Mar 1991. Proceedings. San Diego, CA. 1992.

Conference on Occupational Health Aspects of Advanced Composite Technology in the Aerospace Industry, 5-9 Feb 1989. AAMRL-TR-89-008. Vols I and II, Executive Summary and Proceedings. Wright-Patterson AFB, OH. Mar 1989.

DARCOM/NMC/AFLC/AFSC Commanders Joint Technical Coordinating Group on HAVE NAME (JTCG/HN). *HAVE NAME Guide for Protection of Electrical Equipment from Carbon Fibers*. May 1978.

Faeder, Edward J. and Paul E. Gurba. "Health Effects in the Aerospace Workplace – Some Concerns." SME Conference Proceedings: Composites in Manufacturing 9. Dearborn, MI. 15-18 Jan 1990.

Fire Performance and Suppressibility of Composite Materials. Hughes SBIR Phase II Report HAI 92-1071 DRAFT. 15 Dec 1992.

Fire Safety Aspects of Polymeric Materials, Volume 6: Aircraft: Civil and Military. Report by the National Materials Advisory Board of the National Academy of Sciences. 1977.

Fisher, Karen J. "Is Fire a Barrier to Shipboard Composites?" *Advanced Composites*. Vol 8, No 3: May/Jun 1993.

Gandhi, S. and Richard Lyon. *Health Hazards of Combustion Products from Aircraft Composite Materials,* Draft Manuscript, FAA Technical Center. 1997.

General Advanced Composite Repair Processes Manual. USAF TO1-1-690. McClellan AFB, CA. 1 Aug 1990.

Hetcko, John. "Disposal of Advanced Composite Materials." Defense Division, Brunswick Corporation. Lincoln, NE.

Hubbell, M. Patricia. "Hazard Communication and Composites." McDonnell Douglas Space Systems Company. A3-315-12-1. Huntington Beach, CA 92647.

Kantz, M. "Advanced Polymer Matrix Resins and Constituents: An Overview of Manufacturing, Composition, and Handling." *Applied Industrial Hygiene, Special Issue*. 50(12). P 1-8. 1989.

Mishap Response for Advanced Composites. US Air Force Film. 46th Test Wing Audio-Visual Services, Eglin AFB, FL. Sep 1994.

Morrison, R. General Background on the Filtration Performance of Military Filters. US Army Chemical and Biological Defense Command, Aberdeen Proving Grounds, MD. 1998.

Naval Environmental Health Center, *Advanced Composite Materials*, NEHC-TM91-6. 1991.

Naval Safety Center. *Accident Investigation and Clean up of Aircraft Containing Carbon/Graphite Composite Material Safety Advisory*. Unclassified Telex N03750 from NAS Norfolk, VA. 20 Aug 1993.

Olson, John M. *Aerospace Advanced Composites Interim Technical Mishap Guide*. USAF HQ AFCESA/DF. 22 Mar 1994.

Olson, John M. "Composite Aircraft Mishaps: High Tech Hazards? Part I and II. *Flying Safety Magazine*. Vol 49, No 11 and 12. Nov and Dec 1993.

Olson, John M. *Mishap Risk Control Guidelines for Advanced Aerospace Materials: Environmental,*

Safety, and Health Concerns for Advanced Composites. 28 Oct 1993. USAF Advanced Composites Programs Office, McClellan AFB, CA.

Olson, John M. *Safety, Health, and Environmental Hazards Associated with Composites: A Complete Analysis*. 15 Nov 1992.

"Position Paper on the CORKER Program." Oklahoma City Air Logistics Center. 16 Feb 1993.

Revised HAVE NAME Protection Manual. MP 81-266 MITRE MTR 4654. A.S. Marqulies and D.M. Zasada, Eds. Jun 1981.

Risk Analysis Program Office at Langley Research Center. *Risk to the Public from Carbon Fibers Released in Civil Aircraft Accidents*. NASA SP-448. Washington, DC. 1980.

Safe Handling of Advanced Composite Materials. 2nd Ed. SACMA, Arlington, VA. Jul 1991.

Seibert, John F. *Composite Fiber Hazards*, US Air Force Occupational and Environmental Health Laboratory (AFOEHL) Technical Report 90-226E100178MGA. 1990.

Summary of Medical Evaluation of Boeing Employees Working with Composite Materials Who Have Filed Workers Compensation Claims for Illness. Seattle Medical Care, Association for Independent Practitioners. Seattle, WA.

Thomson, S.A. "Toxicology of Carbon Fibers." *Applied Industrial Hygiene, Special Issue*. 50(12). P 34-36. 1989.

Warnock, Richard. "Engineering Controls and Work Practices for Advanced Composite Repair." *Applied Industrial Hygiene, Special Issue*, 50(12). P 52-53. 1989.

Additional Information Sources for ARFF Personnel

National Fire Protection Association (NFPA) Standards

NFPA 1003, *Standard for Airport Fire Fighter Professional Qualifications*

NFPA 402, *Guide for Aircraft Rescue and Fire Fighting Operations*

NFPA 403, *Standard for Aircraft Rescue and Fire Fighting Services at Airports*

NFPA 405, *Recommended Practice for Recurring Proficiency Training of Aircraft Rescue and Fire Fighting Services*

NFPA 407, *Standard for Aircraft Fuel Servicing*

NFPA 408, *Standard for Aircraft Hand Portable Fire Extinguishers*

NFPA 410, *Standard on Aircraft Hangars*

NFPA 412, *Standard for Evaluating Aircraft Rescue and Fire Fighting Foam Equipment*

NFPA 414, *Standard on Aircraft Rescue and Fire Fighting Vehicles*

NFPA 415, *Standard on Airport Terminal Buildings, Fueling Ramp Drainage, and Loading Walkways*

NFPA 422, *Guide for Aircraft Accident Response*

NFPA 424, *Guide for Airport/Community Emergency Planning*

Federal Aviation Regulations (FARs)

Code of Federal Regulations, Title 14, Part 139, Subpart D:

Sec. 139.315, *Aircraft Rescue and Firefighting: Index Determination*

Sec. 139.317, *Aircraft Rescue and Firefighting: Equipment and Agents*

Sec. 139.319, *Aircraft Rescue and Firefighting: Operational Requirements*

Sec. 139.325, *Airport Emergency Plan*

Federal Aviation Administration (FAA) Advisory Circulars (ACs)

150/5220-12B *Fire Department Responsibility in Protecting Evidence at the Scene of an Aircraft Accident*

150/5220-18B *Airport Safety Self-Inspection*

150/5220-31A *Airport Emergency Plan*

150/5210-21 *Airport Emergency Medical Facilities and Services*

150/5210-5B *Painting, Marking, and Lighting of Vehicles Used on an Airport*

150/5210-6C *Aircraft Fire and Rescue Facilities and Extinguishing Agents*

150/5210-7C *Aircraft Rescue and Firefighting Communications*

150/5210-13A *Water Rescue Plans, Facilities, and Equipment*

150/5210-14A *Airport Fire and Rescue Personnel Protective Clothing*

150/5210-15 *Airport Rescue and Firefighting Station Building Design*

150/5210-17 Chg.1 *Programs for Training of Aircraft Rescue and Fire Fighting Personnel*

150/5210-18 *Systems for Interactive Training of Airport Personnel*

150/5210-19 *Driver's Enhanced Vision System (DEVS)*

150/5220-4B *Water Supply Systems for Aircraft Fire and Rescue Protection*

150/5220-10B *Guide Specification for Water/Foam Aircraft Rescue and Fire Fighting Vehicles*

150/5220-17A *Design Standards for Aircraft Rescue and Fire Fighting Training Facility*

150/5220-19 *Guide Specification for Small Agent Aircraft Rescue and Fire Fighting Vehicles*

150/5230-4 *Aircraft Fuel Storage, Handling, and Dispensing on Airports*

Emergency Response Telephone Numbers

National Response Center	1-800-424-8802
CHEMTREC®	1-800-424-9300
CANUTEC	613-996-6666

Military Shipments:

Explosives/ammunition incidents 703-697-0218

All other dangerous goods incidents 1-800-851-8061

Websites

National Fire Protection Association (NFPA): http://www.nfpa.org

Federal Aviation Administration (FAA): http://www.faa.gov

U.S. Department of Transportation: http://www.dot.gov

Aircraft Rescue and Fire Fighting Working Group (ARFFWG): http://www.arffwg.org

Technical Order 00-105E-9: http://www.robins.af.mil/ti/tilta/documents/to00-105e-9.htm

NOTE: To order Technical Order 00-105E-9, contact:

Thomas L. Stemphoski
HQ AFCESA/CEXF (Fire Protection
139 Barnes Drive, Suite 1
TYNDALL AFB, FL 32403-5319
DSN 523-6150 COM (850) 283-6150

Glossary

NOTE: This glossary is designed to supplement discussions in the text and is not meant to be a comprehensive dictionary of aircraft terms.

A

Abort — The act of terminating a planned aircraft maneuver such as the takeoff or landing. Pilots will normally abort a takeoff if any indication of a possible malfunction exists.

Advanced Aerospace Materials — See *Composite Materials.*

AFFF — See *Aqueous Film Forming Foam.*

Aft/After — The rear or tail section or toward the rear or tail of an aircraft.

Aileron — A movable hinged rear portion of an airplane wing. The primary function of the ailerons is to roll or bank the aircraft in flight.

Air Bill — A shipping document prepared from a bill of lading that accompanies each piece or each lot of air cargo.

Aircraft Accident — An occurrence during the operation of an aircraft in which any person suffers death or serious injury or in which the aircraft receives substantial damage.

Aircraft Arresting System — A device used to engage an aircraft and absorb forward momentum in case of an aborted takeoff and/or landing.

Aircraft Familiarization — Area of ARFF personnel training relating to the various aircraft operated in an airport and the features of these aircraft, including fuel capacity, fuel tank locations, emergency exit locations, operation of emergency exits, passenger seating capacity, etc.

Aircraft Incident — An occurrence, other than an accident associated with the operation of an aircraft, that affects or could affect continued safe operation if not corrected. An incident does not result in serious injury to persons or substantial damage to aircraft.

Airfoil — Any surface, such as an airplane wing, aileron, elevator, rudder, or helicopter rotor, designed to obtain reaction from the air through which it travels. This reaction keeps the aircraft aloft and controls its flight attitude and direction.

Airframe — Major components of an aircraft necessary for flight. These include the fuselage, wings, stabilizers, flight control surfaces, etc. Also refers to a basic model of an aircraft; for example, the Boeing 707 airframe has both civilian and military applications in a variety of configurations.

Airport/Airfield — An area on land or water used or intended to be used for aircraft takeoffs and landings. This includes buildings and facilities. *Aerodrome* is the international term with the same definition.

Airport Emergency Plan (AEP) — A plan formulated by airport authorities to ensure prompt response to all emergencies and other unusual conditions in order to minimize the extent of personal and property damage.

Airport Familiarization — Knowledge of the locations of airport buildings, runways and taxiways, access roads, and surface features, routes, and conditions that may enhance or obstruct the prompt and safe response to accidents/incidents on the airport and those areas surrounding the airport.

Airport Ground Control — The control of aircraft and other vehicular traffic operating in the airport movement area by the airport control tower.

Airport Operations Area (AOA) — The area of an airport where aircraft are expected to operate such as taxiways, runways, and ramps.

Airport Water Distribution System — A system of water mains, piping, valves, hydrants, pumps, etc., under airport authority for the distribution of pressurized water to support ARFF operations on airports.

Airspeed — The speed of an aircraft relative to its surrounding air mass.

Air Surface Detection Equipment (ASDE) — A short-range radar that displays the airport surface. It is used to track and guide surface traffic in low-visibility weather conditions. ASDE may be used to direct radio-equipped emergency vehicles to known accident sites.

Air Traffic Control (ATC) — A service operated by appropriate authority to promote the safe, orderly, and expeditious flow of air traffic.

Approach Lights — A system of lights so arranged to assist pilots in aligning their aircraft with the runway for landing.

Approach Sequence — The order in which two or more aircraft are cleared to approach to land at an airport or while awaiting approach clearance.

Apron/Ramp — A defined area on airports intended to accommodate aircraft for purposes of loading or unloading passengers, mail or cargo, refueling, parking, or maintenance.

Aqueous Film Forming Foam (AFFF) — A synthetic foam concentrate that, when combined with water, is a highly effective extinguishing and blanketing agent on hydrocarbon fuels.

Autorotation — Flight condition in which the lifting rotor of a rotary wing aircraft is driven entirely by action of the air when in flight or, as in the case of a helicopter, after an engine failure.

Auxiliary Power Unit (APU) — A power unit installed in most large aircraft to provide electrical power and pneumatics for ground power, air conditioning, engine start, and backup power in flight. APU also refers to mobile units that are moved from one aircraft to another to provide a power boost during engine startup.

B

Base Leg — The flight path at a right angle to the landing runway off the approach end. The base leg normally extends from the downwind leg to the intersection of the extended runway line. The aircraft must make a 90-degree turn from the base leg before it can begin its final approach.

Below Minimum — Weather conditions below the minimums prescribed by regulation for the particular operation such as takeoff or landing.

Bogie — A tandem arrangement of landing gear wheels with a central strut. The bogie swivels up and down so all wheels stay on the ground as the attitude of the aircraft changes or as the slope of the ground surface changes.

Breakaway/Frangible Fences and Gates — Fences and gates designed and constructed to collapse when impacted by large vehicles to allow rapid access to accident sites. A firefighter's knowledge of their locations is vital in the event of a response off the airport.

Bulkhead — An upright partition that separates one aircraft compartment from another. Bulkheads may strengthen or help give shape to the structure and may be used for the mounting of equipment and accessories.

C

Cabin — Aircraft passenger compartment that may be separated and may contain a cargo area.

Camlock Fastener — A trade name given to a quick-disconnect screw-type fastener, designed to open with a quarter or half turn (similar to Dzus fasteners).

Canopy — Transparent enclosure over the cockpit of some aircraft.

Ceiling — The height above the earth's surface of the lowest layer of clouds reported as "broken" or "overcast," or the vertical visibility into obscuration.

Clearway/Overrun — An area beyond the end of the runway that has been cleared of nonfrangible obstacles and strengthened to allow overruns without serious damage to the aircraft.

Cockpit — The fuselage compartment occupied by pilots while flying the aircraft.

Cockpit Voice Recorder (CVR) — A recording device installed in most large civilian aircraft to record crew conversation and communications and is intended to assist in an accident investigation to determine probable cause of the accident.

Command Post — Command and control point where the incident commander and command staff function and where those in charge of emergency units report to be briefed on their respective assignments.

Composite Materials — Plastics, metals, ceramics, or carbon-fiber materials with built-in strengthening agents. These materials are much lighter and stronger than the metals formerly used for such aircraft components as panels, skin, and flight controls. The newer term is *Advanced Aerospace Materials*.

Control Tower — A unit (facility) established to provide traffic control service for the movement of aircraft and other vehicles in the airport operations area. The control tower contains very sophisticated electronic devices for the control of the flight patterns and airport ground operations.

Controlled Airport — An airport having a control tower in operation. Tower usually, but not always, is staffed by FAA personnel.

Cowl Flaps — Adjustable sections or hinged panels on the engine cowling of reciprocating engines. They are used to control the engine temperature.

Cowling — Removable covering around aircraft engines.

Critical Rescue and Fire Fighting Access Area (CRFFAA) — This is the rectangular area surrounding any given runway. Its width extends 500 feet (150 m) outward from each side of the runway centerline, and its length extends 3,300 feet (1 000 m) beyond each runway end. This is the rectangular area on an airfield where most accidents are expected to occur.

Crosswind Leg — A flight path at right angles to the landing runway off its upwind leg.

D

Deplane — To get out of an aircraft.

Displaced Runway Threshold — The temporary relocation of a runway threshold (beginning or end) due to maintenance or other activity on the runway.

Downwind Leg — The flight path parallel to the landing runway in the direction opposite to landing. The downwind leg normally extends between the crosswind leg and the base leg.

Drag Chute — A parachute device installed on some aircraft that is deployed on landing roll to aid in slowing the aircraft to taxi speed.

Duct — A tube or passage that confines and conducts airflow throughout the aircraft for pressurization, air conditioning, etc.

Dzus Fastener — Trade name given to a half-turn fastener with a slotted head. This type of fastener is used on engine cowlings, cover plates, and access panels throughout the aircraft.

E

Ejection Seat — An aircraft seat capable of being ejected in an emergency to catapult the occupant clear of the aircraft.

Elevator — The hinged, movable control surface at the rear of the horizontal stabilizer. It is attached to the control wheel or stick and is used to control the up-and-down pitch motion of the aircraft.

Emergency Escape Slide — Escape slides connected to aircraft doors and, in some cases, to overwing exits that when deployed will inflate and extend to the ground. Pneumatic in operation, most are automatic by opening the door; some require manual activation, that is, a short pull on a lanyard. Many may be disconnected from the aircraft and used for a flotation device in water.

Emergency Lighting System — A system of interior and exterior low-power incandescent and/or fluorescent lights designed to assist passengers in locating and using aircraft emergency exits, but are not bright enough to assist ARFF personnel in carrying out search and rescue operations.

Empennage — See *Tail*.

Engine Numbers — For identification, engines of multiengine aircraft are numbered consecutively 1, 2, 3, 4, etc., as seen from the pilot's seat. They are numbered left to right across the aircraft even though some may be mounted on the wings or on the tail of the aircraft; for example, in the L-1011 and DC-10, the tail-mounted engine is number 2.

Enplane — To board an aircraft

Exhaust Area — The area behind an engine where hot exhaust gases present a danger to personnel.

F

Final Approach — That portion of the landing pattern in which the aircraft is lined up with the runway and is heading straight in to land.

Firewall — A bulkhead separating an aircraft engine from the aircraft fuselage or wing.

Fixed-Based Operator — An enterprise based on an airport that provides storage, maintenance, or service for aircraft operators.

Flameout — Unintended loss of combustion in turbojet engines resulting in the loss of engine power.

Flameover — Ignition of fire gases that have accumulated at or near the ceiling of an aircraft cabin or other enclosure when the supply of combustion air is increased.

Flame Resistant — Aircraft materials that are not susceptible to combustion to the point of propagating a flame *after* the ignition source is removed.

Flaps — Adjustable airfoils attached to the leading or trailing edges of aircraft wings to improve aerodynamic performance during takeoff and landing. They are normally extended during takeoff, landing, and slow flight.

Flashback — The spontaneous reignition of fuel when the blanket of extinguishing agent breaks down or is compromised through physical disturbance.

Flashover — The stage of a fire at which all surfaces and objects within a space have been heated to their ignition temperature and flame breaks out almost at once over the surface of all objects in the space.

Flash Resistant — Aircraft materials that are not susceptible to burning violently when ignited.

Flight Controls — A general term applied to devices that enable the pilot to control the direction of flight and attitude of the aircraft.

Flight Data Recorder (FDR) — A recording device on large civilian aircraft to record aircraft airspeed, altitude, heading, acceleration, etc., to be used as an aid to accident investigation.

Flight Deck — The cockpit on a large aircraft, separated from the rest of the cabin.

Flight Service Station — A facility from which aeronautical information and related aviation support services are provided to aircraft. This also includes airport and vehicle advisory services for designated uncontrolled airports.

Fore/Forward — The front or nose section of an aircraft or toward that area.

Formers — A frame of wood or metal that is attached to the truss of the fuselage or wing in order to provide the required aerodynamic shape.

Fuel on Board — Amount in pounds (6 to 7 lb per gallon [0.7 kg to 0.8 kg per liter]) of fuel on aircraft remaining.

Fuel Siphoning/Fuel Venting — Unintentional release of fuel from an aircraft caused by overflow, puncture, loose cap, etc.

Fuselage — The main body of an aircraft to which the wings and tail are attached. The fuselage houses the crew, passengers, and cargo.

G

Galley — The food storage and preparation area of large aircraft.

General Aviation — All civil aviation operations other than scheduled air services and nonscheduled operations for remuneration or hire.

Gear Down — Landing gear in down and locked position (have green light in the cockpit).

Grid Map — Map marked either using rectangular coordinates or with azimuthal bearings using polar coordinates; encompasses the airport and should also encompass the emergency response area outside the airport.

Go Around — Maneuver conducted by a pilot whenever a visual approach to a landing cannot be completed.

H

Hazardous Materials/Dangerous Goods — A substance that poses an unreasonable risk to the health and safety of operating or emergency personnel, the public, and/or the environment if it is not properly controlled during handling, storage, manufacture, processing, packaging, use, disposal, or transportation.

High Speed Turnoff/Taxiway — A curved or angled taxiway designed to expedite aircraft turning off the runway after landing.

Hot Refuel/Defuel (Rapid Refuel/Defuel) — Refueling or defueling of an aircraft while the engines are operating.

Hung Gear — One or more of the aircraft landing gear not down and locked (no green-light indication in the cockpit).

Hydraulic System — An aircraft system that transmits power by means of a fluid under pressure.

Hydrazine — A toxic, caustic hypergolic fuel that is a clear, oily liquid with a smell similar to ammonia that poses a health hazard in both the liquid and vapor forms.

Hydroplaning — The condition in which moving aircraft tires are separated from pavement surfaces by steam and/or water, liquid rubber film, resulting in loss of mechanical braking effectiveness.

Hypergolic — A fuel that ignites spontaneously on contact with an oxidizer.

I

Idle Thrust/rpm — An aircraft engine running at the lowest possible speed.

Inboard/Outboard — Refers to location with reference to the centerline of the fuselage; for example, inboard engines are the ones closest to the fuselage, and outboard engines are those farthest away.

Incursion — Any occurrence at an airport involving an aircraft, vehicle, person, or object on the ground that creates a collision hazard or results in a loss of separation with an aircraft taking off, intending to take off, landing, or intending to land.

Inertia Lights — A light mounted in the aircraft structure so that a sharp deceleration, such as a crash situation, will activate the light. It can also be turned on manually and removed from the mounting to be used as a portable flashlight.

Instrument Flight Rules (IFR) — Regulations governing the operation of an aircraft in weather conditions with visibility below the minimum required for flight under visual flight rules.

Instrument Landing — Landing an aircraft by relying only upon instrument data. This may be due to inclement weather or other factors.

Instrument Landing System (ILS) — An electronic navigation system that allows aircraft to approach and land during inclement weather conditions. This area should be avoided if at all possible to avoid breaking guidance beam and interfering with a landing or takeoff.

Intake Area — That area in front of and to the side of a jet engine that might be unsafe for personnel.

Isolation Area — An area used to isolate aircraft carrying hazardous materials or munitions. This area is segregated away from the parking aprons in case an accident or incident should occur.

J

Jet-Assisted Takeoff (JATO) — A rocket or auxiliary jet used to augment normal thrust for takeoffs.

Jet Blast — Wind and/or heat blast created behind an aircraft with engines running.

Jettison — To selectively discard aircraft components such as external fuel tanks or canopies.

Jetway — An enclosed ramp between a terminal and an aircraft for loading and unloading passengers.

L

Landing Roll — The distance from the point of touchdown to the point where the aircraft is brought to a stop or exits the runway.

Leading/Trailing Edge Devices — The forward and rear edges of aircraft wings normally extended for takeoff and landings to provide additional lift at low speeds and to improve aircraft performance.

Longeron (Stringers) — Longitudinal members of the framing of an aircraft fuselage or nacelle, usually continuous across a number of bulkheads or other points of support.

Low Approach — An approach over a runway or heliport where the pilot intentionally does not make contact with the runway.

M

Missed Approach — A maneuver conducted by a pilot whenever an instrument approach cannot be completed into a landing.

Movement Area — The runways, taxiways, and other areas of an airport that are used for taxiing or hover taxiing, air taxiing, and takeoff and landing of aircraft exclusive of loading ramps and aircraft parking areas.

Mutual Aid — Reciprocal aid given one agency by another in times of emergency.

N

Nacelle — The housing of an externally mounted aircraft engine.

National Defense Area (NDA) — The temporary establishment within the United States of "federal areas" for the protection or security of Department of Defense resources. Normally, NDAs are established for emergency situations such as accidents. NDAs may be established, discontinued, or their boundaries changed as necessary to provide protection or security of Department of Defense (DOD) resources.

O

Occupants — Passengers and aircrew aboard an aircraft.

Ordnance — Bombs, rockets, ammunition, and other explosive devices carried on most military aircraft.

Overhead Approach (360 overhead) — A series of standard maneuvers conducted by military aircraft (often in formation) for entry into the airfield traffic pattern prior to landing.

Overrun — In military aviation exclusively, a stabilized or paved area at the end of the runway that is the same width as the runway plus the shoulders. The overrun is used in the event of an emergency that prohibits the aircraft from stopping normally.

P

Petcock — A small faucet or valve for releasing or draining a gas (such as air).

Practical Critical Fire Area (PCA)— Two-thirds of the Theoretical Critical Fire Area (TCA). See also *Theoretical Critical Fire Area.*

Prop or Rotor Wash — Wind blast created behind or around an aircraft with engines running.

R

Radial Engines — Internal-combustion, piston-driven aircraft engines with cylinders arranged in a circle.

Ramp — See *Apron.*

Rapid Response Area — A rectangular area that includes the runway and the surrounding area extending to but not exceeding the airport property line. Its width extends 500 feet (152 m) outward from each side of the runway centerline and its length extends 1650 feet (500 m) beyond each runway end

Reciprocating Engines — Internal-combustion, piston-driven aircraft engines with cylinders arranged in opposition.

Rotor — Rotating airfoil assemblies of helicopters and other rotary-wing aircraft, providing lift.

Rudder — The hinged, movable control surface attached to the rear part of the vertical stabilizer and is used to control the yaw or turning motion of the aircraft.

Runway — A defined rectangular area on a land airport prepared for the takeoff or landing of aircraft along its length.

S

Shipping Papers — See *Air Bill.*

Skin — The outer covering of an aircraft, which includes the covering of wings, fuselage, and control surfaces.

SPAAT (Skin Penetrating Agent Applicator Tool®) — A penetrating nozzle.

Spar — A principal, span-wide structural member of an airfoil or control surface.

Speed brakes — Aerodynamic devices located on the wing or along the rear or underside of the fuselage that can be extended to help slow the aircraft.

Spoilers — Movable panels located on the upper surface of a wing and that raise up into the airflow to increase drag and decrease lift.

Stabilizer — An airfoil on an airplane used to provide stability; that is, the aft horizontal surface to which the elevators are hinged (horizontal stabilizer) and the fixed vertical surface to which the rudder is hinged (vertical stabilizer).

Staging Area — A prearranged, strategically located area where personnel, apparatus, and other equipment can be held in readiness for use during an emergency.

Stopway/Overrun Area — An area beyond the takeoff runway end capable of supporting aircraft that overshoot the runway on aborted takeoff or landing without causing structural damage to the airplane.

Strut — Aircraft structural components designed to absorb or distribute abrupt compression or tension such as the landing gear forces.

T

Tail — Aircraft tail assembly including the vertical and horizontal stabilizers, elevators, and rudders. Also called *Empennage.*

Taxiway — A specially designated and prepared surface on an airport for aircraft to taxi (travel) to and from runways, hangars, etc. In simpler terms, taxiways are the roadways for aircraft movement.

Theoretical Critical Fire Area — The theoretical rectangular area adjacent to an aircraft in which fire must be controlled in order to ensure temporary fuselage integrity and provide an escape route for aircraft occupants.

Threshold — The beginning or end of a runway that is usable for landing or takeoff.

Three-Dimensional Fire — Liquid-fuel fire in which fuel is being discharged from an elevated or pressurized source, creating a pool of fuel on a lower surface.

Thrust — The pushing or pulling force developed by an aircraft engine.

Thrust Reverser — A device or apparatus for diverting jet engine thrust for slowing or stopping the aircraft.

Torching — The burning of fuel at the end of the exhaust pipe or stack of a reciprocating aircraft engine due to excessive richness of the fuel/air mixture.

Traffic Pattern — The traffic flow that is prescribed for aircraft landing at or taking off from an airport. The components of a typical traffic pattern are upwind leg, crosswind leg, downwind leg, base leg, and final approach.

Transportation Area — Location where accident casualties are held after receiving medical care or triage before being transported to medical facilities.

Transport Index — A number placed on the label of a package expressing the maximum allowable radiation level in millirem per hour at one meter (3.3 feet) from the external surface of the package.

Triage — Sorting and classification of accident casualties to determine the priority for medical treatment and transportation.

Triage Tagging — Method used to identify accident casualties as to extent of injury.

Turbojet — A jet engine employing a turbine-driven compressor to compress the intake air, or an aircraft with this type of engine. Also known as a gas turbine.

Turret/Turret Nozzle — A preplumbed master stream appliance on some airport rescue and fire fighting apparatus, capable of sweeping from side to side and designed to deliver large volumes of foam or water.

U

Unannounced Emergency — An emergency that occurs without prior warning.

Uncontrolled Airport — One having no control tower in operation.

Undeclared Hazardous Cargo — Cargo that has not received proper packaging, shipping documentation, or safety precautions required of hazardous shipments.

Upwind Leg — A flight path parallel to the landing runway in the direction of landing.

V

Visual Approach — An approach to landing made by visual reference to the surface.

Visual Flight Rules (VFR) — Rules that govern the procedures for conducting flight operations under conditions of clear visibility.

W

Wake Turbulence — Phenomena resulting from the passage of an aircraft through the atmosphere. The term includes vortices, thrust stream turbulence, jet blast, jet wash, propeller wash, and rotor wash on the ground or in the air.

Wind Sock — Cone-shaped cloth sock located on airports to indicate wind direction and to some extent wind velocity.

Wind Tee — A T-shaped indicator mounted horizontally on a pivot pole to swing freely in the wind. Used as a wind direction indicator or landing direction indicator.

Index

F

W

Z

COMMENT SHEET

DATE ______________________ NAME ____

ADDRESS ____

ORGANIZATION REPRESENTED ____

CHAPTER TITLE ____ NUMBER __________

SECTION/PARAGRAPH/FIGURE __________________________________ PAGE ____________

1. Proposal (include proposed wording or identification of wording to be deleted), OR PROPOSED FIGURE:

2. Statement of Problem and Substantiation for Proposal:

RETURN TO: IFSTA Editor
Fire Protection Publications
Oklahoma State University
930 N. Willis
Stillwater, OK 74078-8045

SIGNATURE ______________________________

Use this sheet to make any suggestions, recommendations, or comments. We need your input to make the manuals as up to date as possible. Your help is appreciated. Use additional pages if necessary.

Your Training Connection.....

The International Fire Service Training Association

We have a free catalog describing hundreds of fire and emergency service training materials available from a convenient single source: the International Fire Service Training Association (IFSTA).

Choose from products including IFSTA manuals, IFSTA study guides, IFSTA curriculum packages, Fire Protection Publications manuals, books from other publishers, software, videos, and NFPA standards.

Contact us by phone, fax, U.S. mail, e-mail, internet web page, or personal visit.

Phone
1-800-654-4055

Fax
405-744-8204

U.S. mail
IFSTA, Fire Protection Publications
Oklahoma State University
930 North Willis
Stillwater, OK 74078-8045

E-mail
editors@osufpp.org

Internet web page
www.ifsta.org

Personal visit
Call if you need directions!